Ives, Herbert Eugene

Airplane Photography

Ives, Herbert Eugene

Airplane Photography

Inktank publishing, 2018

www.inktank-publishing.com

ISBN/EAN: 9783747750001

AIRPLANE PHOTOGRAPHY

BY

HERBERT E. IVES

MAJOR, AVIATION SECTION, SIGNAL OFFICERS RESERVE CORPS, UNITED STATES ARMY; LATELY OFFICER IN CHARGE OF EXPERIMENTAL DEPARTMENT, PHOTOGRAPHIC BRANCH, AIR SERVICE

208 ILLUSTRATIONS

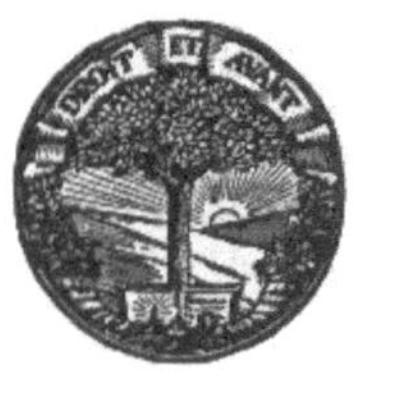

PHILADELPHIA AND LONDON
J. B. LIPPINCOTT COMPANY

TO MY WIFE

A HELPFUL CRITIC, EVEN THOUGH SHE NEITHER PHOTOGRAPHS NOR FLIES

PREFACE

AIRPLANE photography had its birth, and passed through a period of feverish development, in the Great War. Probably to many minds it figures as a purely military activity. Such need not be the case, for the application of aerial photography to mapping and other peace-time problems promises soon to quite overshadow its military origin. It has therefore been the writer's endeavor to treat the subject as far as possible as a problem of scientific photography, emphasizing those general principles which will apply no matter what may be the purpose of making photographs from the air. It is of course inevitable that whoever at the present time attempts a treatise on this newest kind of photography must draw much of his material from war-time experience. If, for this reason, the problems and illustrations of this book are predominantly military, it may be remembered that the demands of war are far more severe than those of peace; and hence the presumption is that an account of how photography has been made successful in the military plane will serve as an excellent guide to meeting the peace-time problems of the near future.

It is assumed that the reader is already fairly conversant with ordinary photography. Considerable space has indeed been devoted to a discussion of the fundamentals of photography, and to scientific methods of study, test, and specification. This has been done because aerial photography strains to the utmost the capacity of the photographic process, and it is necessary that the most advanced methods be understood by those who would secure the best results or contribute to future progress. No pretence is made that the book is an aerial photographic encyclopædia; it is not a manual of instructions; nor is its appeal so popular as it would be were the majority of the illustrations striking aerial photographs of war subjects. It is hoped that the middle course steered has produced a volume which will be informative and inspirational to those who are seriously interested either in the practice of aerial photography or in its development.

The writer is deeply in debt to many people, whose assistance of one sort or another has made this book possible. First of all should be mentioned those officers of the English, French and Italian armies through whose courtesy it is that he can speak at first hand of the photographic practices in these armies at the front. It is due to Lieutenant Colonel R. A. Millikan that the experimental work of which the writer has had charge was initiated in the United States Air Service. To him and to Major C. E. Mendenhall, under whom the work was organized in the Science and Research Division of the Signal Corps, are owing the writer's thanks for the opportunities and support given by them. A similar acknowledgment is made to Lieutenant Colonel J. S. Sullivan, Chief of the Photographic Branch of the Army Air Service, for his interest and encouragement in the compilation of this work, and for the permission accorded to use the air service photographs and drawings which form the majority of the illustrations.

The greatest debt of all, however, is to those officers who have formed the staff of the Experimental Department. To mention them by name: Captain C. A. Proctor, who was charged with our foreign liaison, and who acted as deputy chief during the writer's absence overseas; Captain A. K. Chapman, in charge of the work on optical parts, and later chief of our Rochester Branch; Captain E. F. Kingsbury, who had immediate charge of camera development; Lieutenant J. B. Brinsmade and Mr. R. P. Wentworth, who handled the experimental work on camera mountings and installation; Lieutenant A. H. Nietz, in charge of the Langley Field Laboratory of the Experimental Department; Mr. R. B. Wilsey and Lieutenant J. M. Hammond, who, with Lieutenant Nietz, carried on the experimental work on sensitized materials. A large part of what is new and what is ascribed in the following chapters to "The American Air Servce" is the work of this group of men. Were individual references made, in place of this general and inclusive one, their names would thickly sprinkle these pages. It has been a rare privilege to have associates so able, enthusiastic, and loyal.

THE AUTHOR

NOVEMBER, 1919

CONTENTS

I. INTRODUCTORY

II. THE AIRPLANE CAMERA

III. THE SUSPENSION AND INSTALLATION OF AIRPLANE CAMERAS

IV. SENSITIZED MATERIALS AND CHEMICALS

V. METHODS OF HANDLING PLATES, FILMS AND PAPERS

CONTENTS

VI. PRACTICAL PROBLEMS AND DATA

VII. THE FUTURE OF AERIAL PHOTOGRAPHY

I
INTRODUCTORY

AIRPLANE PHOTOGRAPHY

CHAPTER I

GENERAL SURVEY

Aerial Photography from Balloons and Kites.—Photography from the air had been developed and used to a limited extent before the Great War, but with very few exceptions the work was done from kites, from balloons, and from dirigibles. Aerial photographs of European cities had figured to a small extent in the illustration of guidebooks, and some aerial photographic maps of cities had been made, notably by the Italian dirigible balloon service. Kites had been employed with success to carry cameras for photographing such objects as active volcanoes, whose phenomena could be observed with unique advantage from the air, and whose location was usually far from balloon or dirigible facilities.

As a result of this pre-war work we had achieved some knowledge of real scientific value as to photographic conditions from the air. Notable among these discoveries was the existence of a veil of haze over the landscape when seen from high altitudes, and the consequent need for sensitive emulsions of considerable contrast, and for color-sensitive plates to be used with color filters.

The development of aerial photography would probably however have advanced but little had it depended merely on the balloon or the kite. As camera carriers their limitations are serious. The kite and the captive balloon cannot navigate from place to place in such a way as to permit the rapid or continuous photography of extended areas. The kite suffers because the camera it supports must be manipulated either from the ground or else by some elaborate mechanism, both for pointing and for handling the exposing and plate changing devices. The free balloon is at the mercy of the winds both as to its direction and its speed of travel. The dirigible balloon, as it now exists after its development during the war, is, it is true, not subject to the shortcomings just mentioned. Indeed, in many ways it is perhaps superior to the airplane for photographic purposes, since it affords more space for camera and accessories, and is freer from vibration. It is capable also of much slower motion, and can travel with less danger over forests and inaccessible areas where engine failure would force a plane down to probable disaster. But the smaller types as at present built are not designed to fly so high as the airplane, and the dirigibles, both large and small, are far more expensive in space and maintenance than the plane. For this one reason especially they are not likely to be the most used camera carriers of the aerial photographer of the future. Inasmuch as the photographic problems of the plane are more difficult than those of the dirigible and at the same time broader, the subject matter of this book applies with equal force to photographic procedure for dirigibles.

Development of Airplane Photography in the Great War. —The airplane has totally changed the nature of warfare. It has almost eliminated the element of surprise, by rendering impossible that secrecy which formerly protected the accumulation of stores, or the gathering of forces for the attack, a flanking movement or a "strategic retreat." To the side having command of the air the plans and activities of the enemy are an open book. It is true that more is heard of combats between planes than of the routine task of collecting information, and the public mind is more apt to be impressed by the fighting and bombing aspects of aerial warfare. Nevertheless, the fact remains that the chief use of the airplane in war is *reconnaissance*. The airplane is "the eye of the army."

In the early days of the war, observation was visual. It was the task of the observer in the plane to sketch the outlines of trenches, to count the vehicles in a transport train, to estimate the numbers of marching men, to record the guns in an artillery emplacement and to form an idea of their size. But the capacity of the eye for including and studying all the objects in a large area, particularly when moving at high speed, was soon found to be quite too small to properly utilize the time and opportunities available in the air. Moreover, the constant watching of the sky for the "Hun in the sun" distracted the observer time and time again from attention to the earth below. Very early in the war, therefore, men's minds turned to photography. The all-seeing and recording eye of the camera took the place of the observer in every kind of work except artillery fire control and similar problems which require immediate communication between plane and earth.

The volume of work done by the photographic sections of the military air service steadily increased until toward the end of the war it became truly enormous. The aerial negatives made per month in the British service alone mounted into the scores of thousands, and the prints distributed in the same period numbered in the neighborhood of a million. The task of interpreting aerial photographs became a highly specialized study. An entirely new activity —that of making photographic mosaic maps and of maintaining them correct from day to day—usurped first place among topographic problems. By the close of the war

2

scarcely a single military operation was undertaken without the preliminary of aerial photographic information. Photography was depended on to discover the objectives for artillery and bombing, and to record the results of the subsequent "shoots" and bomb explosions. The exact configurations of front, second, third line and communicating trenches, the machine gun and mortar positions, the "pill boxes," the organized shell holes, the listening posts, and the barbed wire, were all revealed, studied and attacked entirely on the evidence of the airplane camera. Toward the end of the war important troop movements were possible only under the cover of darkness, while the development of high intensity flashlights threatened to expose even these to pitiless publicity.

Limitations to Airplane Photography Set by War Conditions.—The ability of the pilot to take the modern high-powered plane over any chosen point at any desired altitude in almost any condition of wind or weather gives to the plane an essential advantage over the photographic kites and balloons of pre-war days. There are, however, certain disadvantages in the use of the plane which must be overcome in the design of the photographic apparatus and in the method of its use. Some few of these disadvantages are inherent in the plane itself; for instance, the necessity for high speed in order to remain in the air, and the vibration due to the constantly running engine. Others are peculiar to war conditions, and their elimination in planes for peace-time photography will give great opportunities for the development of aerial photography as a science.

Chief among the war-time limitations is that of economy of space and weight. A war plane must carry a certain equipment of guns, radio-telegraphic apparatus and other instruments, all of which must be readily accessible. Many planes have duplicate controls in the rear cockpit to enable the observer to bring the plane to earth in case of accident to the pilot. Armament and controls demand space which must be subtracted from quarters already cramped, so that in most designs of planes the photographic outfit must be accommodated in locations and spaces wretchedly inadequate for it. Economy in weight is pushed to the last extreme, for every ounce saved means increased ceiling and radius of action, a greater bombing load, more ammunition, or fuel for a longer flight. Hence comes the constant pressure to limit the weight of photographic and other apparatus, even though the tasks required of the camera constantly call for larger rather than smaller equipment.

To another military necessity is due in great measure the forced development of aerial photographic apparatus in the direction of automatic operation. The practice of entrusting the actual taking of the pictures to observers with no photographic knowledge, whose function was merely to "press the button" at the proper time, necessitated cameras as simple in operation as possible. The multiplicity of tasks assigned to the observer, and in particular the ever vital one of watching for enemy aircraft, made the development of largely or wholly automatic cameras the war-time ideal of all aerial photographic services. Whether the freeing of the observer from other tasks will relegate the necessarily complex and expensive automatic camera to strictly military use remains to be seen.

CHAPTER II

THE AIRPLANE CONSIDERED AS A CAMERA PLATFORM

An essential part of the equipment of either the aerial photographer or the designer of aerial photographic apparatus is a working knowledge of the principles and construction of the airplane, and considerable actual experience in the air.

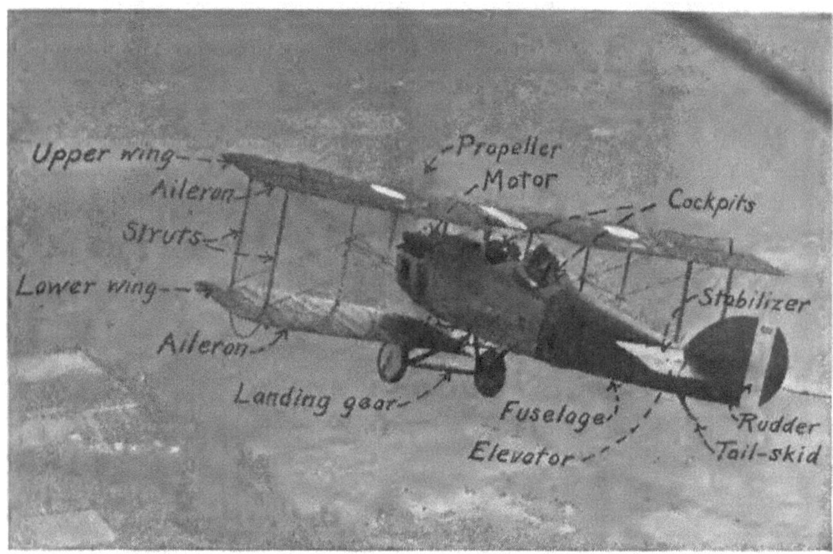

Fig. 1.—The elements of the plane.

Conditions and requirements in the flying plane are far different from those of the shop bench or photographic studio. As a preliminary to undertaking any work on airplane instruments a good text-book on the principles of flight should be studied. Such general ideas as are necessary for understanding the purly photographic problems are, however, outlined in the next paragraphs.

Construction of the Airplane.—The modern airplane (Fig. 1) consists of one or more *planes*, much longer across than in the direction of flight (*aspect ratio*). These are inclined slightly upward toward the direction of travel, and their rapid motion through the air, due to the pull of the *propeller* driven by the *motor*, causes them to rise from the earth, carrying the *fuselage* or body of the airplane. In the fuselage are carried the pilot, observer, and any other load. Wheels below the fuselage forming the *under-carriage* or *landing gear* serve to support the body when running along the ground in taking off or landing. The pilot, sitting in one of the *cockpits*, has in front of him the *controls*, by means of which the motion of the plane is guided (Figs. 2 and 3). These consist of the engine controls—the *contacts* for the ignition, the *throttle*, the oil and gasoline supply, air pressure, etc., and the steering controls—the *rudder bar*, the *stick* and the *stabilizer control*. The rudder bar, operated by the feet, controls both the *rudder* of the plane, which turns the plane to right or left in the air, and the *tail skid*, for steering on the ground. The *stick* is a vertical column in front of the pilot which, when pushed forward or back, depresses or raises the *elevator* and makes the machine dive or climb. Thrown to either side it operates the *ailerons* or wing tips, which cause the plane to roll about its fore and aft axis. The stabilizer control is usually a wheel at the side of the cockpit, whose turning varies the angle of incidence of the small stabilizing plane in front of the elevator, to correct the balance of the plane under different conditions of loading.

The *fuselage* consists usually of a light hollow framework of spruce or ash, divided into a series of bays or compartments by upright members, connecting the *longerons*, which are the four corner members, running fore and aft, of the plane. Diagonally across the sides and faces of these bays

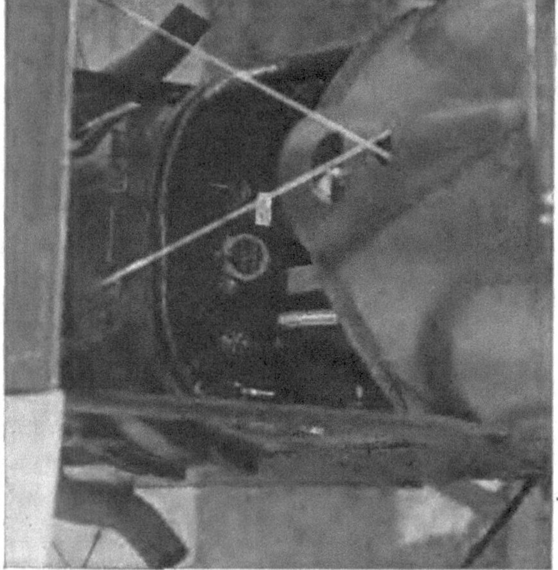

FIG. 2.—Forward cockpit of DeHaviland 4, showing instrument board.

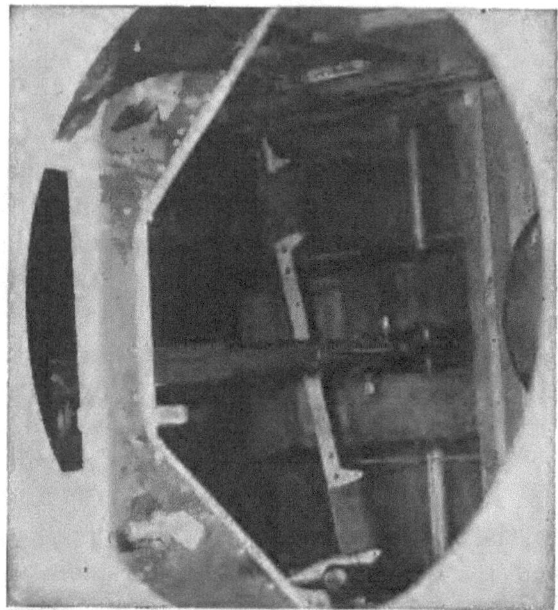

FIG. 3.—Rear cockpit of DeHaviland 4, showing rear "stick" and rudder bar.

are stretched taut piano wires, and the whole structure is covered with canvas or linen frabic. Cross-wires and fabric are omitted from the top of one or more bays to permit their being used as cockpits for pilot and observer. In later designs of planes the wire and fabric construction has been superseded by ply-wood veneer, thereby strengthening the fuselage so that many of the diagonal bracing wires on the inside

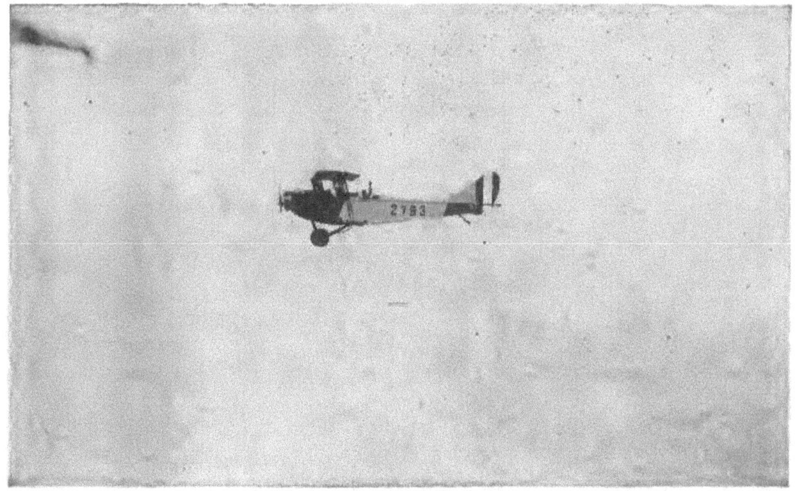

Fig. 4.—Biplane in flight.

are dispensed with. This greatly increases the accessibility of the spaces in which cameras and other apparatus must be carried.

The fuselage differs greatly in cross-section shape and in roominess according to the type of engine. In the majority of English and American planes, with their vertical cylinder or V type engines, the fuselage is narrow and rectangular in cross-section. In many French planes, radial or rotary engines are used and the fuselage is correspondingly almost circular, and so is much more spacious than the English and American planes of similar power. The shape and size of the plane body has an important bearing on the question of camera installation.

Types of Planes.—The most common type of plane is the *biplane* (Fig. 4), with its two planes, connected by struts and wires, set not directly over each other, but *staggered*, usually with the upper plane leading. *Monoplanes* were in favor in the early days of aviation, and *triplanes* have been

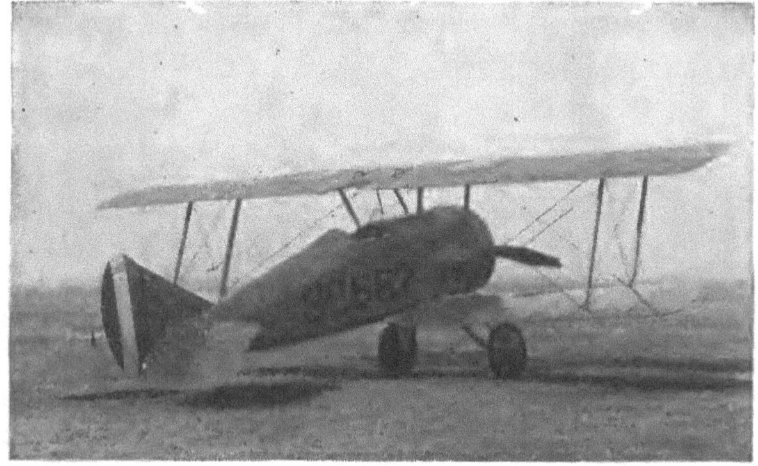

Fig. 5.—A single-seater.

used to some extent. According to the position of the propeller planes are classified as *tractors* or *pushers*, tractors being at present the more common form. Planes are further classified as *single-seaters* (Fig. 5), *two-seaters*, and *three-seaters*. These motor and passenger methods of classification are now proving inadequate with the rapid development of planes carrying two, three, and even more motors, divided between pusher and tractor operation, and carrying increasingly large numbers of passengers. Aside from structure, planes may be further classified according to their uses, as *scout, combat, reconnaissance, bombing,* etc. Planes equipped with floats or pontoons for alighting on the water are called *seaplanes* (Fig. 182), and those in which the fuselage is boat-shaped, to permit of floating directly on the water, are *flying boats* (Fig. 183).

The Plane in the Air.—The first flight of the photographic observer or of the instrument expert who is to work upon airplane instruments is very profitably made as a "joy ride," to familiarize him with conditions in the air. His experience will be somewhat as follows:

The plane is brought out of the hangar, carefully gone over by the mechanics, and the engine "warmed up." The pilot minutely inspects all parts of the "ship," then climbs up into the front cockpit. He wears helmet and goggles, and if the weather is cold or if he expects to fly high, a heavy wool-lined coat or suit, with thick gloves and moccasins, or an electrically heated suit. The passenger, likewise attired, climbs into the rear cockpit and straps himself into the seat. He finds himself sitting rather low down, with the sides of the cockpit nearly on a level with his eyes. To either side of his knees and feet are taut wires leading from the controls to the elevator, stabilizer, tail skid and rudder. If the machine is dual control, the stick is between his knees, the rudder bar before his feet. None of these must he let his body touch, so in the ordinary two-seater his quarters are badly cramped.

At the word "contact" the mechanics swing the propeller, and, sometimes only after several trials, the motor starts, with a roar and a rush of wind in the passenger's face. After a moment's slow running it is speeded up, the intermittent roar becomes a continuous note, the plane shakes and strains, while the mechanics hold down the tail to prevent a premature take-off. When the engine is properly warmed up it is throttled to a low speed, the chocks under the wheels are removed, the mechanics hold one end of the lower wing so that the plane swings around toward the field. It then "taxis" out to a favorable position facing into the wind with a clear stretch of field before it. After a careful look around to see that no other planes are landing, taking off, or in the air near by, the pilot opens out the engine, the roar increases its pitch, the plane moves slowly along the ground, gathers speed and rises smoothly into the air. Near the ground the air is apt to be "bumpy," the plane may drop or rise abruptly, or tilt to either side. The pilot instantly corrects these deviations, and the plane continues to climb until steadier air is reached.

At first the passenger's chief impressions are apt to be the deafening noise of the motor, the heavy vibration, the terrific wind in his face. If he raises his hand above the edge of the cockpit he realizes the magnitude of wind resistance at the speed of the plane, and hence the importance of the *stream-line section* of all struts and projecting parts.

When he reaches the desired altitude the pilot levels off the plane and ceases to climb. Now his task is to maintain the plane on an even keel by means of the controls, correcting as soon as he notes it, any tendency to "pitch," to "roll" or to "yaw" off the course. The resultant path is one which approximates to level straight flying to a degree conditioned by the steadiness of the air and the skill of the pilot. If he is not skilful or quick in his reactions he may keep the plane on its level course only by alternately climbing and gliding, by flying with first one wing down and then the other, by pointing to the right and then to the left. The skilled photographic pilot will hold a plane level in both directions to within a few degrees, but he will do this easily only if the plane is properly balanced. If the load on the plane is such as to move the center of gravity too far forward with respect to the *center of lift* the plane will be *nose-heavy*, if the load is too far back it will be *tail-heavy*. Either of these conditions can be corrected, at some cost in efficiency, by changing the inclination of the stabilizer. When the plane reaches high altitudes in rare air, where it can go no further, it is said to have reached its *ceiling*. It here travels level only by pointing its wings upward in the climbing position, so that the fuselage is no longer parallel to the direction of flight. An understanding of these pecularities of the plane in flight is of prime importance in photographic map making, where the camera should be accurately vertical at all times.

The direction and velocity of the wind must be carefully considered by the pilot in making any predetermined course or objective. The progress of the plane due to the pull of the propeller is primarily with reference to the air. If this is in motion the plane's *ground speed* and direction will be altered accordingly. In flying with or against the wind the ground speed is the sum or difference, respectively, of the plane's *air speed* (determined by an air speed indicator) and the speed of the wind. If the predetermined course lies more or less across the wind the plane must be pointed into the wind, in which case its travel, with respect to the earth, is not in the line of its fore and aft axis. The effect of "crabbing," as it is called, on photographic calculations is discussed later (Figs. 136 and 138).

When the plane has reached the end of its straight course and starts to turn, its level position is for the moment entirely given up in the operation of *banking* (Fig. 6). Just as the tracks on the curve of a railroad are raised on the outer side to oppose the tendency of the train to slip outward, so the plane must be tilted, by means of the ailerons, toward the inside of the turn. A point to be clearly kept in mind about a bank is that if correctly made a plumb line inside the fuselage will continue to hang

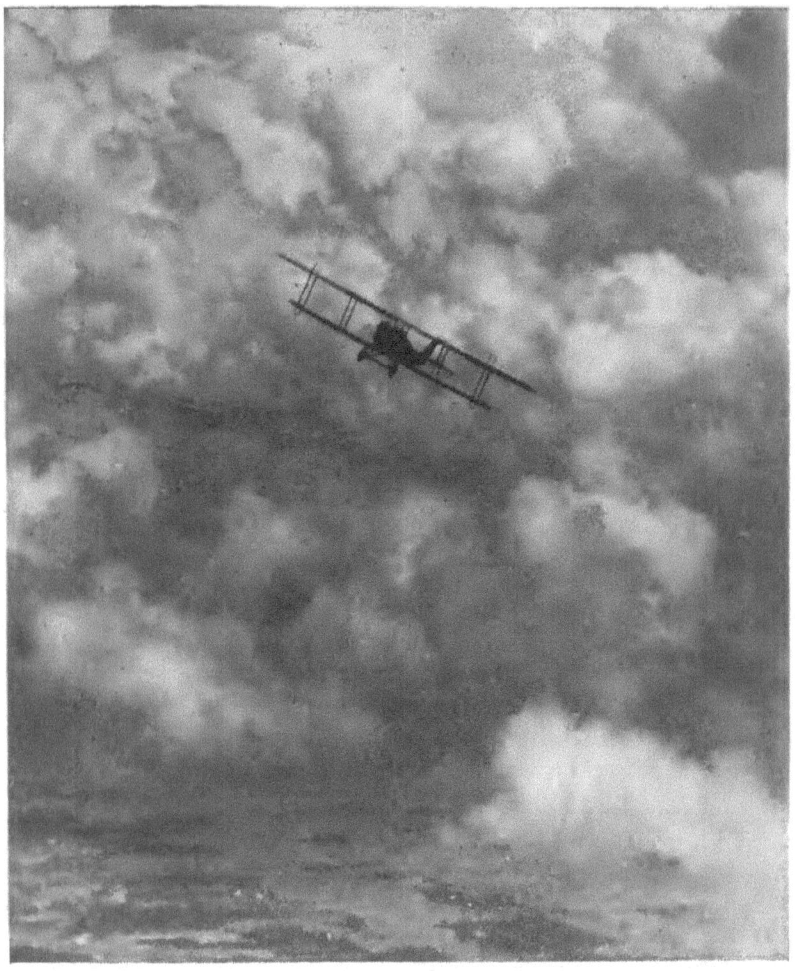

Fig. 6.—Banking.

vertical *with respect to the floor of the plane, and not with respect to the earth,* for the force acting on it is the combination of gravity and the acceleration outward due to the turn. Only some form of gyroscopically controlled pointer, keeping its direction in space, will indicate the inclination of the plane with respect to the true vertical. If the banking is insufficient the plane will *side slip* outward or *skid;* if too great, it will side slip inward.

As part of the "joy ride" the pilot may do a few "stunts," such as a "stall," a "loop," a " tail spin," or an "Immelman." From the photographic standpoint these are of interest in so far as they bear on the question of holding the camera in place in the plane. The thing to be noted here is that (particularly in the loop), if these maneuvers are properly performed, there is little tendency toward relative motion between plane and apparatus. In a perfect loop it would, for instance, be unnecessary, due to the centrifugal force outward, for the observer to strap himself in. It is, however, unwise to place implicit confidence in the perfection of the pilot's aerial gymnastics. No apparatus should be left entirely free, although, for the reason given, comparatively light fastenings are usually sufficient.

When nearing the landing field the pilot will throttle down the engine and commence to glide. If he is at a considerable altitude he may come down a large part of the distance in a rapid spiral. As the earth is approached the air pressure increases rapidly, and the passenger, if correctly instructed, will open his mouth and swallow frequently to equalize the air pressure on his ear drums. Just before the ground is reached the plane is leveled off, it loses speed, and, if the landing is perfect, touches and runs along the ground without bouncing or bumping. Frequently, however, the impact of the tail is sufficiently hard to cause it to bump badly, with a consequent considerable danger to apparatus of any weight or delicacy. This is especially apt to occur in hastily chosen and poorly leveled fields such as must often be utilized in war or in cross-country flying.

Appearance of the Earth from the Plane.—The view from the ordinary two-seater is greatly restricted by the engine in front and by the planes to either side and below (Figs. 7, 8, and 9). By craning his neck over the side, or by looking down through an opening in the floor, the passenger has an opportunity to learn the general appearance of the subject he is later to devote his attention to photographing. Perhaps the most striking impression he receives will be that of the *flatness* of the earth, both in the sense of absence of relief and in the sense of absence of extremes of light and shade. The absence of relief is due to the fact that at ordinary flying heights the elevations of natural objects are too small for the natural separation of the eyes to give any stereoscopic effect. The absence of extremes of light and shade is in part due to the fact that the natural surfaces of earth, grass and forest present no great range of brightness; in part to the small relative areas of the parts in shadow; in considerable part to the layer of atmospheric haze which lies as an illuminated veil between the observer and the earth at altitudes of 2000 meters and over (Figs. 10 and 11). Due to the combination of these factors the earth below presents the appearance of a delicate pastel.

As the gaze is directed away from the territory directly below, the thickness of atmosphere to be pierced rapidly increases, until toward the horizon (which lies level with the observer here as on the ground) all detail is apt to be obliterated to such an extent that only on very clear days can the horizon itself be definitely found or be distinguished from low lying haze or clouds (Fig. 4).

Airplane Instruments.—Mounted on boards in front of the pilot and observer are various instruments to indicate

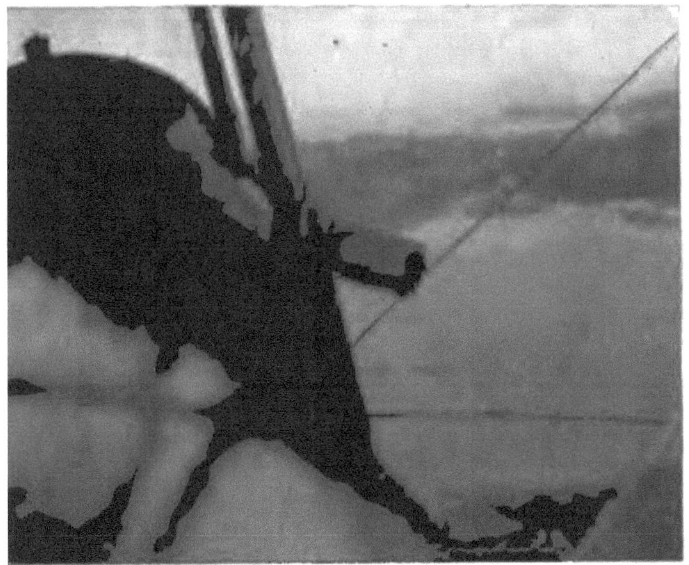

Fig. 7.—The view ahead.

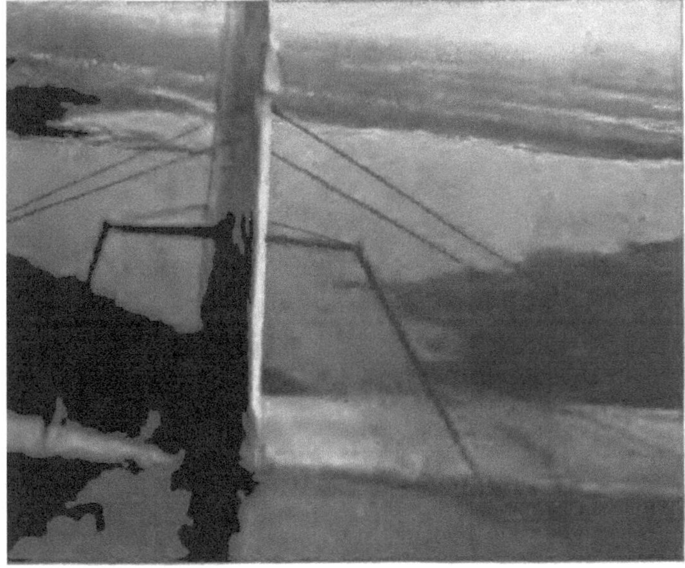

Fig. 8.—The view astern.

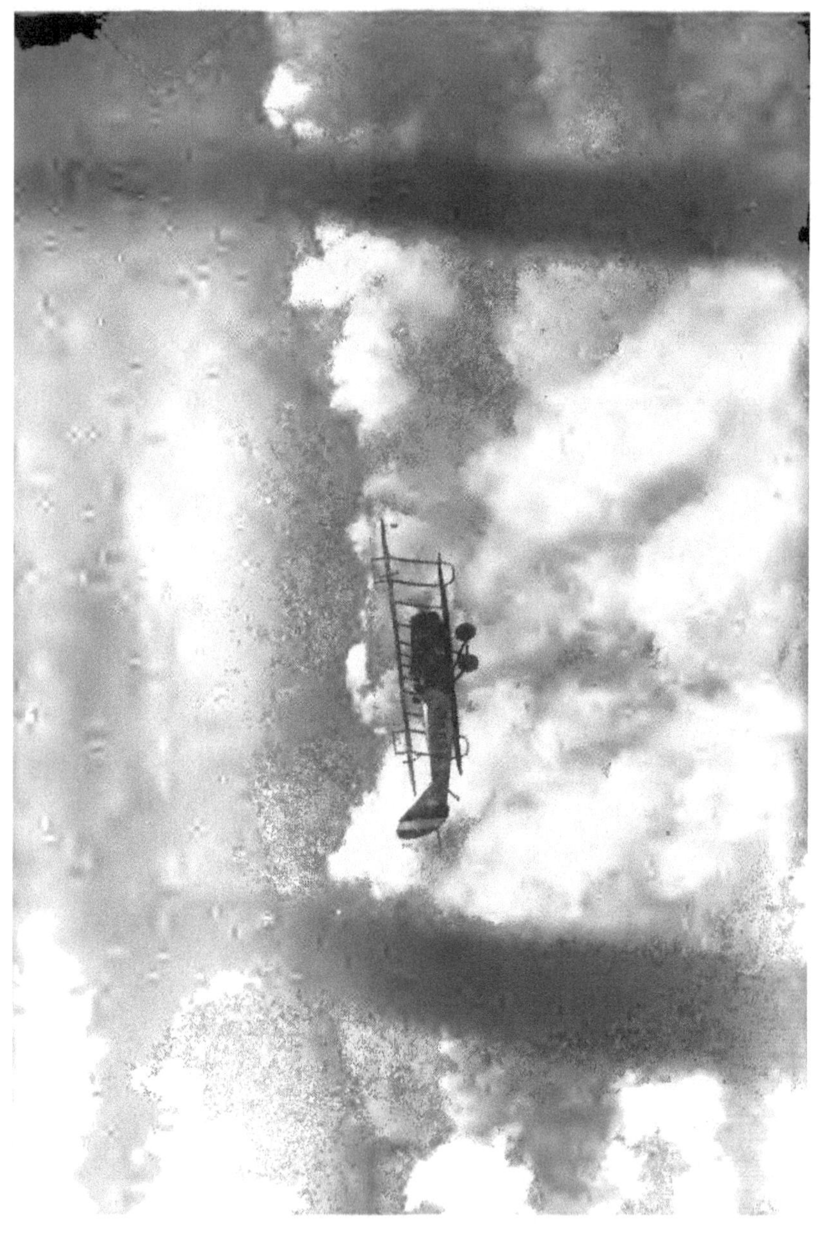

the performance of engine and plane (Fig. 2). Those of interest to the photographic observer are the *compass*, the *altimeter*, the *air speed indicator*, the *inclinometers*.

The *compass* is usually a special airplane compass, with its "card" immersed in a damping liquid. Like most of the direction indicating instruments on a plane its indications

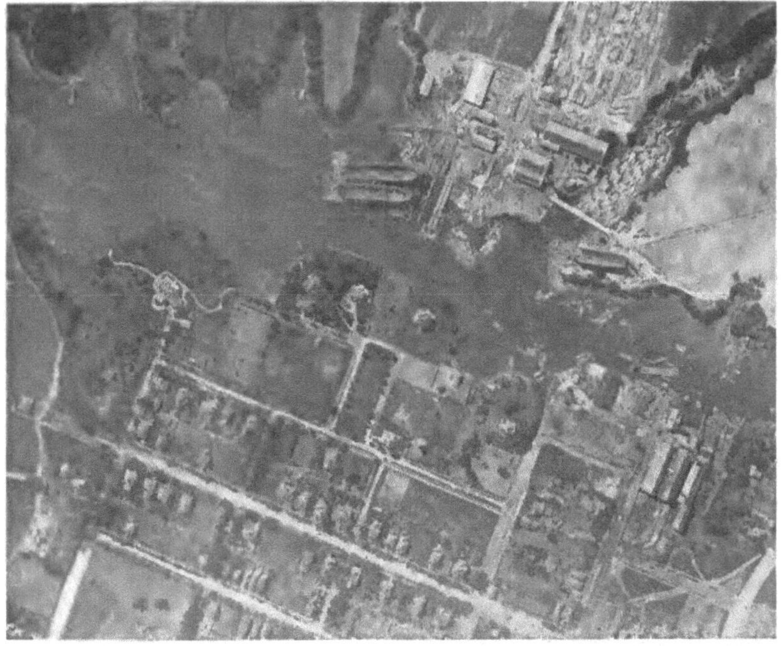

FIG. 10.—Appearance of the earth from a low altitude—3000 feet or less.

are only of significance when the plane is pursuing a steady course. On turns or rapid changes of direction of any sort perturbations prevent accurate reading.

The *altimeter* is of the common aneroid barometer type. On American instruments it is usually graduated to read in 100-foot steps. While somewhat sluggish, it is quite satisfactory for all ordinary determinations of altitude in photo-

3

graphic work. Were primary map making to be undertaken, where the scale was determinable only from the altitude and focal length of the lens, the ordinary altimeter is hardly accurate enough.

The *air speed indicator* consists of a combination of Venturi and Pitot tubes, producing a difference of pressure

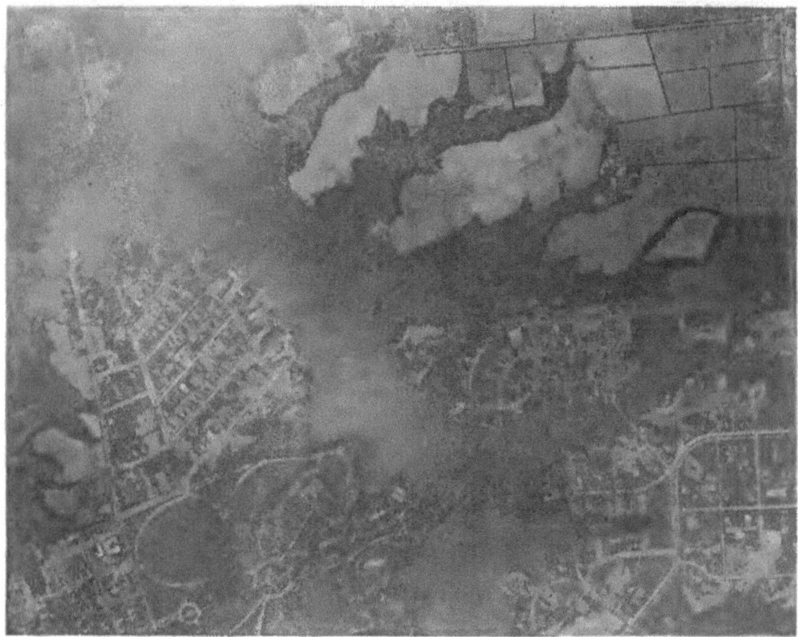

Fig. 11.—Appearance of the earth from a high altitude—10,000 feet or more.

when in motion through the air which is measured on a scale calibrated in air speed. This instrument is important for determining, in combination with wind speed, the *ground speed* of the plane, on the basis of which is calculated the interval between exposures to secure overlapping photographs. Its accuracy is well above that necessary for the purpose.

Inclinometers for showing the lateral and fore and aft angle of the plane with the horizontal, are occasionally used, and have also been incorporated in cameras. The important point to remember about these instruments is that they are controlled not alone by gravity but as well by the acceleration of the plane in any direction. They consequently indicate correctly only when the plane is flying straight. On a bank the lateral indicator continues to indicate "vertical" if the bank is properly calculated for the turn.

II
THE AIRPLANE CAMERA

CHAPTER III

THE CAMERA—GENERAL CONSIDERATIONS

Chief Uses of an Airplane Camera.—The kinds of camera suitable for airplane use and the manner in which they must differ from cameras for use on the ground are determined by consideration of the nature of the work they must do. Four kinds of pictures constitute the ordinary demands upon the aerial photographer. These are single objectives or *pin points*, *mosaic maps* of strips of territory or large areas, *oblique views*, and *stereoscopic views*. Each of these presents its own peculiar problems influencing camera design.

Pinpoints consist of such objects as gun emplacements, railway stations, ammunition dumps, and other objects of which photographs of considerable magnification are desired for study. Here the instrumental requirements are *sufficient focal length of lens* to secure an image of adequate size; *means for pointing* the camera accurately; enough *shutter speed* to counterbalance the speed of the plane; sufficiently *wide lens aperture* to give adequate exposure with the shutter speed required; *means of supporting* the camera to protect it from the vibration of the plane.

Mosaic maps are built up from a large number of photographs of adjacent areas. In addition to the above requirements, mosaic maps demand *lenses free from distortion* and covering as large a plate as possible, in order to keep to a minimum the number of pictures needed to cover a given area; means for keeping the camera *accurately vertical*, and means for *changing the plates or films* and resetting the shutter rapidly enough to avoid gaps between successive pictures. At lowaltitudes and high ground speedsthe interval between exposures becomes a matter of only a few seconds.

Oblique views are made at angles of from 12 to 35 degrees from the horizontal, usually from comparatively low altitudes. They have been found to be particularly suitable for the use of men who have no training in photographic interpretation, being more like the pictures with which the men are familiar. Distributed among the infantry before an attack, they have proved indispensable aids to the proper knowledge of the ground to be covered. The additional requirement here is for *high shutter speed* to elminate the effect of the relatively very rapid movement of the foreground.

Stereoscopic views are among the most useful of all airplane pictures. They are made from successive exposures, the separation of the points of view being obtained not by two lenses at the distance of the eyes apart, but by the motion of the plane. For this purpose the views should overlap by at least 60 per cent; this,therefore,requires a very *short interval between exposures.* For stereo-oblique views this may mean that they are taken at intervals as short as one or two seconds.

Chief Differences between Ground and Air Cameras.—Certain definite differences are thus seen to stand out between airplane cameras and the ordinary kind. It is essential that the apparatus for use in the air shall have high lens and shutter speed, means for rapid changing of plates, and anti-vibration suspension. Without these features a camera is of little use for aerial work. These requirements lead inevitably to greater complexity of design. One simplification over ground cameras, however, is brought about by the fact that all exposures are made on objects beyond the practical infinity point of the lens; consequently, all cameras are fixed focus. This fixed focus feature is a positive advantage in construction, since it permits of the simple rigid box form, desirable and necessary to withstand the strains due to the weight of the lens and the stresses from the plane. But with the abandonment of all provision for focussing in the air must go special care that the material used in constructing the camera body is as little subject as possible to expansion and contraction with temperature, since there is often a drop of 30 to 40 degrees Centigrade from ground to upper air. The effect of change of temperature on focus will be treated in the discussion of lenses.

In addition to these differences, we must keep in mind certain requirements which are conditioned by the nature and place of aerial navigation. Thus all mechanical devices which will fail to function at the low temperatures and pressures met at high altitudes are entirely unsuitable. Experience has shown, too, that we must avoid all mechanism depending primarily on *springs* and on the action of *gravity*. Vibration, and the motion of the plane in all three dimensions, conspire to render mechanical motions unreliable when actuated by these agencies. All plate changing, shutter setting, and exposing operations should be as nearly as possible positively controlled motions. Because of the cold of the upper air all knobs, levers and catches must be made extra large and easy to handle with heavy gloves. Circular knurled heads to such parts as shutter setting movements are to be avoided in favor of bat-wing keys or levers. Grooves for the reception of magazines must be as large and smooth as possible, and guides to facilitate the magazines' introduction should be provided (Fig. 50). No releases or adjustments which depend upon hearing or upon a delicate sense of touch are feasible in airplane apparatus. Wherever possible, large *visible* indicators of the stage of the cycle of operations should be provided. Loose parts are to be shunned, as they are invariably lost in service. Complete operating instructions should be placed on the apparatus wherever possible, to minimize the confusion due to changing and uninstructed personnel.

The Elements of the Airplane Camera.—Disregarding its means of suspension, the airplane camera proper consists essentially of *lens, camera body, shutter,* and *plate or film holding and changing box.*

In certain of the aerial cameras developed early in the war all of these elements were built together in a common enclosure. Later it was generally recognized that a unit system of interchangeable parts is preferable. In the case of the lens there arose various requirements for focal length, from 25 to 120 centimeters, according to the work to be done. Rather than use an entirely different camera for each different kind of work, it is better to have lenses of various focal lengths, mounted in tubes or cones, all built to attach to the same camera body. In the case of the shutter it is desirable to be able to repair or calibrate periodically. By making the shutter a removable unit, the provision of a few spares does away with the need for putting the whole camera out of commission. Similar considerations hold with reference to other parts.

A further material advantage that comes from making airplane cameras in sections is the greater ease with which they are inserted in the plane, usually through the openings between diagonal cross-wires. It is in fact only by virtue of this possibility of breaking up into small elements that some of the larger cameras could be inserted in the common types of reconnaissance plane. Illustrations of the building up of cameras from separate removable elements are given in the detailed discussion of the individual types.

Types of Airplane Cameras.—During the course of

the war airplane cameras have been classified on various bases, in different services. In the French service, where the de Maria type of camera was standardized early in the war, the usual classification was based on focal length; thus the standard cameras were spoken of as the 26, the 50 and the 120 (centimeter). A further distinction was then made according to the size of plate, this being originally 13×18 centimeters for the 26 centimeter, and 18×24 centimeters for the larger cameras. In the English service the 4×5 inch plate was used almost exclusively, and their various types of cameras were known by serial letters—C, E, L, etc. Both these modes of classification became inadequate with the ultimate agreement to standardize on the 18×24 centimeter size for all plates, and to carry lenses of all focal lengths in interchangeable elements.

For purposes of description and discussion, it is most convenient to classify cameras according to their method of operation and the sensitive material employed. On this basis we may distinguish among cameras using *plates* three kinds—*non-automatic cameras*, *semi-automatic cameras*, and *automatic cameras.* We may similarly discuss *film cameras*, but having treated the plate cameras comprehensively, it will be found that the discussion of all types of film camera can be handled most conveniently by studying the differences in construction and operation introduced by the characteristics of film as compared to plates.

CHAPTER IV

LENSES FOR AERIAL PHOTOGRAPHY

General Considerations.—The design and selection of lenses for aerial photography present on the whole no problems not already encountered in photography of the more familiar sort. Indeed, the lens problem in the airplane camera is in some particulars more simple than in the ground camera. For instance, there is no demand for depth of focus—all objects photographed are well beyond the usually assumed "infinity focus" of 2000 times the lens diameter. Such strictly scientific problems of design as pertain to aerial photographic lenses are ones of degree rather than of kind. Larger aperture, greater covering power, smaller distortion, more exquisite definition—these always will be in demand, and each progressive improvement will be reflected in advances in the art of aerial photography. But many lens designs perfected before the war were admirably suited, without any change at all, for aerial cameras.

Of the utmost seriousness, however, with the Allies, was the problem of securing lenses of the desired types in sufficient numbers. The manufacture of the many varieties of optical glass essential to modern photographic lenses was almost exclusively a German industry, which had to be learned and inaugurated in Allied countries since 1914. In consequence of this entirely practical problem of quantity production without the glasses for which lens formulæ were at hand, some new lens designs were produced. Whether any of these possess merits which will lead them to be preferred over pre-war designs, when the latter can again be manufactured, remains to be seen.

While the glass problem was still unsolved, aerial cameras had to be equipped with whatever lenses could be secured by requisition from pre-war importation and manufacture, and later, with lenses designed to utilize those glasses whose manufacture had been mastered in the allied countries. It is important that the historical aspect of this matter be well understood by the student of aerial photographic methods, for the use of these odd-lot lenses reacted on the whole design of aerial cameras and on the methods of aerial photography, particularly in England and the United States. Almost without exception the available lenses were of short focus, considered from the aerial photographic standpoint; that is, they lay between eight and twelve inches. This set a limit to the size of the airplane camera, quite irrespective of the demands made by the nature of the photographic problem. Lenses of these focal lengths produced images which, for the usual heights of flying, were generally considered too small, and which were, therefore, almost always subsequently enlarged. Such was the English practice, which was followed in the training of aerial photographers in America, where exactly similar conditions held at the start with respect to available lenses. French glass and lens manufacturers did succeed in supplying lenses of longer focus (50 centimeters), in numbers sufficient for their own service, although never with any certainty for their allies. The French, therefore, almost from the start, built their cameras with lenses of long focus, and made contact prints from their negatives.

Practices adopted under pressure of an emergency to meet temporary practical limitations often come to dominate the whole situation. This is particularly true of aerial photography in the British and American services. The small apparatus built around the stop-gap short focus lenses fixed the plane designer's idea of an airplane camera, and the space it should occupy. This was directly reflected in the designs of the English planes, and the American planes copied after them. Meanwhile the American photographic service in France associated itself with the French service, adopting its methods and apparatus, and using French planes whose designs were not being followed in American construction. The task of harmonizing the photographic practice as taught in America, following English lines, with French practice as followed in the theater of war, and of adapting planes built on English designs so that they could carry French apparatus, was a formidable one, not likely to be soon forgotten by any who had a part in it.

Photographic Lens Characteristics.—Whole volumes have been written on the photographic lens, and on the optical science utilized and indeed brought into being by its problems. Such works should be consulted by those who intend to make a serious study of the design of lenses for aerial use. No more can be attempted, no more indeed is relevant here, than an outline review of the chief characteristics and errors of photographic lenses, considering them with special reference to aerial needs.

The modern photographic lens is, broadly speaking, a development of the simple convex or converging lens. Its function is the same: to form a real image of objects placed before it. But the difference in performance between the simple lens and the modern photographic objective is enormous. The simple lens forms a clear image only close to its axis, for light of a single color, and as long as its aperture is kept quite small as compared to the distance at which the image is formed. The photographic lens, on the other hand, is called upon to produce a clear image with light of a wide range of spectral composition, sharply defined over a flat surface of large area, and it must do this with an aperture that is large in comparison with the focal length, whereby the amount of light falling on the image surface shall be a maximum. This ideal is approximated to a really extraordinary degree by the scientific combination and arrangement of lens elements made from special kinds of glass in the best photographic lenses of the anastigmat type. The result is of necessity a set of compromises, whereby the

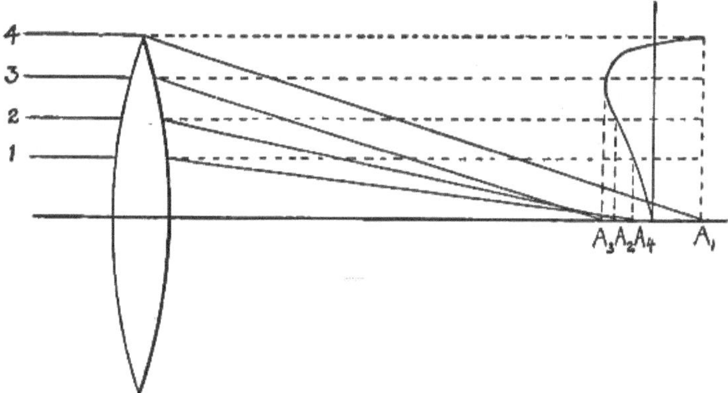

FIG. 12.—Diagrammatic representation of spherical aberration.

outstanding errors are reduced to a size judged permissible in view of the work the lens is to do. These errors or *aberrations* are briefly reviewed below, in order that the reader may readily grasp the terms in which the performance and tolerances in aerial lenses are described.

Spherical Aberration and Coma.—Suppose we focus on a screen, by means of a simple convex lens the image of a distant point of light. Suppose for simplicity that this image is located on the axis of the lens and that light of only one color is used, such as yellow. It will be found that the smallest image that can be obtained is not a point, but a small disc. This is due to the fact that the rays of light passing through the outer portions of the lens are bent more than those passing through the lens in the region near the center. This effect is shown in Fig. 12 by the usual mode of representing it graphically. Here the figures 1, 2, 3, 4, represent distances from the axis of the lens, and the letters A_1, A_2, A_3, A_4, the points of convergence of the rays from 1, 2, 3, 4, etc. These distances projected upward on to the produced lens points form a curve which shows at a glance the extent and direction of the error due to each part of the lens. This information is of value where the lens is fitted with an adjustable diafram. With some types of correction sharper definition may be obtained by reducing the aperture. With others, however, diaframing impairs definition, by destroying the balance between under and over correction which averages to make a good image. In aerial lenses it is not customary to use diaframs, as all the light possible is desired. Consequently the reduction of spherical aberration must be accomplished by proper choice of lens elements and their arrangement.

Off the axis of the lens the image of a point source takes on an irregular shape, due to oblique spherical aberration or *coma*.

Chromatic Aberration.—Because of the inherent properties of the glass of which it is made, a simple collective lens does not behave in the same way with respect to light of different colors. If one attempts, with such a lens, to focus upon a screen the image of a distant white light, it will be found that the blue rays will not focus at the same point as the red rays, but will come together nearer the lens. Modern photographic lenses are compounded of two or more kinds of glass in such a way as to largely eliminate this defect, the presence of which is detrimental to good definition. Such lenses are called achromatic, and the property of a lens by virtue of which this defect is eliminated is called its *chromatic correction.*

Chromatic correction is never perfect, but two colors of the spectrum can be brought to a focus in the same plane, and to a certain extent the departure of other colors from this plane can be controlled. Off the axis of the lens outstanding chromatic aberration results in a difference in the size of images of different colors, known as *lateral chromatism.*

Like spherical aberration, chromatic aberration is a contributing factor to the size of the image of a point source, which determines the defining power of a lens. It is, however, an error whose effect is to some extent dependent on the kind of sensitive plate used. Two lenses may give images of the same size (in so far as it is governed by chromatic aberration), if a plate of narrow spectral sensitiveness is used, while giving images of different size on panchromatic plates of more extended color sensibility. The choice of the region of the spectrum for which chromatic correction is to be made is thus governed by the color of the photographically effective light. While in ordinary photography the blue of the spectrum is most important, in aerial work where color filters are habitually used with isochromatic plates the green is most important, and color correction centered about this region constitutes a real difference of design peculiar to aerial lenses. Similarly the general use of deep orange or red filters with red sensitive plates, for heavy mist penetration, would call for a shift of correction to that part of the spectrum.

Astigmatism and Covering Power.—Suppose the lens forms at some point off its axis an image of a cross. Suppose one of the elements of the cross to be on a radius from the center of the field, the other element parallel to a tangent.

4

The rays forming the images of these two elements of the cross are subject to somewhat different treatment in their passage through the lens. The curvature of the lens surfaces is on the whole greater with respect to the rays from the radial element than to those from the tangential element. They are therefore refracted more strongly and come to a focus nearer the lens. The arms of the cross are consequently not all in focus at once. This error, termed *astigmatism,* is rather well shown in Fig. 15, where the images of the outlying concentric circles are sharp in the radial, but blurred in the tangential direction.

Astigmatism can be largely compensated for, and its character controlled. The most usual correction brings the two images in focus together both at the axis, and on a circle at some distance out. This second locus of coincidence may or may not be in the same plane as the first, depending on which disposition produces the best average correction. The mean between the two foci determines the focal plane of the lens, which is in general somewhat curved. The *covering power* of a lens is given by the size of the field which is sufficiently flat and free from astigmatism for the purpose for which the lens is used. This is largely determined by the astigmatism, but the other aberrations are also important.

Illumination.—The amount of light concentrated by the lens on each elementary area of the image determines its brightness or illumination. The ideal image would, of course, be equally bright over its whole area of good definition, and for lenses of narrow angle this is approximately true. But when it is desired to cover a wide angle the question of illumination becomes serious. The relationship between angle from the axis and illumination is that illumination is proportional to the fourth power of the cosine of the angle. This relationship is shown in the following table:

Angle	Image brightness
0°	100 per cent.
10°	94.1 per cent.
20°	78.0 per cent.
30°	56.2 per cent.
40°	34.4 per cent.
50°	17.1 per cent.

If the field of view is 60°, which corresponds to an 18×24 centimeter plate with a lens of 25 centimeter focus, the brightness is only 56 per cent., and the necessary exposure at the edge approximately 1.8 times that at the center. This effect is shown in Fig. 15. It is very noticeable if the exposure is so short as to place the outlying areas in the under-exposure period.

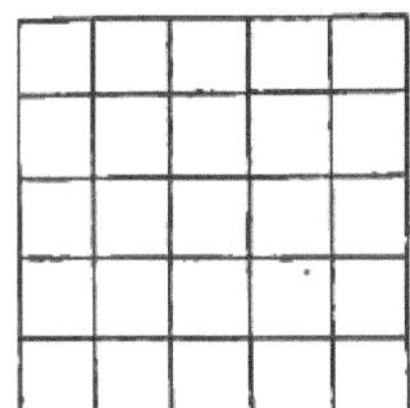
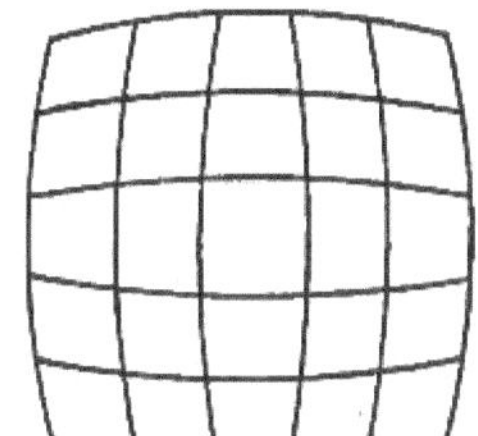
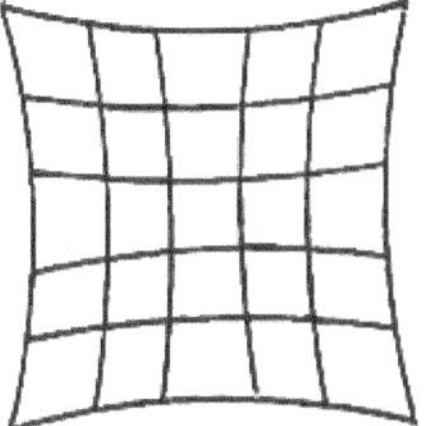

Fig. 13.—Barrel and pin-cushion distortion.

Distortion.—Sometimes a lens is relatively free from all the aberrations, mentioned above, so that it gives sharp, clear images on the plate, yet these images may not be exactly similar to the objects themselves as regards their geometrical proportions; in other words, the image will show distortion. Lens distortion assumes two typical forms, illustrated in Fig. 13, which shows the result of photographing a square net-work with lenses suffering in the one case from "barrel" distortion and in the other from "pin-cushion" distortion. In the first the corners are drawn in relative to the sides; in the latter case the sides are drawn in with respect to the corners. Either sort is a serious matter in precision photography, such as aerial photographic mapping aspires to become. It must be reduced to a minimum and its amount must be accurately known if negatives are to be measured for the precise location of photographed objects. In general symmetrical lenses give less distortion than the unsymmetrical (Fig. 14).

Lens Testing and Tolerances for Aerial Work.—Simple and rapid comparative tests of lenses may be made by

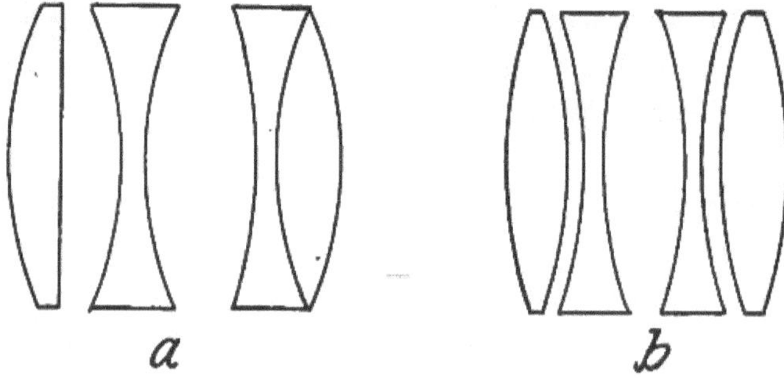

FIG. 14.—Arrangement of elements in two lenses suitable for aerial work: *a*, Zeiss Tessar; two simple and one cemented components (unsymmetrical); *b*, Hawk-eye Aerial; two positive elements of heavy barium crown, two negative of barium flint, uncemented (symmetrical).

photographing a *test chart*, consisting of a large flat surface on which are drawn various combinations of geometrical figures—lines, squares, circles, etc.—calculated to show up any failures of defining power. For testing aerial lenses the chart should be as large as possible, so that it may be photographed at a distance great enough for the performance of the lens to be truly representative of its behavior on an object at infinite distance. This means in practice a chart of 4 or 5 meters side, to be photographed at a distance 20 to 30 times the focal length of the lens.

A typical photograph of such a chart is shown in Fig. 15. It reveals at a glance the more conspicuous lens errors.

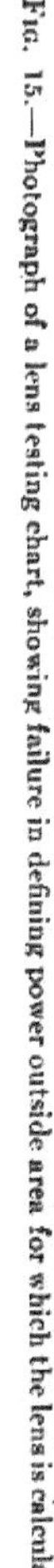

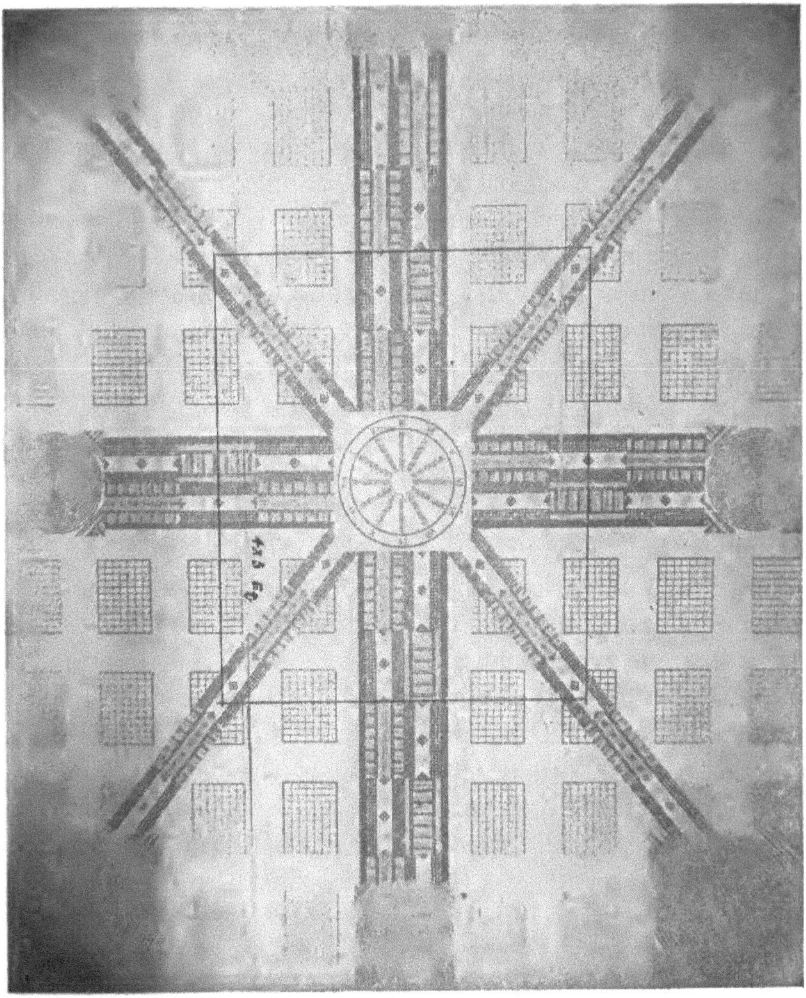

Fig. 15.—Photograph of a lens testing chart, showing failure in defining power outside area for which the lens is calculated.

At the sides and corners the concentric circles show the lens's astigmatism, by the clear definition of the lines radial to the center of the field and their blurring in the tangential direction. The falling off in illumination with increasing distance from the center is also exhibited; and the blurring of all detail outside the rectangle for which the lens was calculated shows that spherical, chromatic, and other aberrations have become prohibitively large.

But the only complete test of a lens is the quantitative measurement of errors made on an optical bench. A point source of light, which may at will be made of any color of the spectrum, is used as the object and its image formed by the lens in a position where it can be accurately measured for location, size, and shape by a microscope. A chart giving the results of such a test is shown in Fig. 16. In the upper left-hand corner is shown the position of the focus for the different colors of the spectrum. Below this is recorded the lateral chromatism at 21 degrees, in terms of the difference in focus for a red and a blue ray. Below this again comes the distortion, or shift of the image from its proper position, for various angles (plotted at the extreme right) from the lens axis. To the right of this is the image size, at each angle, and finally, to the right of the diagram, are plotted the distances of the two astigmatic foci from the focal plane, together with the mean of the two foci, which practically determines the shape of the field.

An important point to notice is that these data are uniformly plotted in terms of a lens of 100 millimeters focal length irrespective of the actual focal length of the lens measured. Thus this particular chart is for a 50 centimeter lens but would be plotted on the same scale for a 25 or a 100 centimeter lens. Underlying this practice is the assumption that all the characteristics of lenses of the same design and aperture are directly proportional to their focal length. If this were so, then a 50 centimeter lens would give double the size of image that a 25 centimeter does, and so on. As a matter of fact, test shows that the size of the image does not increase so rapidly as the focal length; so that while the image size for a 25 centimeter lens would be, say,

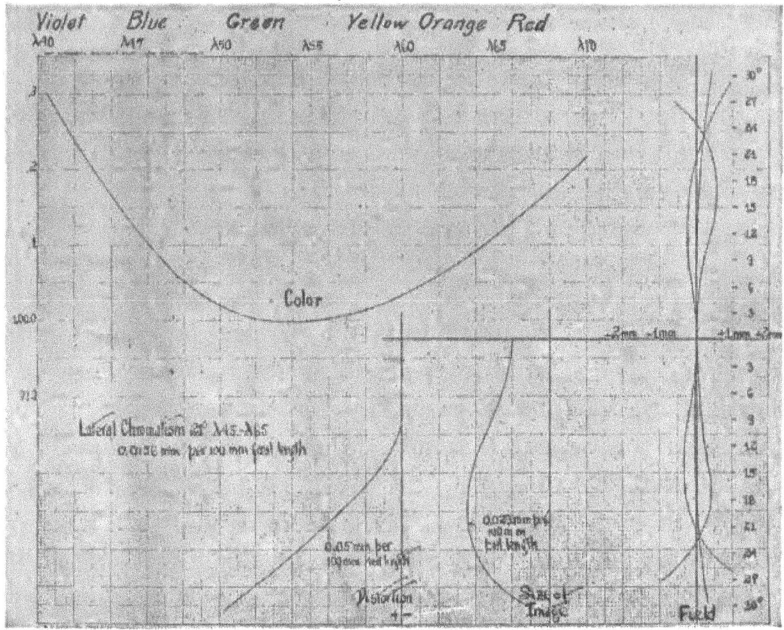

Fig. 16.—Chart recording measurements of lens characteristics.

.05 millimeters per 100 millimeters focal length, it will be only .03 or .04 millimeters per 100 millimeters focal length for a 50 centimeter lens. The actual size of a point image will therefore be greater, though not proportionately greater.

The chart presents tests on a good quality lens, and so gives a good idea of the permissible magnitude of the various errors. In many ways the most important figure is that for image size, including as it does the result of all the aberrations. In the example given, this varies from .075 to .15 mm. actual size. For the same type of lens of 25 centimeters focus this range will be from .05 to .10 mm. Since these are commonly used focal lengths, a good average figure for image size, commonly used in aerial photographic calculations, is 1/10 mm. In regard to astigmatic tolerances, the two astigmatic foci should not be separated at any point by more than 6 to 7 millimeters, and the mean of these should not deviate from the true flat field by more than ½ millimeter, in each case the figures being based on the conventional 100 millimeters focal length. Distortion should not be over .08 millimeter at 18° or .20 millimeter at 24° from the axis (per 100 millimeters focal length).

Lens Aperture.— In the simple lens the aperture is merely the diameter. In compound lenses the aperture is not the linear opening but the effective opening of an internal diafram. Photographically, however, aperture has come to have a more extensive meaning. While in the telescope the actual diameter of an objective is perhaps the most important figure, and in the microscope the focal length, in photography the really important feature is the amount of light or illumination. This is determined by lens opening and focal length together; specifically, by the ratio of the lens area to the focal length. The common system of representing photographic lens aperture is by the ratio of focal length to lens diameter, the lens being assumed to be circular. Thus F/5 (often written F.5) indicates that the diameter is one-fifth the focal length.

Two points are to be constantly borne in mind in connection with this system of representation. First, all lenses of the same aperture (as so represented) give the same illumination of the plate (except for differences due to loss of light by absorption and reflection in the lens system).

This follows simply from the fact that the illumination of the plate is directly proportional to the square of the lens diameter, and inversely as the square of the focal length. Secondly, the illumination of the plate is inversely as the square of the numerical part of the expression for aperture. That is, lenses of aperture F/4.5 and F/6 give images of relative brightness $\left(\frac{6}{4.5}\right)^2 = 1.78$.

What lens aperture, and therefore what image brightness, is feasible, is determined chiefly by the angular field that must be covered with any given excellence of definition. The largest aperture ordinarily used for work requiring good definition and flat field free from distortion is F/4.5. Anastigmatic lenses of this aperture cover an angle of 16° to 18° from the axis satisfactorily, which corresponds to an 18×24 centimeter plate with a lens of 50 centimeters focus. Lenses with aperture as large as F/3.5 were used to some extentin German hand cameras of 25centimeters focal length, with plates of 9×12 centimeters. English and American lenses of this latter focal length were commonly of aperture F/4.5, designed to cover a 4×5 inch plate.

As a general rule the greater the focal length the smaller the aperture—a relationship primarily due to the difficulty of securing optical glass in large pieces. Thus while 50 centimeter lenses of aperture F/4.5 are reasonably easy to manufacture, the practicable aperture for quantity production is F/6, and for 120 centimeter lenses, F/10. This means that a very great sacrifice of illumination must be faced to secure these greater focal lengths. As is to be expected from the state of the optical glass industry, the German lenses were of generally larger aperture for the same focal lengths than were those of the Allies. Besides the F/3.5 lenses already mentioned, their 50 centimeter lenses were commonly of aperture F/4.8, their 120 centimeter lenses of aperture F/7, or of about double the illuminating power of the French lenses of the same size.

Demands for large covering power also result in smaller aperture. The 26 centimeter lenses used on French hand cameras utilizing 13×18 centimeter plates were commonly of aperture F/6 or F/5.6. The lens of largest covering power decided on for use in the American service was of 12 inch focus, to be used with an 18×24 centimeter plate (extreme angle 26°); the largest satisfactory aperture for this lens is F/5.6.

Ordinarily the question of aperture is closely connected with that of diaframs, whereby the lens aperture may be reduced at will. Diaframs have been very little used in aerial photography. All the aperture that can be obtained and more is needed to secure adequate photographic action with the short exposures required under the conditions of rapid motion and vibration peculiar to the airplane. Any excess of light, over the minimum necessary to secure proper photographic action, is far better offset by increase of shutter speed or by introduction of a color filter. For this reason American aerial lenses were made without diaframs. In the German cameras, however, adjustable diaframs are provided (Fig. 43), controlled from the top of the camera by a rack and pinion. In the camera most used in the Italian service an adjustable diafram is provided, but this is occasioned by the employment of a between-the-lens shutter of fixed speed, so that the only way exposure can be regulated is by aperture variation, a method which has little to recommend it.

The Question of Focal Length.—In aerial photography the lens is invariably used at fixed, infinity, focus. Under these conditions the simple relationship holds that the size of the image is directly proportional to the focal length and inversely proportional to the altitude. If any chosen scale is desired for the picture the choice of focal length is determined by the height at which it is necessary to fly. This at least would be the case were there no limitation to the practicable focal length—which means camera size—and were one limited to the original size of the picture as taken; that is, were the process of enlargement not available. But the possibility of using the enlarging process brings in other questions: Is the defining power of a short focus lens as good in proportion to its focal length as that of a long focus lens? If so a perfect enlargement from a negative made by a short focus lens would be identical with a contact print from a negative made with a lens of longer focus. Is defining power lost in the enlarging process with its necessary employment of a lens which has its own errors of definition and which must be accurately focussed?

Certain factors which enter into comparisons of this sort in other lines of work, such as astronomical photography, play little part here. These are, first, the optical resolving power of the lens, which is conditioned by the phenomena of diffraction, and is directly as the diameter; and, second, the size of the grain of the plate emulsion. The first of these does not enter directly, because the size of a point image on the axis of the lens, due merely to diffraction, is very much less than that given by any photographic lens which has been calculated to give definition over a large field, instead of the minute field of the telescope. Yet it may contribute toward somewhat better definition with a long focus lens because of the actually larger diameter of such lenses. The second factor is not important, because, as will be seen later, the resolving power of the plates suitable for aerial photography is considerably greater than that of the lens. The emulsion grain is in fact only a quarter or a fifth the size of the image as given by a 25 centimeter lens, and enlargements of more than two or three times are rarely wanted.

A series of experiments was made for the U. S. Air Service to test out these questions, using a number of representative lenses of all focal lengths, both at their working apertures and at identical apertures for all. With regard to lens defining power, as shown by the size of a point image, the answer has already been reported in a previous section. Lenses of long focus give a relatively smaller image than lenses of the same design of short focus. In regard to the whole process of making a small negative and enlarging it, the loss of definition is quite marked, as compared to the pictures of the same scale made by contact printing from negatives taken with longer focus lenses.

This answer is clear-cut only for lenses calculated to give the same angular field. Thus a 10 inch lens covering a 4×5 inch plate has about the same angle as a 50 centimeter lens for an 18×24 centimeter plate. When, however, it comes to the longer foci, such as 120 centimeters, the practical limitation to plate size (18×24 cm.) has been passed, and the angular field is less than half that of the 50 centimeter lens. The 120 centimeter lens need only be designed for this small angle, with consequent greater opportunities for reduction of spherical aberration. It is therefore an open question whether a 50 centimeter lens designed to cover a plate of linear dimensions $\frac{50}{120}$ times that used with the regular 50 centimeter lens could not be produced of such quality that it would yield enlargements equal to contacts from a 120 centimeter lens. If so, lenses of larger aperture could be used, and a considerable saving in space requirements effected.

Focal lengths during the Great War were decided by the nature of the military detail which was to be revealed and by the altitudes to which flying was restricted in military operations. In the first three years of the war the development of defences against aircraft forced planes to mount steadily higher, so that the original three or four thousand feet were pushed to 15,000, 18,000, and even higher. Lenses of long focus were in demand, leading ultimately to the use of some of as much as 120 centimeters (Fig. 41). In the last months of the war the resumption of open fighting made minute recording of trench details of less weight, while the preponderance of allied air strength permitted lower flying. In consequence, lenses of shorter focus and wider angle came to the fore, suitable for quick reconnaissance of the main features of new country. At the close of the war the following focal lengths were standard in the U. S. Air Service, and may be considered as well-suited for military needs. Peace may develop quite different requirements.

Focal length	Aperture	Plate size
10 inch	F/4.5	4×5 inch
26 cm.	F/6	13×18 cm.
12 inch	F/5.6	18×24 cm.
20 inch	F/6.3 to F/4.5	18×24 cm.
48 inch	F/10 to F/8	18×24 cm,

The question of the use of *telephoto lenses* in place of lenses of long focus is frequently raised. Lenses of this type combine a diverging (concave) element with the normal converging system, whereby the effect of a long focus is secured without an equivalent lens-to-plate distance. This reduction in "back focus" may be from a quarter to a half. Were it possible to obtain the same definition with telephoto lenses as with lenses of the same equivalent focus, they would indeed be eminently suitable for aerial work because of their economy of length. But experience thus far has shown that the performance of telephoto lenses, as to definition and freedom from distortion, is distinctly inferior, so that it is best to hold to the long focus lens of the ordinary type.

Lenses Suitable for Aerial Photography.—Among the very large number of modern anastigmat lenses many were found suitable for airplane cameras and were used extensively in the war. A partial list follows: The Cooke Aviar, The Carl Zeiss Tessar, the Goerz Dogmar, the Hawkeye Aerial, the Bausch and Lomb Series Ic and IIb Tessars, the Aldis Triplet, the Berthiot Olor.

The Question of Plate Size and Shape.—Plate size is determined by a number of considerations, scientific and practical. If the type of lens is fixed by requirements as to definition, then the dimensions of the plate are limited by the covering power. From the standpoint of economy of flights and of ease of recognizing the locality represented in a negative, by its inclusion of known points, lenses of as wide angle as possible should be used. If the focus is long, this means large plates, which are bulky and heavy. If the finest rendering of detail is not required a smaller scale may be employed, utilizing short focus lenses and correspondingly smaller plates. Thus a six inch focus lens on a 4×5 inch plate would be as good from the standpoint of angular field as a 12 inch on an 8×10 inch plate. This is apt to be the condition with respect to most peace-time aerial photography, which may be expected to free itself quickly from the huge plates and cameras of war origin.

For work in which great freedom from distortion of any sort is imperative, small plates will be necessary, for two reasons. One is that the characteristic lens distortions are largely confined to the outlying portions of the field. The other is that a wide angle of view inevitably means that all objects of any elevation at the edge of the picture are shown partly in face as well as in plan, which prevents satisfactory joining of successive views (Fig. 128). In making a mosaic map of a city, if a wide angle lens is employed with large plates, the buildings lying along the junctions of the prints can be matched up only for one level. If this is the ground level, as it would be to keep the scale of the map correct, the roofs will have to be sacrificed. In extreme cases a house at the edge of a junction may even show merely as a front and rear, with no roof, while in any case the abrupt change at these edges from seeing one side of all objects to seeing the opposite side is not pleasing.

The table in a preceding section gives the relation of plate size to focal length found best on the whole for military needs. Deviations from these proportions in both directions are met with. In the English service the LB camera, which uses 4×5 inch plates, is equipped with lenses of various focal lengths, up to 20 inches. The German practice, as well as the Italian, was almost uniform use of 13×18 centimeter plates for all focal lengths. Toward the end of the war, however, some German cameras of 50 centimeter focal length were in use employing plates 24×30 centimeters.

It will be recognized that these plate sizes are chosen from those in common use before the war. A simiiar observation holds with even greater force on the question of plate *shape.* Current plate shapes have been chosen chiefly with reference to securing pleasing or artistic effects with the common types of pictures taken on the ground. These shapes are not necessarily the best for aerial photography. Indeed the whole question of plate shape should be taken up from the beginning, with direct reference to the problems of aerial photography and photographic apparatus.

A few illustrations will make this clear, taking Fig. 17

as a basis. If it is desired to do spotting (the photography of single objectives), the best plate shape would be circular, for that shape utilizes the entire covering area of the lens. If it is desired to make successive overlapping pictures, either for mapping, or for the production of stereoscopic pairs, a rectangular shape is indicated. If the process of plate changing is difficult or slow, it is advisable, in order to give maximum time for this operation, to have the long side of the rectangle parallel to the line of flight (indicated by the arrow). If economy of flights is a consideration, as in making a mosaic map of a large area, it is advantageous to have as wide a plate as the covering power of the lens will permit. Reference to Fig. 17 shows that this means a plate of small dimensions in the direction of flight. If the changing of plates or film is quick and easy, the maximum use of the lens's covering power is made by such a rectangle whose long side approximates the dimensions of the lens field diameter. This is in fact the choice made in the German film mapping camera (Figs. 61 and 63), whose picture is 6×24 centimeters. An objection to this from the pictorial side, lies in the many junction lines cutting up the mosaic. Another objection, if the plane does not hold a steady course, is the failure to make overlaps on a turn. (Fig. 62.) Here as everywhere the problem is to decide on the most practical compromise between all requirements.

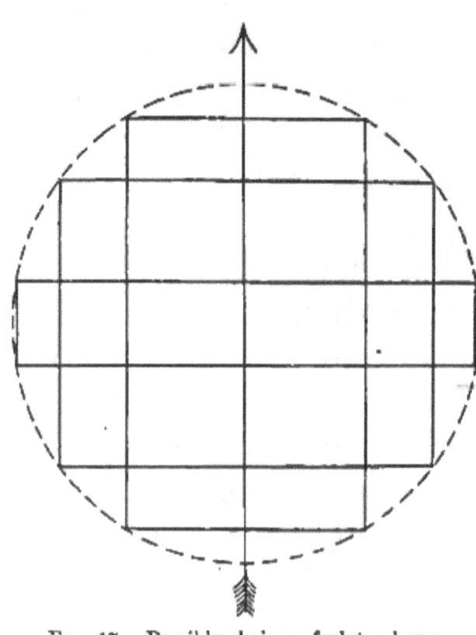

Fig. 17.—Possible choices of plate shape.

Focussing.—The process of focussing aerial cameras was at first deemed a mystery, though undeservedly so. A belief was long current that "ground" focus and "air" focus differ. In other words, that a camera focussed upon a distant object on the ground would not be in focus for an object the same distance below the camera when in the plane. Belief in this mysterious difference went so far that certain instruction books describe in detail the process of focussing a camera by trial exposures from the air.

Careful laboratory tests performed for the U. S. Air Service showed that neither low temperature nor low pressure, such as would be met at high altitudes, alter the focus of any ordinary lens by a significant amount, and that the possible contraction of the camera body was of negligible effect on the focus (not more than $\frac{1}{200}$ per cent. per degree centigrade with a metal camera). In complete harmony with these tests has been the experience that if the ground focussing is done carefully, by accurate means, then the air focus is correct. The whole matter thus becomes one of precision focussing.

The best method, applicable if the air is steady, is to focus by *parallax.* The ground glass focussing screen is marked in the center with a pencilled cross. Over this is mounted, with Canada balsam, a thin microscope cover-glass. The camera is directed on an object a mile or more away, and the image formed by the lens is examined by a magnifying glass through the virtual hole formed by the affixed cover-glass. With the pencil line in focus the head is moved from side to side. If the image and pencil mark coincide they will move together as the head is moved. If the image moves away from the pencil mark and in the *same* direction as the eye moves, the image is too near the lens. If the image moves away in the *opposite* direction to

5

the motion of the eye, it is too far from the lens. In either case the focus is to be corrected accordingly.

In place of a distant object, which may waver with the motion of the air, we may use an image placed at infinity by optical means. The *collimator*, an instrument for doing this, consists of a test object (lines, circles, etc.) placed accurately at the focus of a telescope objective. The camera lens is placed against this and focussed by parallax, as with a distant object. Collimators are employed in camera factories, and should be part of the equipment of base laboratories where repairing and overhauling of cameras is done.

Lens Mounts.—All that is required for the mounting of an aerial camera lens is a rigid platform, with provision for enough motion of the lens to adjust its focus accurately. As already explained, the lens works at fixed, infinity, focus, and therefore needs no adjustment during use. It must be held far more rigidly than would be possible by the bellows, which is an almost invariable adjunct of focussing cameras. The use of ordinary types of hand cameras on a plane is rarely successful just because of the bellows, which is strained and rattled by the rush of wind.

The lens mountings thus far used have been simple affairs. In the French cameras the lens is merely screwed into a flange which in turn is fastened by screws to a platform in the camera body. Adjustment for focussing is not provided; instead, the flange is raised on thin metal rings or washers, cut of such thickness by trial as to bring the lens to focus, once and for all.

The U. S. Air Service method of mounting is to provide the lens barrel with a long thread, which screws into a flange that in turn is mounted on a platform in the camera cone, by means of thumb-screws. The lens is focussed by screwing in and out, and then clamped by a screw through the side, bearing on the thread. The whole mount may be quickly removed by loosening the thumb-screws, and once focussed in one cone, can be transferred to another similar, machine-made cone without change of focus. Fig. 18 shows a 20 inch lens mounted in this manner. The photograph shows as well the ring on the front of the lens by means of which circular color filters may be held in place.

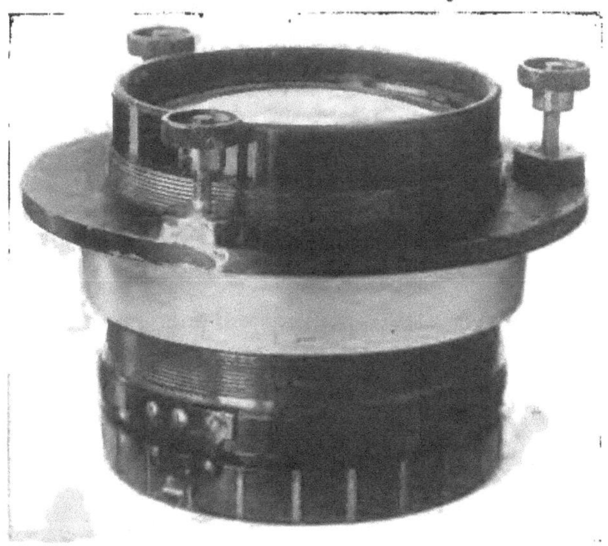

Fig. 18.—50 centimeter F/6 lens in U. S. standard mount, showing color filter retaining ring and catch.

This ring screws down on the filter, and the catch is dropped into the nearest vertical groove to the tight position.

A somewhat different and better method of tightening the lens in the flange, when focussed, has been adopted in the English lens mount, which is in general similar to the American. The threaded part of the flange is split by a slot cut parallel to the flange base, and a screw is run into the flange from the front, through the split portion. By tightening this screw, which is always accessible, the split part of the flange is squeezed together, thus rigidly holding the lens barrel.

CHAPTER V

THE SHUTTER

Permissible Exposure in Airplane Photography.—A definite limitation to the length of exposure in airplane cameras is set by the motion of the plane. If we represent the speed of the plane by S, the altitude of the plane by A, and the focal length of the lens by F, we obtain at once from the diagram (Fig. 19), that s, the rate of movement of the image on the plate, is given by the relation,

$$\frac{s}{S}=\frac{F}{A}$$

If we call the permissible movement d, then the permissible exposure time, t, is given by the relation—

$$t=\frac{d}{s}=\frac{Ad}{FS}$$

As a representative numerical case, expressing all quantities in centimeters and in centimeters per second, let $F=50$, $S=\frac{20{,}000{,}000,}{3600}$ (200 kilometers per hour), and $A=300{,}000$, then

$$s=\frac{50\times 20{,}000{,}000}{300{,}000\times 3600}=.9 \text{ centimeters}$$

If we take for the permissible undetectable movement, .01 centimeter, which is, as has been shown, a reasonable figure for lens defining power, we have, then, that the *longest permissible exposure is .011 second*—in round numbers, one-hundredth.

In flying with a slow plane, or in flying against the wind, the exposure can sometimes be increased to as much as double this length. Diminishing F would similarly extend the allowable exposure, but the ratio of F to A approximates to a constant in actual practice; in other words, a certain resolution and size of image have been found desirable. If flying is forced higher, a longer focus lens is used; if lower flying is possible, a lens of shorter focus. This relationship

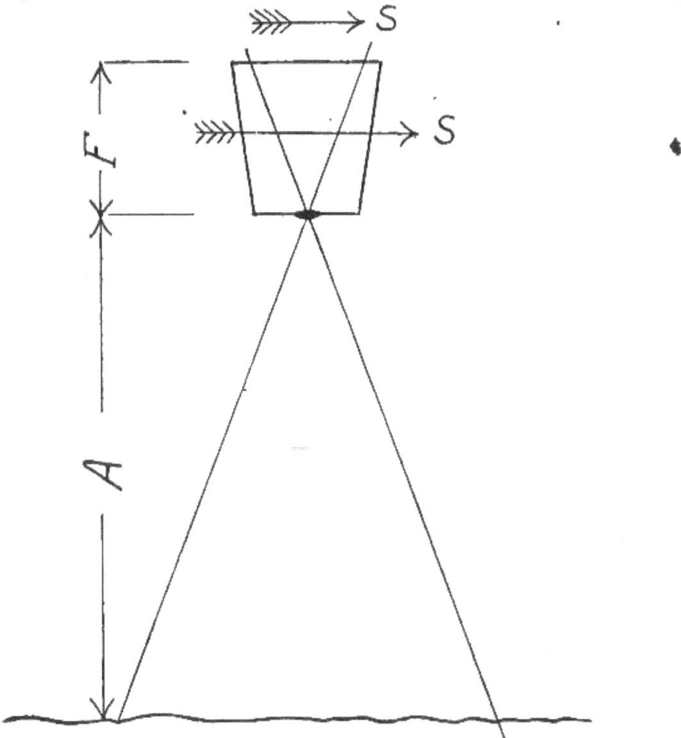

Fig. 19.—Relative motion of plane and photographic image.

has, of course, been derived from war-time experience. Probably much of the prospective peace-time mapping work will impose substantially easier requirements as to definition and will thus allow longer exposures.

For low oblique views the longest exposure is much less. Taking 45 degrees as a representative angle for the foreground, and 500 meters as a representative height, the value of t becomes $\frac{1}{600}$.

These figures will illustrate two important points: they show how severe is the limitation as to exposure, with the consequent heavy demand on lens and sensitive material speed; and they show how important it is to secure a shutter with the maximum light-giving power for a specified length of exposure. This leads to a study of the characteristics as to efficiency of the two common types of shutter, namely, *shutters at or between the lens,* and *focal-plane shutters.*

Characteristics of Shutters Located at the Lens.—Of the various shutters located at the lens the most common is the type that is clumsily but descriptively termed the "between-the-lens" shutter. This is composed of thin hard rubber or metal leaves or sectors which overlap and which are pulled open to make the exposure. It may require two operations, one for setting and one for exposing, or it may, as in some makes, set and expose by a single motion. Clock escapements, or some form of frictional resistance, are depended on to control the interval between opening and closing. This shutter is the one almost universally employed on small hand cameras and on all lenses up to about two inches diameter. It gives speeds sometimes marked as high as $\frac{1}{300}$ second, although usually not over $\frac{1}{100}$ on actual test.

Between-the-lens shutters have been used to some extent on the shorter focus (up to 25 centimeter) aerial cameras, notably in the Italian service. They suffer, however, from two limitations. In the first place we have not yet solved the mechanical problems met with in trying to make the shutter of large size (as for 50 centimeter $F/6$ lenses) at the same time to give high speeds. In the second place the efficiency of the type is low because a large part of the exposure time is occupied by the opening and closing of the sectors.

If we define the *efficiency* of a shutter as the ratio of the amount of light it transmits during the exposure to the amount of light it would transmit were it wide open during the whole period, then the efficiency of the ordinary between-the-lens shutter is of the order of 60 per cent. This means 1.6 times the motion of the image for the same photographic action that we should have with a perfect shutter. The accompanying photographic record (Fig. 20) of the opening and closing process of this type of shutter clearly illustrates its deficiencies.

Characteristics of the Focal-Plane Shutter.—Long before the days of aerial photography the problem of a high-

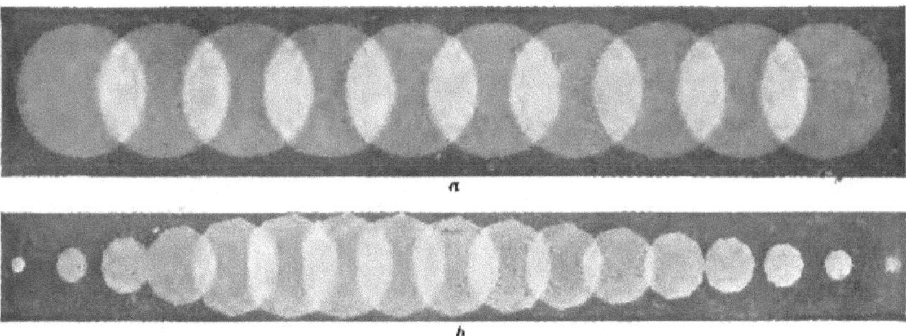

FIG. 20.—Effective lens opening at equal intervals of time: (*a*) during focal plane shutter exposure; (*b*) during between-the-lens shutter exposure.

efficiency high-speed shutter for photographing moving objects on the ground—railway trains or racing automobiles—had already led to the development of the *focal-plane shutter.* This is a type peculiarly adapted to the problems of the airplane camera. It consists essentially of a curtain, running at high speed close to the photographic plate, the exposure being given by a narrow rectangular slot.

If the focal-plane shutter is in virtual contact with the sensitive surface the efficiency, as defined above, is 100 per cent., since the whole cone of rays from the lens illuminates the plate during the whole time of exposure. But if the curtain is not carried close to the plate the efficiency falls off rapidly with distance, especially so for small apertures of the slot.

The efficiency of the focal-plane shutter may be calculated as follows: Let the focal length of the lens be F, its diameter

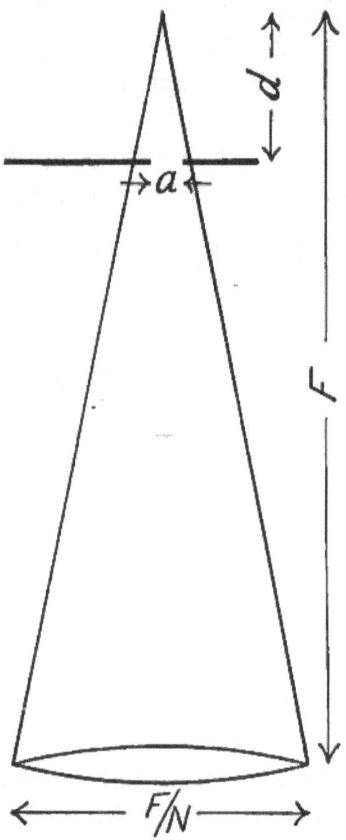

Fig. 21.—Calculation of focal plane shutter efficiency.

be F/N, the width of the slot be a, and the distance from plate to curtain d (Fig. 21). Now if the curtain is moving at a uniform speed, the time taken for the slot to traverse the whole cone of rays, from the instant it enters till the instant it leaves, will be directly proportional to

$$\frac{d}{F}\left(\frac{F}{N}\right)+a=\frac{d}{N}+a$$

If the curtain were in contact with the plate the time taken for the same amount of light to reach the sensitive surface would be proportional to a. Again defining shutter efficiency as the ratio of the light transmitted to what would have been transmitted were the shutter fully open for the total time of exposure, the efficiency, E, is given at once by the expression—

$$E=\frac{a}{\frac{d}{N}+a}$$

As an example let the lens aperture be $F/6$, so that $N=6$; let $d=1$, and $a=1$, then $E=\frac{6}{7}$. In the French de Maria

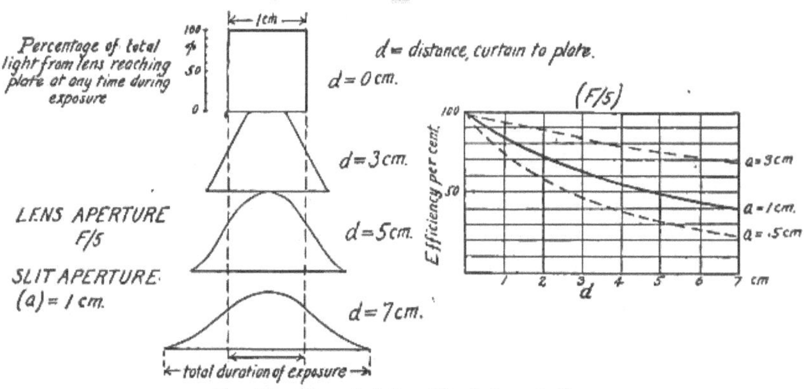

FIG. 22.—Characteristics of focal plane shutter.

cameras, where $d=4$ centimeters, $E=60$ per cent. for the aperture assumed, which is representative. Fig. 22 exhibits diagrammatically the chief characteristics of the focal plane shutter.

In view of the necessity for *some* distance between shutter and plate it is obviously important to keep a as large as possible, depending for the requisite shutter speed on the velocity of the curtain. Large aperture and high curtain speed are also found to be desirable when we consider the distortion produced by the focal-plane shutter.

Distortions Produced by the Focal-plane Shutter.—While the time of exposure of any point on the plate can, with the focal-plane shutter, easily be made $\frac{1}{100}$ second or less, the whole period during which the shutter is moving is much greater than this. For instance, a 1 centimeter opening which gives $\frac{1}{100}$ second exposure takes $\frac{1}{10}$ second to move across a 10 centimeter plate, or nearly $\frac{1}{5}$ second for an 18 centimeter plate. With a moving airplane this means that the point of view at the end of the exposure has moved forward compared to that at the beginning, by the amount of motion of the plane in the interval. If the shutter moves in the direction of motion of the plane the image will be magnified; if in the opposite direction, it will be compressed along the axis of motion. The amount of this distortion is calculated as follows:

Let the velocity of the plane be V, and that of the shutter be v. Let the focal length of the camera be F, and the altitude A. If the camera were stationary, a plate of length l would receive on its surface an image corresponding to a distance $\frac{A}{F} \times l$ on the ground. Due to the motion of the shutter the end of the exposure occurs at a time $\frac{l}{v}$ after the start. In this time the plane has moved a distance $V \times \frac{l}{v}$; hence the point photographed at the end of the shutter travel is $\frac{Vl}{v}$ within or beyond the original space covered by the plate, depending on the direction of motion of the curtain. The distortion, D, is given by the ratio of this distance to the length corresponding to the normal stationary field of view:

$$D = \frac{\frac{V}{v} \times l}{\frac{A}{F} \times l} = \frac{VF}{vA}$$

When $V = 200$ kilometers per hour, $v = 100$ centimeters per second, $F = 50$ centimeters, $A = 3000$ meters, we have—

$$D = \frac{20{,}000{,}000 \times 50}{3600 \times 100 \times 300{,}000} = \text{approximately } \frac{1}{100}$$

Or if the actual distance error on the ground is desired,

$$\frac{Vl}{v} = 10.8 \text{ meters}$$

As a percentage error this one per cent. is small compared with other uncertainties, such as film shrinkage or the error of level of the camera. As an absolute error in surveying, thirty feet is, of course, excessive.

The distortion is diminished for any specified shutter speed by making the speed of travel of the curtain as large as possible and by correspondingly increasing the aperture. In connection with film cameras, another solution which has been suggested is to move the film continuously during the exposure in the direction of the plane's motion. The requisite speed of the film v' to eliminate distortion is given by the relation:

$$\frac{v'}{V} = \frac{F}{A}$$

For the values of V, F, and A used above, $v' = .92$ centimeters per second. This speed is clearly that which holds the image stationary on the film—a fact which suggests another object for such movement, namely, to permit of longer exposures.

The effect of focal plane distortion may be averaged out in the making of strip maps, if the shutter is constructed so as to move in opposite directions on successive exposures. The first picture will be magnified, the second compressed, and so on, but a strip formed of accurately juxtaposed pictures will be substantially accurate in over-all length. Such a shutter is embodied in one of the German film cameras (Fig. 61).

Distortion of the kind above discussed is absent with between-the-lens shutters, which may conceivably be improved in efficiency and in feasible size. If so they would merit serious consideration for aerial mapping.

Methods and Apparatus for Testing Shutter Performance.—With a focal-plane shutter the desirable qualities in performance are three in number: (1) *Adequate speed range*, which may be taken as from $\frac{1}{50}$ to $\frac{1}{500}$ second for aerial work, (2) *good efficiency*, which has already been treated, and (3) *uniformity of speed* during its travel across the plate. Before the advent of aerial photography little attention was paid to speed uniformity, differences of 50 per cent. in initial and final speed being common in focal-plane shutters, and but little noticed in ordinary landscape work because of the natural variation of brightness from sky to ground. In the making of aerial mosaic maps the non-uniformity of density across the plate results in a most offensive series of abrupt changes of tone at the junction points of the successive prints (Fig. 140), an effect which must be minimized by manipulation of the printing light.

Instruments for testing the speed and uniformity of action of focal-plane shutters are an essential part of any laboratory for developing or testing photographic apparatus and some simple device for setting and checking shutter speed should be available in the field. Every such speed tester must contain some form of time counting element—pendulum, tuning fork or clockwork. Elaborate shutter testers, suitable for determining all the characteristics of all types of shutter, have been developed and used in certain of the photographic research laboratories. For the study and setting of focal-plane shutters (whose efficiency need not be measured, as it can be simply calculated from linear dimensions), the following simple kinds of apparatus are adequate:

Clock dial type of shutter tester. This consists essentially of a black clock dial carrying a white pointer which makes its complete revolution in one second or less. If this dial is photographed by the camera under test, the width of the

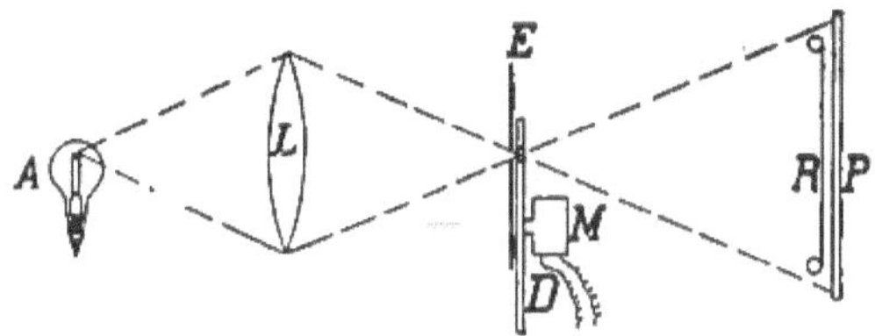

FIG. 23.—Apparatus for testing focal plane shutter speed throughout the travel of the curtain.

sector traced during the exposure by the moving pointer shows the time interval. If the dial is photographed at several points on the plate—beginning, middle and end of the shutter travel—the complete characteristics of the shutter can be determined.

Interrupted light type of shutter tester. For the study of uniformity of shutter action alone the apparatus shown in Fig. 23 may be employed. *A* is a high intensity light source, such as an arc or a gas filled tungsten lamp. *L* is a convex lens, focussing an image of the light source on a small aperture in the screen *E*. *D* is a sector disc which, driven by the motor *M*, interrupts the transmitted light with a frequency determined by the number of openings of the sector and by the speed of rotation, which must be measured by a tachometer. The light diverging from the aperture in E falls upon the shutter S, which for this test is reduced to a narrow slit of one millimeter or less. Passing through the shutter opening the light falls upon the photographic plate P. The principle is simple: If the light is uninterrupted, the plate P is exposed at all points; due to the interruptions, a series of parallel lines of photographic action result, and

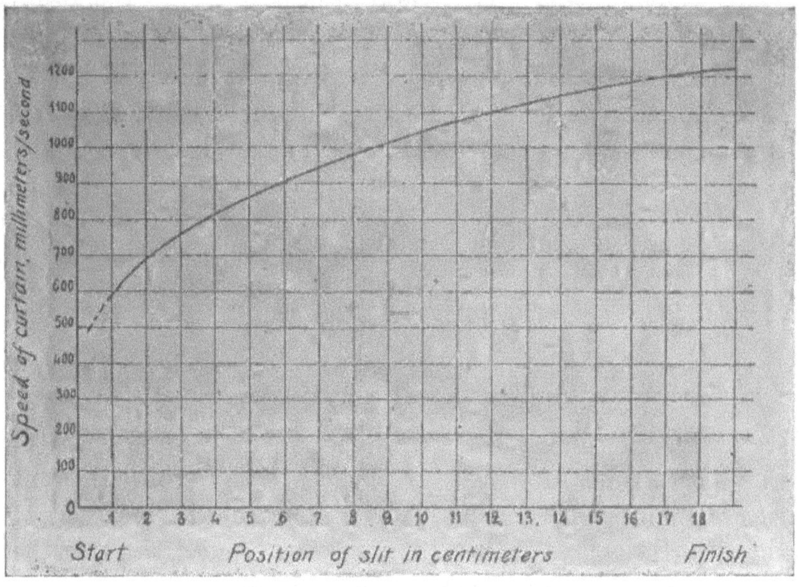

Fig. 24.—Performance of Klopcic shutter.

their distance apart gives a measure of the speed of the shutter at any chosen point in its travel. A performance curve of the French Klopcic shutter is shown in Fig. 24. The variation in speed lies over a range of two to one. So serious is this defect in these shutters that diaframs are sometimes inserted in the French cameras to cut off part of the light from the lens on the most exposed end of the plate. This expedient produces uniformity of photographic action, but does not overcome the movement of the image, which is one of the chief faults of excessive exposure.

A more complete apparatus, adapted both to absolute speed determinations and to the study of uniformity of action, is that worked out and used in the United States Air Service (Fig. 25). At A is a high intensity light source, an image of which is focussed by the lens L_1 upon a slit E, in front of which stands a tuning fork T, of period 1024 or 2048 per second. The light diverging from the slit is received by a second lens, L_2 which is arranged either to focus the slit image upon the shutter curtain or to render the rays

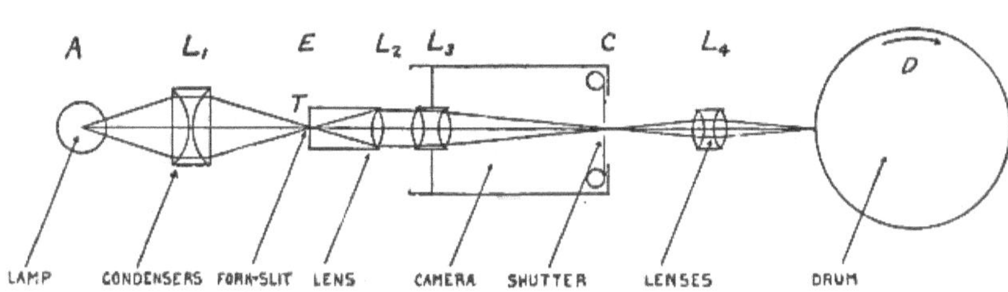

Fig. 25.—Optical system of shutter tester for Air Service, U. S. Army.

parallel, so that an entire camera may be inserted. In the latter case the camera lens L_3 serves to focus the slit image on the curtain C. After passing through the curtain aperture the light is focussed by the lens L_4 on the rotatable drum D, which carries a strip of sensitive film.

The operation of testing a shutter consists in focussing the slit image on the portion of the shutter whose performance is required, striking the tuning fork to set it vibrating, rotating the drum rapidly and setting off the shutter. There is thus obtained on the sensitive film an exposed strip resembling in appearance the edge of a saw, the number of teeth showing the time interval in vibrations of the tuning fork. Three exposures usually give all the points necessary for a practical knowledge of the shutter's uniformity of action. A point of some importance, learned from numerous shutter tests, is that a focal-plane shutter should be tested in the position in which it is to be used. Aerial camera shutters should be tested in the horizontal position.

Types of Focal-plane Shutters.—A variety of means have been utilized for securing the necessary variation in speed in focal-plane shutters. Their success is to be measured by the actual speed range and by the uniformity of speed attained. In aerial cameras at present in use we find *variable tension* of the curtain spring, the aperture being fixed; *variable opening* with fixed tension; *multiple curtain openings* with fixed spring tension; and combinations of two or all of these methods of speed control. The problem of covering the aperture during the operation of winding up or setting the shutter has led to further elaborations of shutter mechanism. These take the form of *lens or shutter flaps, auxiliary curtains*, and shutters of the *self-capping* type. Shutters embodying all these features are briefly described below.

Representative Shutters.—The Folmer variable tension shutter is used on the United States Air Service hand-held and hand-operated plate camera and on some of the film cameras. It consists of a fixed aperture curtain wound on a curtain roller in which the spring can be set to various tensions, numbered 1 to 10. The range of speeds attainable is at best about three to one, or from $\frac{1}{100}$ to $\frac{1}{300}$ second, considerably shorter than the range indicated as desirable. Its uniformity of travel is variable with the tension, as shown by representative performance curves in Fig. 30. Lacking any self-capping feature the shutter is provided either with an auxiliary curtain, or in the hand-held camera with flaps in front of the lens, opened by the exposing lever before the curtain is released (Fig. 39). This shutter is made a removable unit in the 18×24 centimeter hand-operated camera, but is built into the hand-held and film cameras.

The Ica shutter used on the standard German aerial cameras is a good example of the multiple slit curtain (Fig. 26). Four fixed aperture slits are provided, with a single

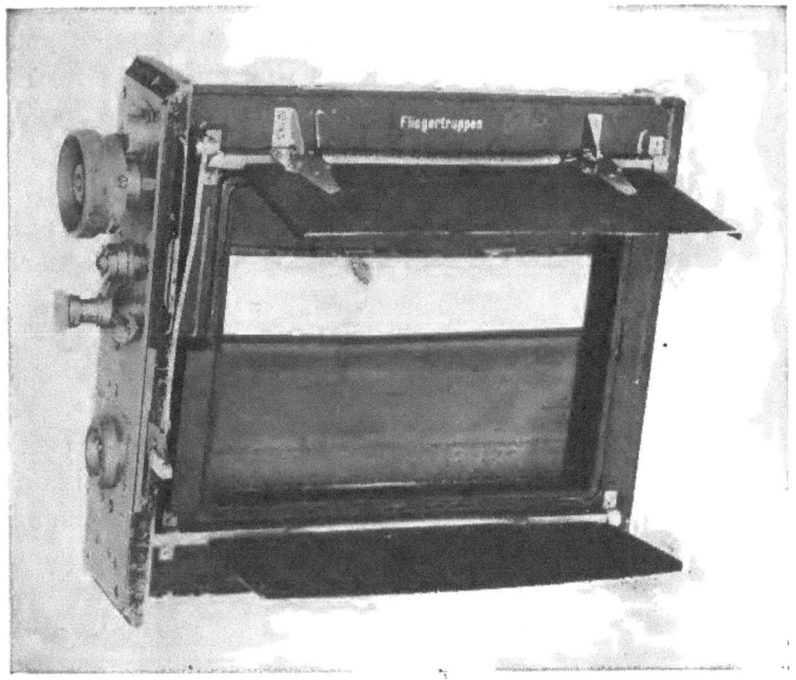

Fig. 26.—Removable four-slit shutter of German (Ica) camera, showing flaps.

tension, the openings roughly in the ratio 1, $\frac{1}{2}$, $\frac{1}{4}$, $\frac{1}{8}$, which when the spring tension is properly adjusted give exposures of $\frac{1}{90}$, $\frac{1}{180}$, $\frac{1}{375}$, $\frac{1}{750}$ second. To pass from one exposure time to another the setting milled head is wound up to successively higher steps or else exposed one or more times without resetting, depending on the direction it is desired to go. Capping during setting, or during exposure, in order to change the opening, is provided for by a pair of flaps on the shutter unit, which open into the camera body. The mechanical work on these shutters is of excellent quality, the curtain running with exceptional smoothness. Provision is made for adjusting the tension until the marked speeds are attained; this is presumably done in a repair laboratory to which the shutter only need be sent, as it is a removable unit. Tests made on one of these shutters wound to its highest tension are shown in Fig. 30. The marked speeds are not attained, and there is considerable lack of uniformity from start to finish of the travel.

L camera variable-aperture shutter. The shutter of the L type camera (Fig. 27) is representative of one of the most primitive methods of varying aperture. The two jaws of the slit are held together by a long cord passing completely around the aperture, fastened permanently at one end and attached at its other end by a sliding clasp or saddle. As this saddle is forced in one direction the slit is closed, in the opposite direction the cord becomes slack, and after the shutter is released once or twice the slit assumes a wider opening. A chronic trouble is the breaking of the cords. Its opening can be changed only after the plate magazine is removed.

U. S. Air Service variable-aperture shutter. This shutter is incorporated in the American deRam and in other late American cameras (Fig. 28). Its characteristic feature is the introduction of an idler, whose distance from the main curtain roller can be varied. Tapes whereby the following curtain is attached to the spring roller pass over this idler, and by changing its position the aperture or distance between the two curtain elements is altered over a large range. Tests of this shutter are shown in Fig. 30. A speed of $\frac{1}{50}$ second is provided for by a slit width of five centimeters, and the highest speed is fixed only by the practical limit of

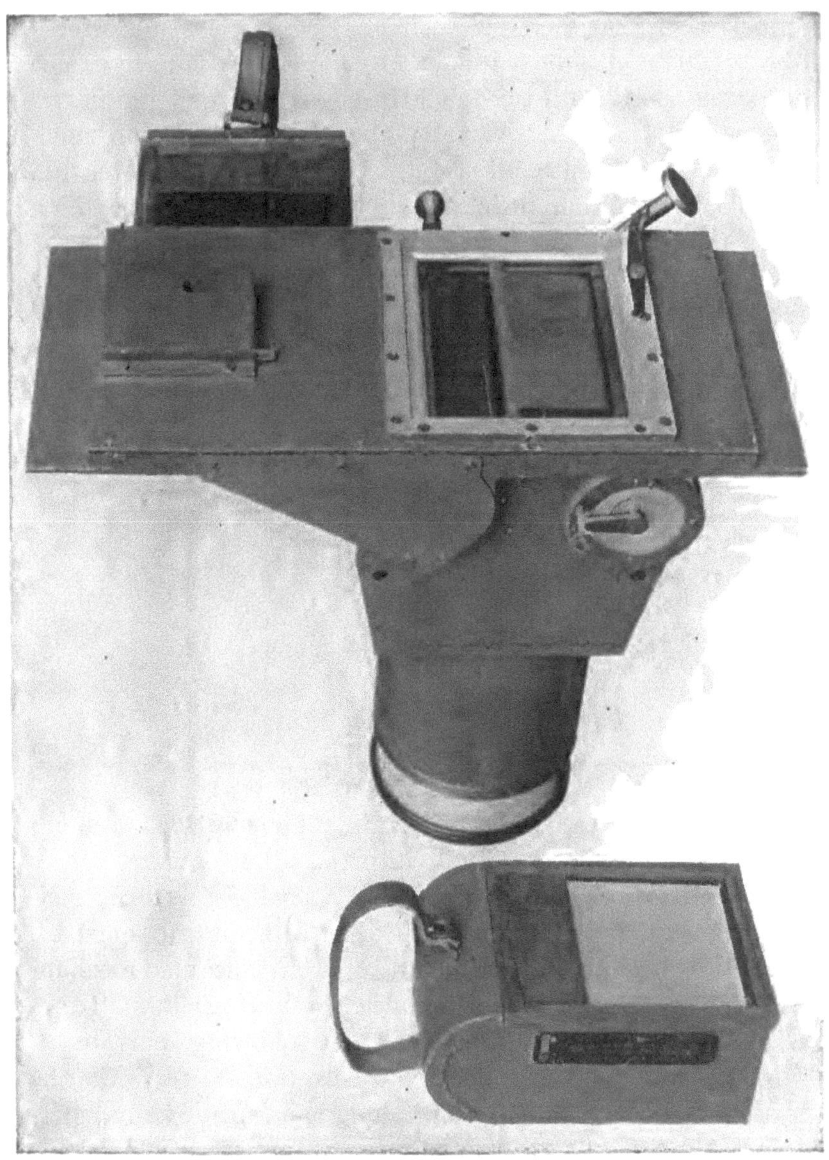

FIG. 27.—"L" type camera showing open negative magazines and shutter mechanism.

approach of the jaws. Experiment shows great uniformity of rate of travel to be attainable by combining careful choice of spring length and tension with good workmanship in the mechanical features. Variable-aperture fixed-tension shutters have a definite advantage over the variable-tension type in that they can utilize for all speeds that tension which gives uniform action. The capping feature of this shutter is provided in the American deRam by flaps, in the automatic

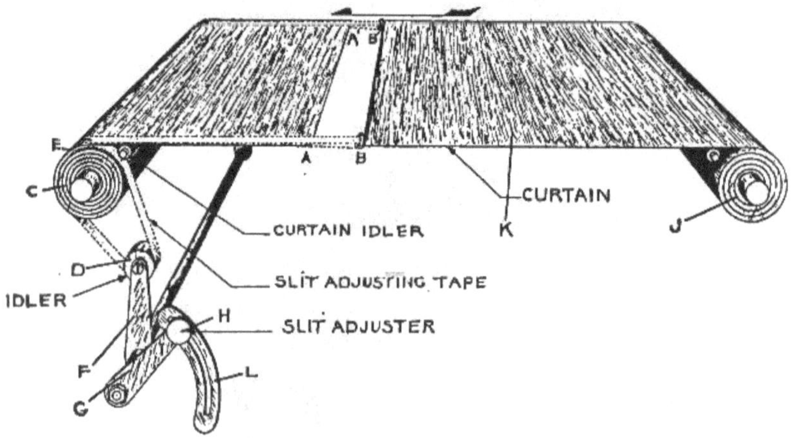

Fig. 28.—Variable aperture curtain developed in U. S. Air Service, and used in American deRam, and "K" type automatic film cameras.

film camera by an auxiliary curtain. The shutter is removable in the deRam, but built into the other camera.

The Klopcic variable-tension, variable-aperture, self-capping shutter is an example of an attempt to meet all shutter requirements with an entirely self-contained mechanism. It is shown diagrammatically in Fig. 29. Tapes G_1, G_2 are used to connect the following curtain B directly to the spring roller T, at a fixed distance, while the leading curtain, A, may be slid along the tapes by small friction buckles, C_1, C_2, auxiliary springs R_1, R_2 serving to keep it taut in any position. When the shutter is being set the buckles are arrested against stops while the winding-up continues for what is to be the following half of the curtain in exposing. When released the curtain moves across with an aperture fixed by the point of setting of the buckle stops. At the end of the travel the buckles are arrested by other stops, while the following portion of the curtain con-

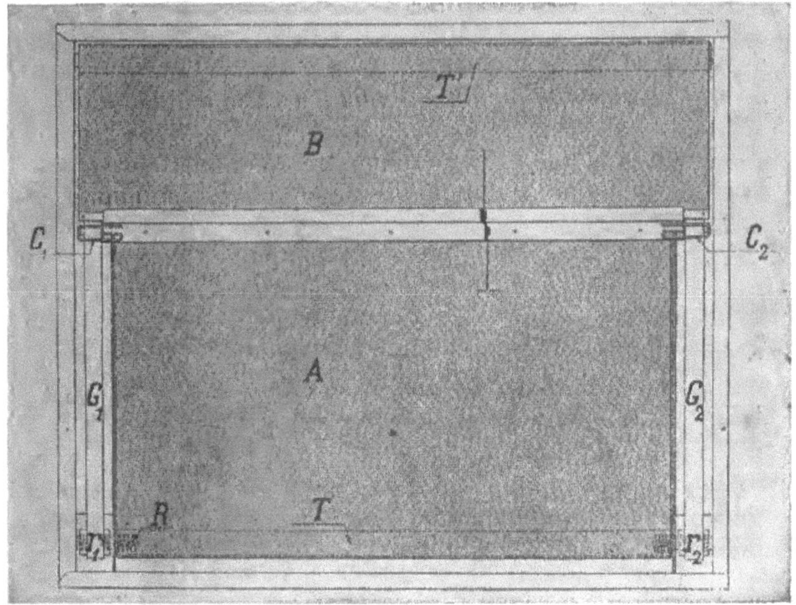

FIG. 29.—Mechanism of Klopcic variable aperture self-capping shutter.

tinues its travel to the end. On re-winding, therefore, the aperture is closed. Variable tension as well as variable aperture is provided, although little used. In the French cameras a lens flap is also inserted behind the lens, but this is not needed if the self-capping feature functions properly. On the hand cameras this flap is said to be necessary in order to prevent a curious kind of accident: if the camera is held on the knee, pointing upward, an image of the sun may be formed on the curtain and burn a hole through it.

The performance of the French shutter in respect to uniformity has already been shown in Fig. 24. It leaves very much to be desired. Besides non-uniformity of action during its travel it exhibits another common defect of variable-tension shutters, namely, the curtain must be released several times after a change of tension before the new speed is established (Fig. 30, tensions 5 and 5′).

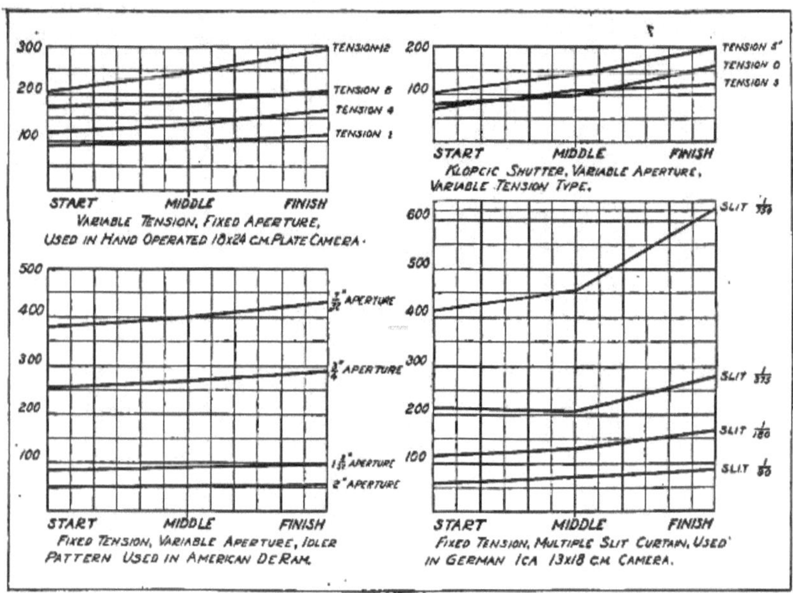

Fig. 30.—Performances of various shutters used on aerial cameras. Speeds expressed in reciprocals of fractional parts of one second.

The French shutter as made for the de Maria cameras is a removable unit. The small size (13×18 cm.) sets by the straight pull of a projecting pin, the larger (18×24 cm.) by winding up a milled head. The former is the more convenient motion for an aerial camera. Care must be taken with either type that the motion of setting is not stopped when the first resistance is encountered; this occurs when the tape buckles strike their stop and the slit begins to open.

CHAPTER VI

PLATE-HOLDERS AND MAGAZINES

In the earlier days of airplane photography the ordinary plate-holder or double dark slide was used to some extent, but it is ill-suited to the purpose because of the considerable time and attention required for its operation. It has nevertheless the merit of adding little to the length of the camera, and it works in any position. For these reasons it has remained in occasional use for the taking of oblique views with long focus cameras in a cramped fuselage.

Next in order of progress rank the simple box magazines, for holding a dozen, eighteen or twenty-four plates, as used in the English C, E, and L type cameras. These are little more than boxes with sliding lids which when open permit the introduction or removal of the plates. Figs. 45 and 46 illustrate the magazine of this type as made for the English C and E cameras. It is constructed of wood, grooved to fit tracks on the camera, and is furnished with a sliding door or lid hinged in the middle to fold down out of the way when open. The eighteen plates are carried in metal sheaths, both to provide opaque screens between them, and to protect them from injury in the mechanism of the camera. Fig. 27 shows the all-metal magazine made for the American model L camera. This differs from the English in material of construction, plate capacity (24 instead of 18) and manner of operating the slide, which is built up of three thicknesses of phosphor bronze and draws out through metal guides bent into semicircular form. A snap catch holds this slide at either end of its travel. The leather strap introduced in the American model for carrying and handling is a distinct improvement. These magazines contain no springs or other mechanism, as the cameras with which they are used depend upon the action of gravity for emptying the upper (feeding) magazine, and filling the lower (receiving) one.

Next in order of complexity may be ranked the *bag magazine* (Figs. 31 and 44). In this the exposed plate is pulled out of the magazine proper by a metal slide or rod into a leather bag. The rod is then pushed back, the plate in its metal sheath is grasped through the leather bag, lifted

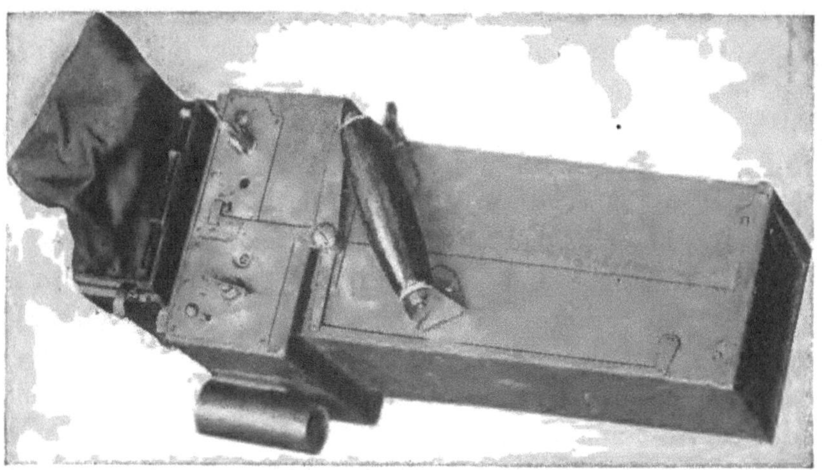

Fig. 31.—Aerial hand camera (U. S. type A-2).

to the back of the magazine, and forced in behind the other plates. The number of plates exposed is indicated either by numbers on the backs of the sheaths, visible through a red glazed opening in the back, or else by a counter actuated by the metal slide rod. Usually twelve are carried in a magazine. For aerial work the common design of this magazine as used for ground work must be modified by providing extra large easily grasped hooks both on the draw rod and on the dark slide, which must be drawn before making the first exposure and replaced after the last. The small rings and grips of the standard commercial magazine are almost impossible to handle through heavy gloves.

The next type of magazine is represented by three designs, the *Gaumont* and *deMaria*, used very generally by the French during the war, and the *Ernemann*, used almost universally in the German air service (Figs. 32, 40 and 42). In all of these the operation of plate changing is the same: the end of the magazine is pulled out and thrust back, a more simple operation than the bag manipulation just described. The internal workings are different according to size. In the smaller French magazines (13×18 cm.) the camera is first pointed upward, all the plates are drawn out except the one to be changed, and this, with the aid of springs, drops to the bottom, after which the other plates push back over it. The plates pull out in the direction of their long dimension. In the larger French magazine (18×24 cm.) only the exposed plate pulls out. The pull is in the direction of the shorter dimension of the plate, which is lifted up by heavy springs and slides back over the top of the pile. In the Ernemann magazine only six plates are carried, which there is good reason to believe represent the maximum feasible number, judging by the reports of jambs and breakages in the twelve-plate French magazines. In all of these magazines laminated wood slides pull out and in at each operation, and while satisfactory if made and operated in one climate, experience indicates that if made in America and sent abroad swelling of the wood may be expected to prevent their successful operation.

Alternative forms of magazine, somewhat more practical from the standpoint of manufacture and export, are several designs embodying *two compartments* (Fig. 32). In the most simple of these the plates are moved, immediately before or after exposing, from the unexposed to the exposed side. Illustrative of this type are the Folmer designs, in which the to-and-fro motion is imparted by a rack geared to a pinion

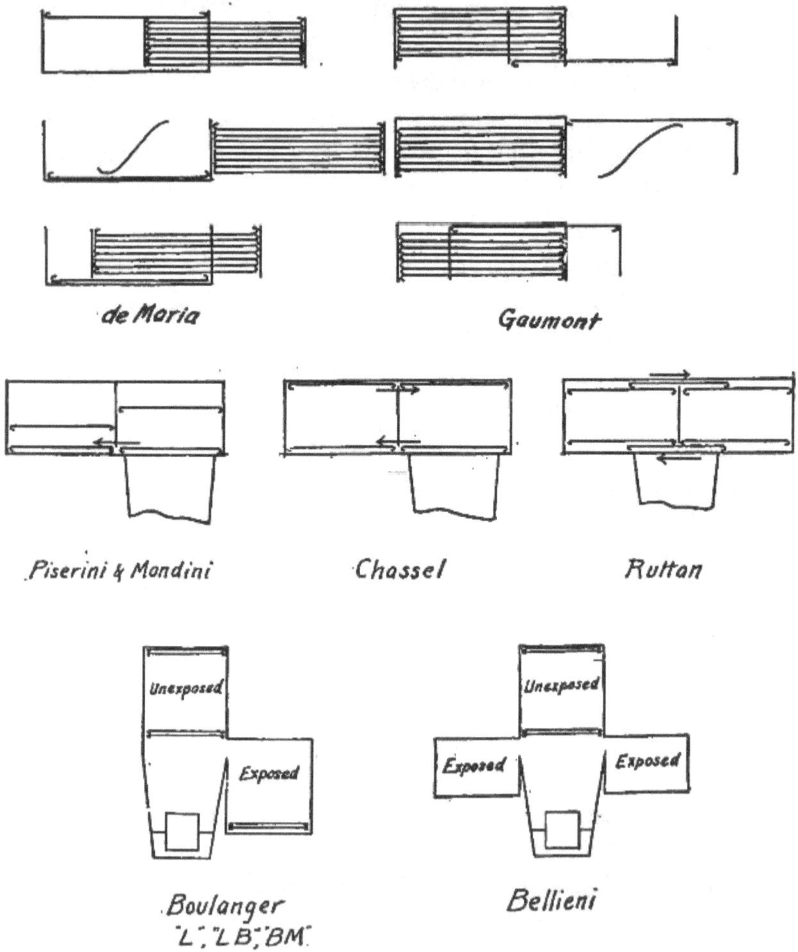

Fig. 32.—Various plate magazines used on aerial cameras.

actuated either by a lever, in the hand camera, or by the power drive, in the automatic design (Figs. 33 and 53). Another illustration is afforded by the Piserini and Mondini

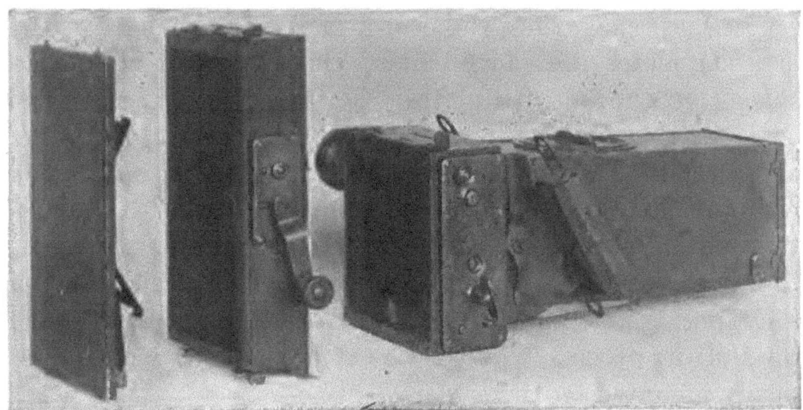

Fig. 33.—U. S. Air Service hand camera, with two-compartment magazine.

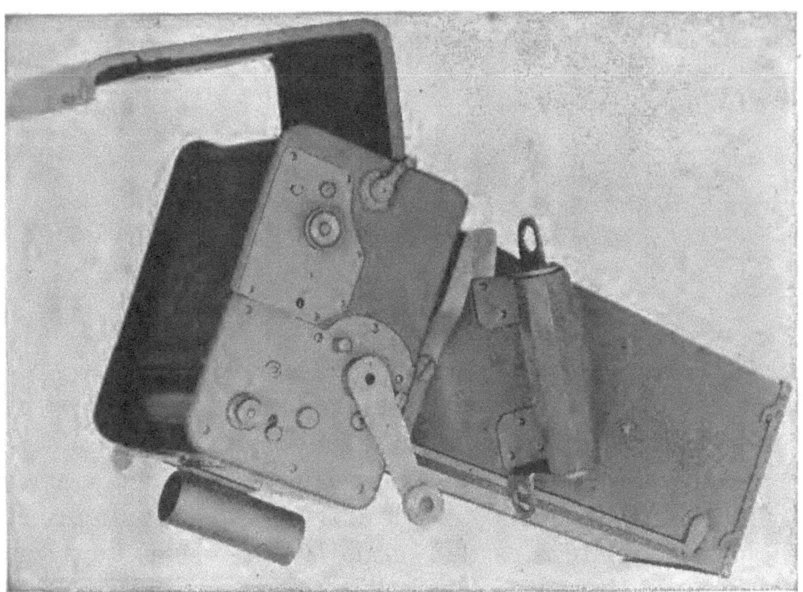

Fig. 34.—Film type hand camera.

magazine, in which the operation of changing is performed by a back-and-forth motion of a hand-grip, which also sets the camera shutter (Fig. 47).

In these magazines the center of gravity changes as the exposed plates are moved over, and only half the inside space is occupied with plates. These objections are overcome in the Chassel form, where both compartments are always full. Transfer of the bottom exposed plate from one compartment to the other is compensated for by the simultaneous shift of the top plate in the receiving compartment, to the feeding side. In a modification of this idea by Ruttan the exposing position is when the plates are half-way through the shifting process, whereby the magazine may be symmetrically mounted on the camera body.

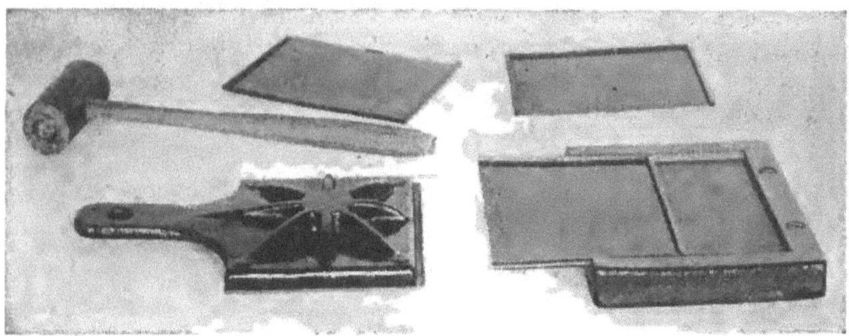

Fig. 35.—Apparatus for straightening plate sheaths.

Other more complicated magazines have been designed, some of which are shown in the diagrammatic *ensembles* of Figs. 32 and 48. In the Jacquelin, the main body of plates is raised while the bottom (exposed) plate is folded against the side. The main body of plates then drops back to place; the exposed plate is carried on upward and folds down on the back of the pile. The Bellieni magazine system uses three, a central feeding one and two below for receiving, one on each side of the camera body. These were made solely for attachment to captured German cameras. In the Fournieux magazine the plates are carried in an interior rotating box. The plate to be exposed is dropped off the front of the pile, down to the focal plane, and after exposure is picked up and placed at the back of the pile, which has turned over in the meanwhile. The deRam rotating magazine is described in connection with the camera of which it is an essential part (Fig. 52).

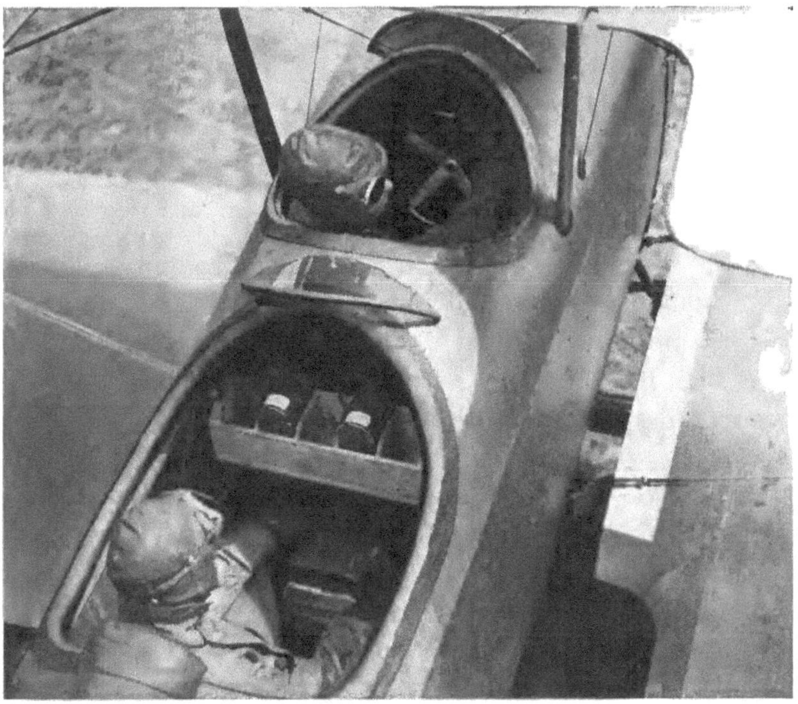

Fig. 36.—Training plane equipped for photography, showing "L" camera in floor mount and magazine rack forward of the observer.

For the protection of the plates during their manipulation, and in the camera, all plate magazines thus far developed carry them in thin *metal sheaths.* These add greatly both to the weight and to the time necessary to handle the plates, but no means have as yet been found for dispensing with them. Fig. 35 shows a representative sheath or septum, as used in the L camera. On three sides the edge is bent up and turned over, forming a ledge for the plate to press against. The fourth side is left open for inserting the plate, which is then held in by a small upward projecting lip, and kept close against the ledges by narrow springs at the sides. To insert or remove the plate the projecting lip is depressed, either by springing the sheath by pressure from the sides or by using an appropriate tool.

Care of sheaths. Unless systematically taken care of, plate sheaths become bent or dented They are then a menace to camera operation, catching or jamming in the plate changing process, breaking plates and damaging camera mechanisms. In order to maintain them flat and true, steel forms are necessary on which the sheaths may be hammered to shape with a mallet (Fig. 35).

Magazine racks. Reconnaissance and mapping call for a number of exposures much greater than the capacity of one 12, 18, or 24 plate magazine. Additional magazines must therefore be carried. These should be in racks convenient to the observer (Fig. 36), securely held yet capable of quick removal and insertion. In the rack designed to carry two of the metal magazines for the American L Camera, the magazines slide into loose grooves formed by a metal lip. They are prevented from slipping out by a spring catch, past which they slide when inserted but which is instantly thrown aside by pressure of the thumb as the hand grasps the magazine handle for removal.

CHAPTER VII

HAND-HELD CAMERAS FOR AERIAL WORK

Field of Use.—The first cameras to be used for aerial photography were hand-held ones of ordinary commercial types. Indeed the idea is still prevalent that to obtain aerial photographs the aviator merely leans over the side with the folding pocket camera of the department store show window and presses the button. But the Great War had not lasted long before the ordinary bellows focussing hand camera was replaced by the rigid-body fixed-focus form, equipped with handles or pistol grip for better holding in the high wind made by the plane's progress through the air. Even this phase of aerial photography was comparatively short-lived. The need for cameras of great focal length, and the need for apparatus demanding the minimum of the pilot's or observer's attention, both tended to relegate hand-held cameras to second place, so that they were comparatively little used in the later periods of the war.

Yet for certain purposes they have great value. They can be used in any plane for taking oblique views, and for taking verticals, in any plane in which an opening for unobstructed view can be made in the floor of the observer's cockpit. They can be quickly pointed in any desired direction, thus reducing to a minimum the necessary maneuvering of the plane, a real advantage when under attack by "Archies" or in working under adverse weather conditions.

For peace-time mapping work the hand-held camera, when equipped with spirit-levels on top, and when worked by a skilful operator, possesses some advantages over anything short of an automatically stabilized camera. For experimental testing of plates, filters and various accessories, the ready accessibility of all its parts makes the hand-held camera the easiest and most satisfactory of instruments.

The limitations of the hand-held camera lie in its necessary restriction to small plate sizes and short focal lengths, and in the fact that it must occupy the entire attention of the observer while pictures are being taken—the latter a serious objection only in war-time.

Essential Characteristics.—In addition to the general requirements as to lens, shutter and magazine, common to all aerial cameras, the hand camera must meet the special problems introduced by holding in the hands, especially over the top of the plane's cockpit. An exceptionally good system of handles or grips must be provided whereby the camera can be pointed when pictures are taken, and held while plates are being changed and the shutter set. The weight and balance of the camera must be correct within narrow limits; the wind resistance must be as small as possible; the shutter release must be arranged so as to give no jerk or tilt to the camera in exposing.

As to the method of holding the camera, a favorite at first among military men was the pistol grip, with a trigger shutter release (Fig. 37). Because of the size and weight of the camera the pistol grip alone was an inadequate means of support and additional handles on the side or bottom had to be provided for the left hand. Small (8×12 cm.) pistol grip cameras were used to some extent by the Germans (Fig. 42), and a number of 4×5 inch experimental cameras of this type were built for the American Air Service (Fig. 37). But the grasp obtained with such a design is not so good as is obtained with handles on each side or with flat straps to go over the hands. The camera balances best with the handles in the plane of the center of gravity. As to weight, no set rules are laid down, but experience has shown that a fairly heavy camera—as heavy as is convenient to handle—will hold steadier than a light one. The American 4×5 inch cameras described below weigh with their magazines in the neighborhood of twelve pounds.

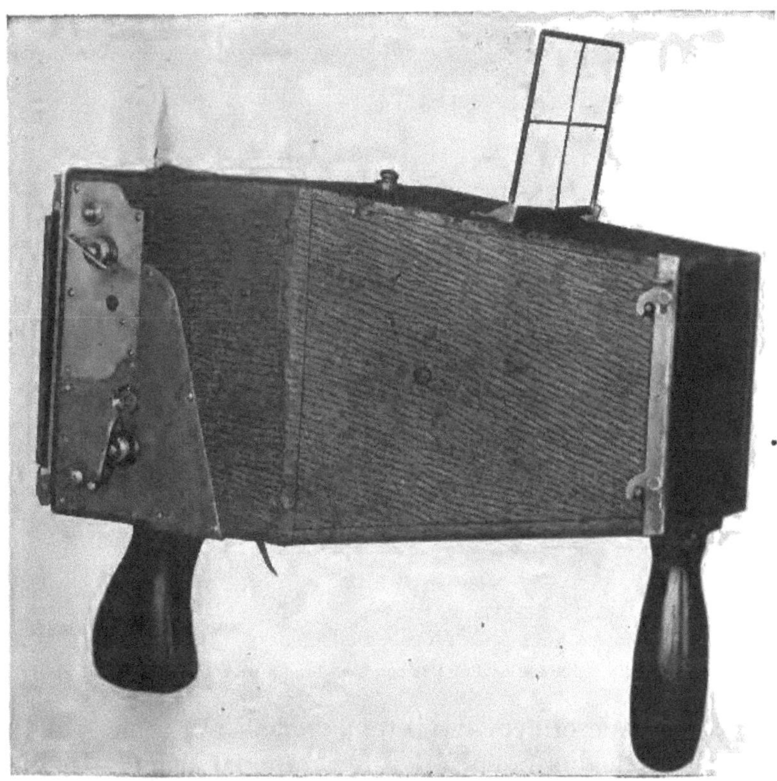

Fig. 37.—Pistol-grip aerial hand camera.

Representative Types of Hand-held Cameras.—French and German hand-held cameras are essentially smaller editions of their standard long-focus cameras, and a description of them will apply to a considerable extent to the large

7

cameras to be discussed in a later chapter. The English and American hand-held cameras are generally quite different in type from the large ones, which are used attached to the plane.

The French hand-held camera uses 13×18 centimeter plates, carried in a deMaria magazine, and has a lens of 26 centimeters focus. The shutter is the Klopcic self-capping

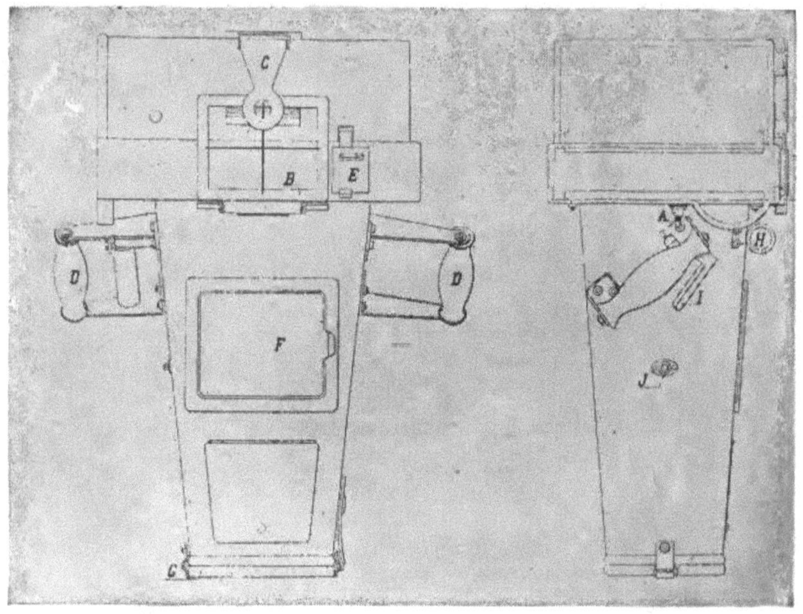

FIG. 38.—Diagram of French (deMaria) 26 cm. focus hand camera, using 13x18 cm. plates.

type already described, and is removable. The camera body, built of sheet aluminum, takes a pyramidal shape. In Fig. 38, *A* is the shutter release and *B* the rectangular sight, of which *C* is the rear or eye sight. The complete sight may be placed either on the top or on the bottom of the camera. At *D* are the handles, sloping forward from top to bottom; *E* is a catch for holding the magazine; *F* is a door for reaching the back of the lens and the lens flap; *G* is a snap clasp for holding the front door of the camera closed; *H* is a ring for attaching a strap to go around the observer's neck; *I* is the lever which opens the flap behind the lens and releases the focal-plane shutter; *J* is a snap catch for holding the front door of the camera open.

The operations with this camera are three in number. Starting immediately after the exposure, the camera is pointed lens upward and the plate changed by pulling the inner body of the magazine out and then in; next the shutter is set; then the camera is pointed, and finally exposed by a gentle pull on the exposing lever.

The English hand-held camera (Fig. 186). This differs from the French in the size of plate (4×5 inch), in the shape of the camera body, which is circular, and in the type of shutter, which is fixed-tension variable-opening. In the longer focus camera (10 to 12 inch) the shutter is self-capping, and the aperture is controlled by a thumb-screw at the side. In the smaller (6 inch) a lens flap is provided in front of the lens and the shutter aperture is varied by a sliding saddle and cord. The handles of the camera are placed vertical, instead of sloping as in the French. The shutter is released by a thumb-actuated lever. Double dark slides are used, as the multiple plate magazine has not found favor in the English service.

The German hand-held camera (Fig. 42). The German hand-held camera is, like their whole series, built of canvas-covered wood, the body having an octagonal cross-section. It is equipped with the Ica shutter and uses the Ernemann six plate (13×18 cm.) magazine. The excellent system of grips by which the camera is held and pointed is an especially commendable feature. On the right-hand side is a handle similar to the French type, but carefully shaped to fit the hand. The left-hand grip consists of a long, rounded block of wood running diagonally from top to bottom of the side, with a deep groove on the forward side for the finger tips, while over the hand is stretched a leather strap, the whole aim being to provide an absolutely sure and comfortable hold on the camera during the plate changing and shutter setting operations.

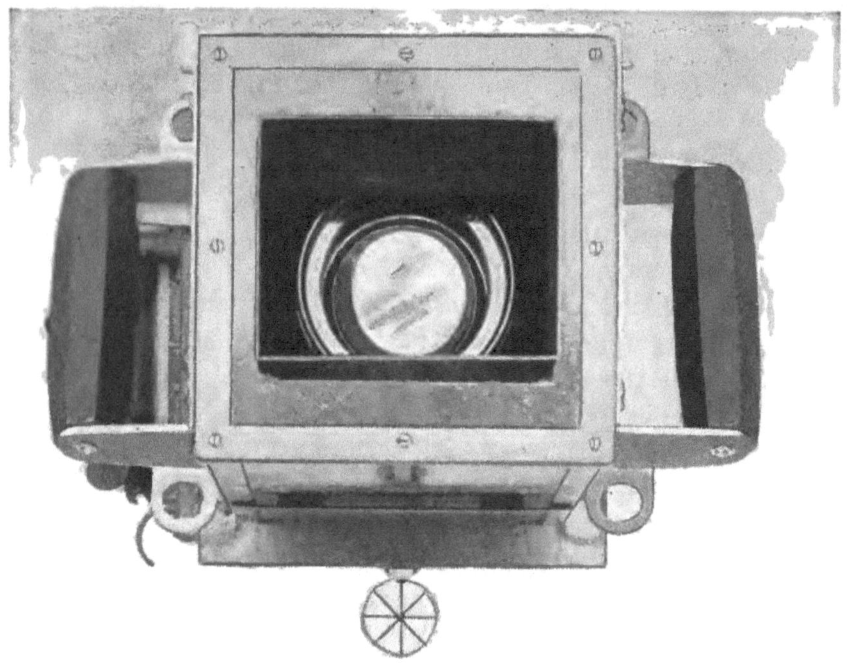

FIG. 39.—Front view of U. S. aerial hand camera, showing lens flaps partly open, and details of tube sight.

United States Air Service hand cameras. The hand camera developed for the United States Air Service and manufactured by the Eastman Kodak Co. is made in three models, using the bag magazine, a two-compartment magazine, and roll film, respectively. The shutter is of the fixed (one or two) aperture variable tension type, built into the camera. A distinctive feature is the double lens flap, in front of the lens actuated by the thumb pressure shutter release (Fig. 39). In the bag magazine camera the shutter is set, as a separate operation, by a wing handle, and a similar handle controls the tension adjustment. In the two-compartment type (Fig. 33) the shutter wind-up is geared to the plate changing lever, so that but one operation is necessary to prepare the camera for exposure. In the film type (Fig. 34) a single lever motion sets the shutter and winds up the film ready for the next exposure. After the last exposure of all the film is wound backward on its own (feeding) roller before removing from the camera. The film is held flat by a closely fitting metal plate behind, and by guides at the edges in front, an arrangement which with small sizes works fairly well although the exquisite sharpness of focus attainable with plates is not to be expected. The saving in weight made possible by the use of film in place of plates in metal sheaths is about three pounds per dozen exposures.

In all these cameras the sight—a tube with front and back cross wires—is placed at the bottom. This position has been found the most convenient for airplane work, as it necessitates the observer raising himself but little above the cockpit, a matter of prime importance when the tremendous drive of the wind is taken into account.

CHAPTER VIII

NON-AUTOMATIC AERIAL PLATE CAMERAS

The ideal of every military photographic service has been an automatic or at least a semi-automatic camera, in order to reduce the observer's work to a minimum. Yet as a matter of fact almost all the aerial photography of the Great War was done with entirely hand-operated cameras. The primary reason for this was that no entirely satisfactory automatic cameras were developed, cameras at once simple to install and reliable when operated. Even the propeller-drive semi-automatic L type of the British Air Service was very commonly operated by hand, for many of the pilots and observers regarded the propeller merely as another part to go wrong.

Any automatic mechanism in the airplane must work well in spite of vibration, three dimensional movements, and great range of temperature. The requirements were well recognized when the war closed, but had not yet been met. Careful study of the conditions and needs by competent designers of automatic machinery may be expected to result at an early date in reliable cameras of the automatic type, but the description below of hand-operated cameras really covers practically all the cameras found satisfactory in actual warfare.

General Characteristics of Hand-operated Cameras.—As distinguished from the hand-held cameras the larger hand-operated cameras are characterized by the greater focal length of their lenses, the size of plate employed, and the manner of holding—by some form of anti-vibration mounting attached directly to the fuselage.

Except for the early English C and E type cameras which called for 10 inch lenses and 4×5 inch plates, the general practice at the close of the war by agreement between the French, English and American Air Services, was for the use of 18×24 centimeter plates and for lenses with focal lengths of approximately 25, 50 and 120 centimeters. The English also made use of a 14 inch (35 centimeter) lens, and never made a regular practice of anything larger than 50 centimeters. The Germans and Italians restricted themselves to the 13×18 centimeter size of plate, while a lens of 70 centimeters focal length was standardized with the Germans, in addition to the 25, 50, and 120 centimeter.

The particular focal length was determined by the nature of the photographic mission. Where large areas were to be covered at low altitudes or without the demand for exquisite detail, the shorter focus lenses suffice. The most commonly used lens in the French Service was the 50 centimeter, while the 120 was employed when high flying was necessary or when minute detail was required. As already mentioned, the common practice was to keep cameras of all focal lengths available, but the ideal at the close of the war was to have the camera nose and lens a detachable unit, so that any focal length desired could be secured with the same camera body.

The standard French camera. The hand-held form of French camera has already been described. The cameras for larger plate sizes and longer focus lenses differ only in the addition of a Bowden-wire distance release for the shutter and in the use of the Gaumont magazine which operates without the necessity of pointing the exposed side of the magazine upward. Fig. 40 illustrates the 50 centimeter camera, and Fig. 41 the 120.

The German Ica cameras. These are larger editions of the light wood hand camera already described, but with the

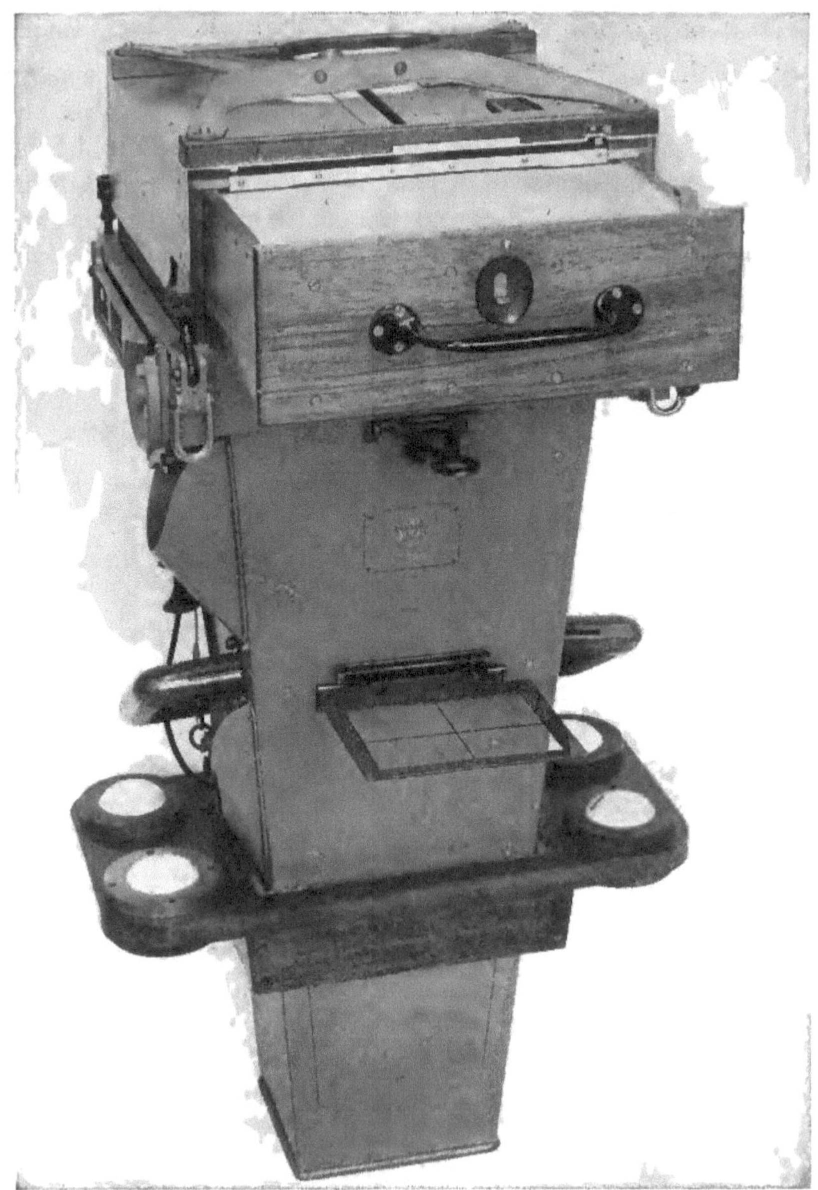

FIG. 40.—50 centimeter deMaria hand operated camera on tennis ball mounting.

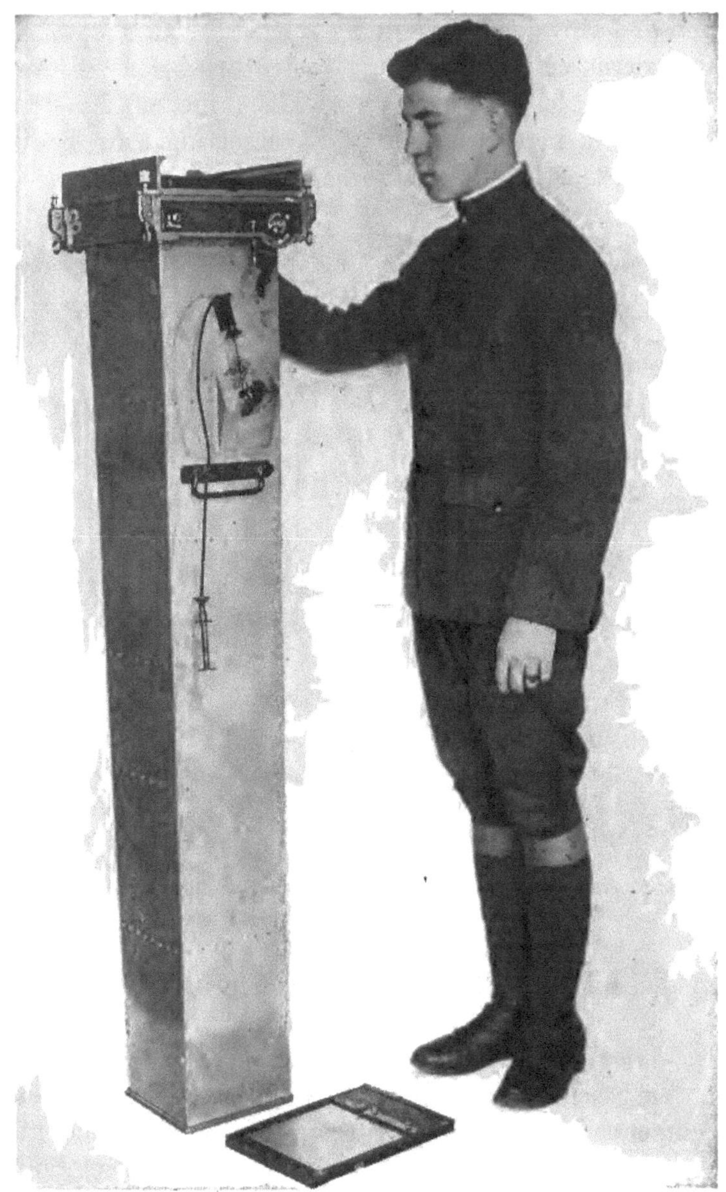

Fig. 41.—120 centimeter deMaria camera.

addition of a Bowden-wire shutter release. The body of the larger cameras carries a distinctive feature in the distance control of the lens diafram, worked by means of a lever which actuates racks, pinions and connecting rods leading to the lens. On the side of the camera body a shallow box

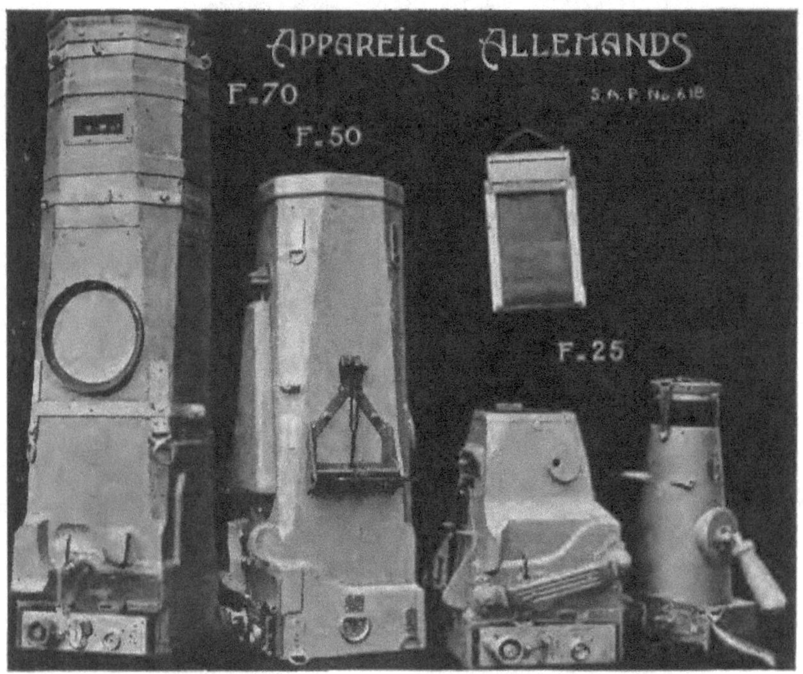

Fig. 42.—German aerial cameras.

is provided for carrying the color filter in its bayonet joint mount to fit on the lens (Figs. 42 and 43).

The hand-operated bag-magazine camera of the United States Air Service (Type M) is similar to the small hand-held camera, but differs in three respects: a removable shutter (of the variable-tension fixed-aperture type) embodying an auxiliary curtain for capping during the setting opera-

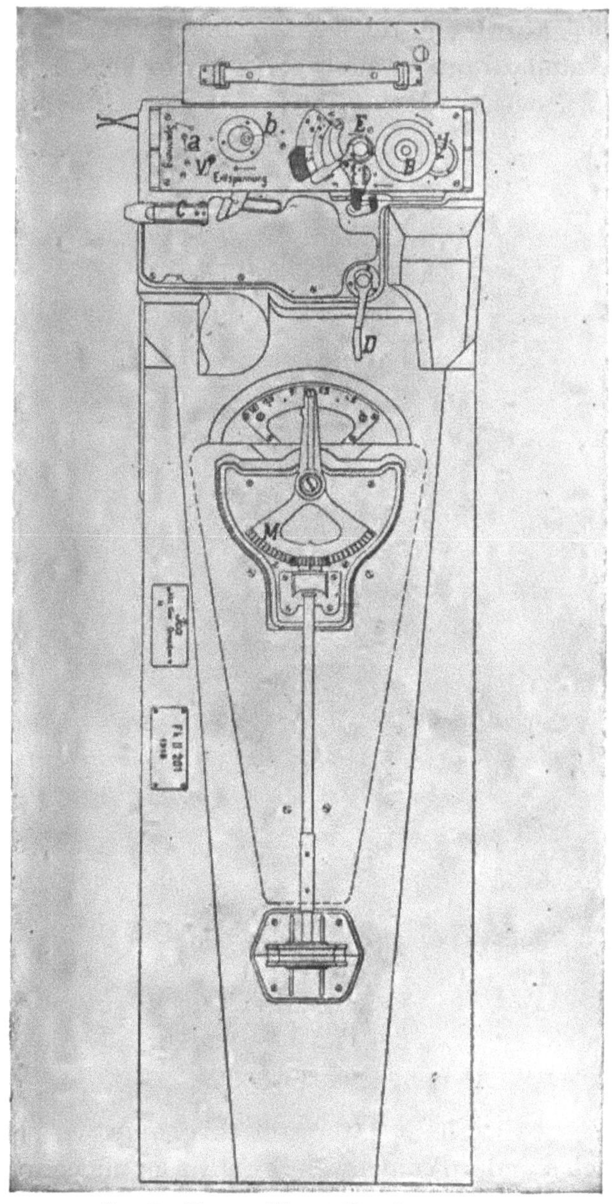

FIG. 48.—Diagram of German 50 centimeter camera.

tion; a Bowden-wire shutter release; and the employment of a set of standard interchangeable cones to hold lenses of several focal lengths. The 20 inch and 10 inch cones are

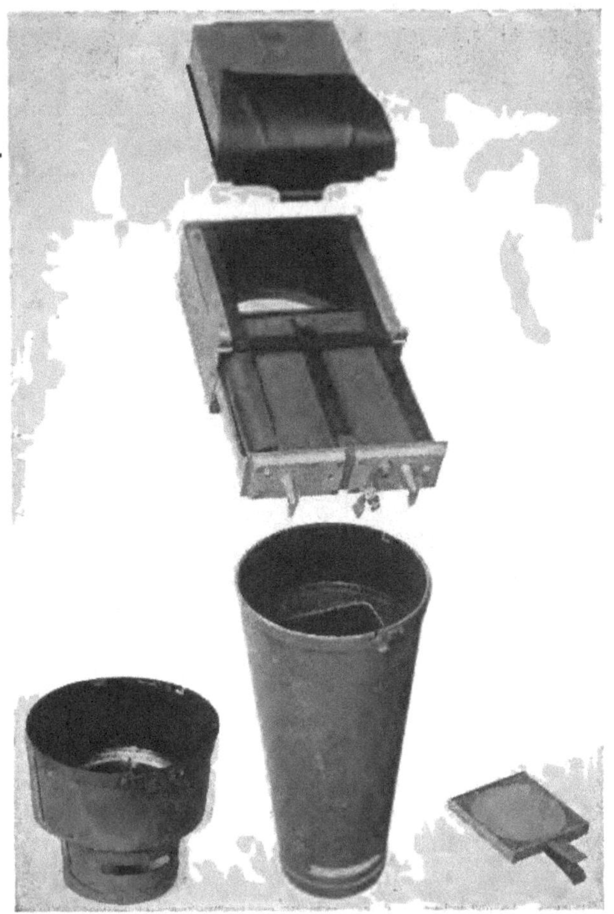

Fig. 44.—U. S. hand-operated aerial camera (type M) with 10 and 20 inch cones.

shown in Fig. 44. The operation of this camera is similar to the French standard cameras, but not so simple because of the number of motions required in manipulating the bag.

Its chief objection for war work lies in fact in the magazine, which should be superseded by a two-compartment or other satisfactory type of plate changing chamber. The camera alone, with 20 inch cone, weighs approximately 40 pounds; the loaded magazine, with its plates in metal sheaths, 15 pounds.

The English C and E type cameras. The C and E type cameras have now chiefly an historic interest. They were

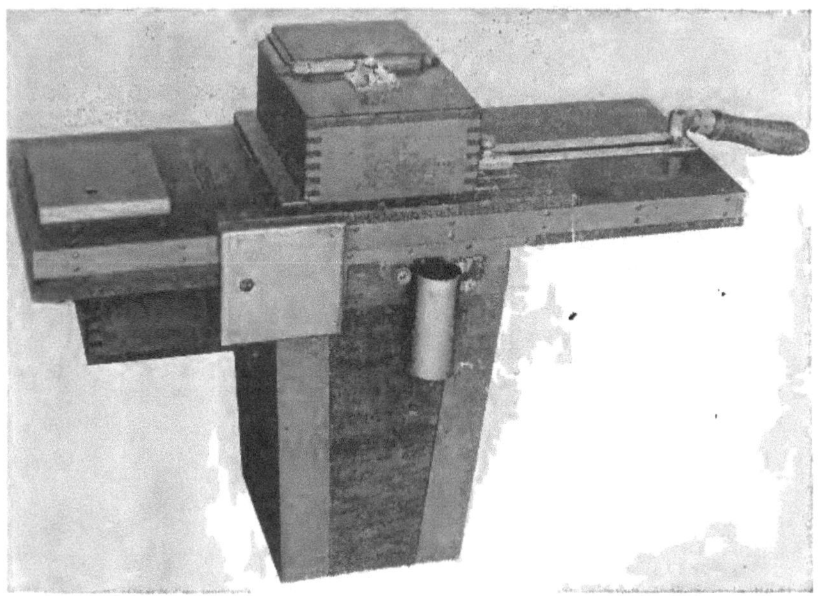

Fig. 45.—English C type aerial camera.

the first used in the English service, fixed to the fuselage, and were later used in training work in England and in the United States. They were never built for plates larger than 4×5 inch nor for lenses of more than 12 inch focus, a limitation set by the lenses available at the time of their design.

In several respects the mode of operation of the two types is the same. The unexposed plates are held in a magazine lying above the camera, in the axis of the lens (Fig. 32).

After exposure the bottom plate is carried to one side and allowed to fall by the action of gravity into the receiving magazine. In the C type (Fig. 45) an opaque slide is drawn between the lens and the (variable-opening) shutter during the setting operation. During the exposure period this slide projects into a compartment on the opposite side of the

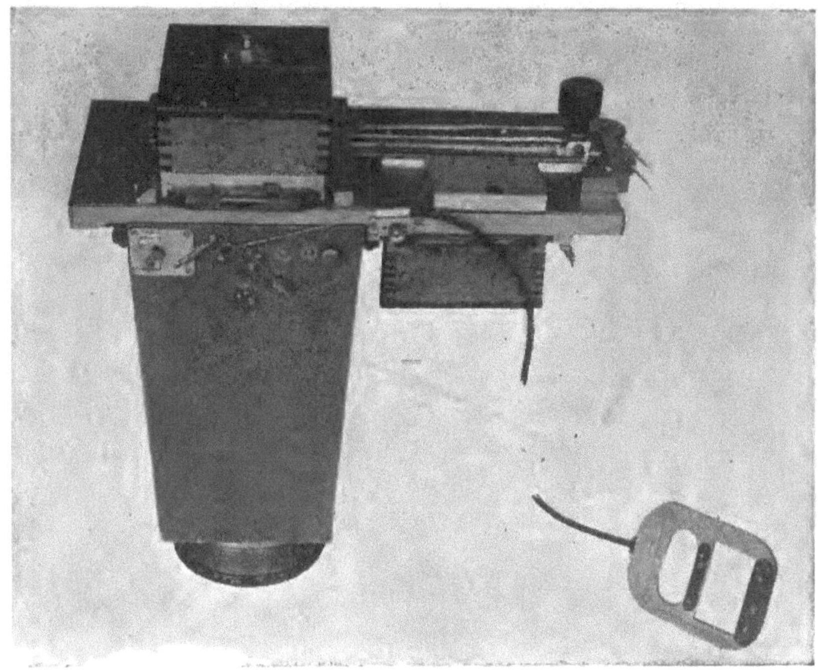

Fig. 46.—English type "E" hand-operated plate camera.

camera from the receiving magazine, thus making the camera mechanism three plates wide. In the E type (Fig. 46), a flap over the lens makes it possible to dispense with the sliding screen, and reduces the camera to about the width of two plates. In the C type the plates are changed by a handle on top of the camera; in the E type provision is made for distance control by cords, and for shutter release by a Bowden wire. In both cameras the operation of plate changing also sets the shutter, a definite advance over the two preparatory motions in the French apparatus. The C type was constructed of wood, the E of metal.

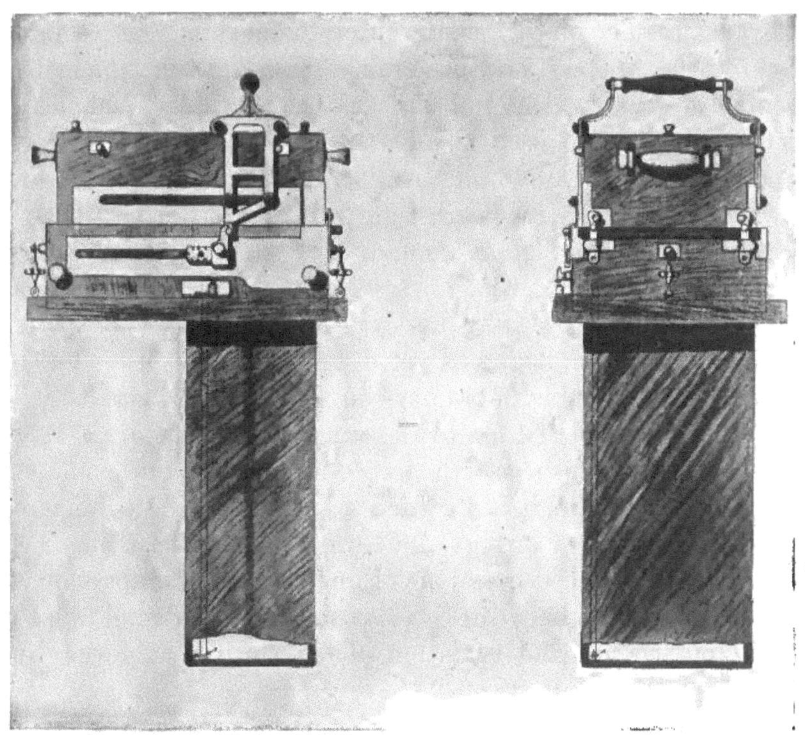

FIG. 47.—Italian (Piserini and Mondini) two compartment magazine hand-operated camera.

Italian two-compartment magazine camera. A camera designed by Piserini and Mondini was used to some extent by the Italian service toward the close of the war (Fig. 47). This has the desirable feature just noted in the C and E cameras: the operations of plate changing and shutter setting are performed in a single motion. Unlike those cameras, however, the plates are changed from one compartment to another of the magazine already described, without dependence on gravity, by an entirely positive shifting action. The setting of the self-capping focal-plane shutter is accomplished by a projecting finger engaging the shutter mechanism. Cameras of this general type, built for 18 by 24 centimeter plates, with interchangeable lens cones, removable shutters, and preferably magazines in which the center of gravity does not shift as the plates are changed, represent the next step in advance of the French practice, and may indeed prove all that is necessary or desirable in camera complexity for peace-time photography from the air.

The standard Italian camera and similar types. The camera (Lamperti) which the Italian Air Service used almost exclusively during the war exemplifies a type quite different from anything as yet described (Figs. 48 and 49). Plates to the number of twenty-four (13×18 cm.) are loaded into a chamber at the top of the camera. Each plate is held in a septum furnished with projecting lugs at one end. A lever acting through a Bowden wire, exposes the bottom plate, which then swings downward about these lugs as pivots, and is forced by a pair of fingers into a compartment at the side. The between-the-lens shutter has a single speed of 1/150 second, and variation of exposure is achieved by altering the lens aperture.

The great advantage of this camera is its simplicity, a single motion performing all the operations. Its disadvantages are its dependence on gravity for operation, its between-the-lens shutter, the limitation set to the number of exposures, and the necessity for removing the whole camera to take out the plates for developing. In actual practice the camera has worked out well. The better light found in the Italian as contrasted with the northern theatre of war makes the between-the-lens shutter at high speed adequate,

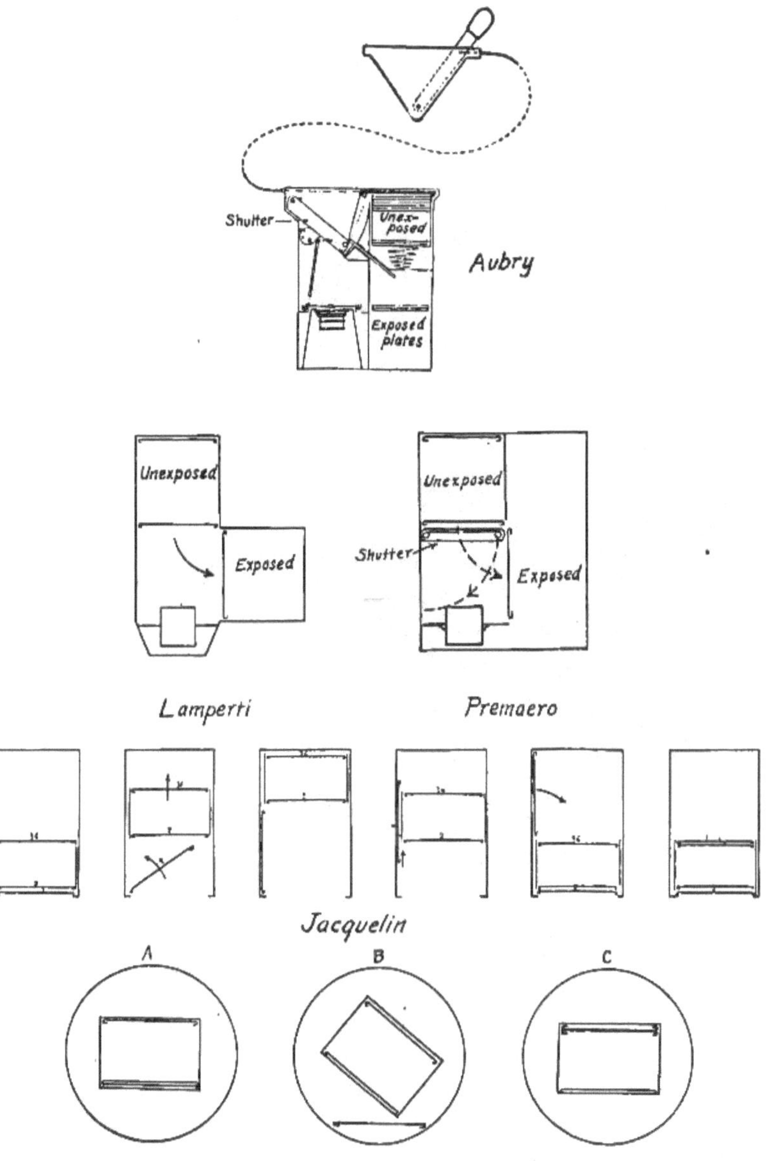

Fig. 48.—Various plate changing devices.

while the limitation to the number of exposures has been met by carrying several complete cameras in each plane. Because of the Bowden-wire operation these cameras need not be accessible to the observer or pilot, so that the practice

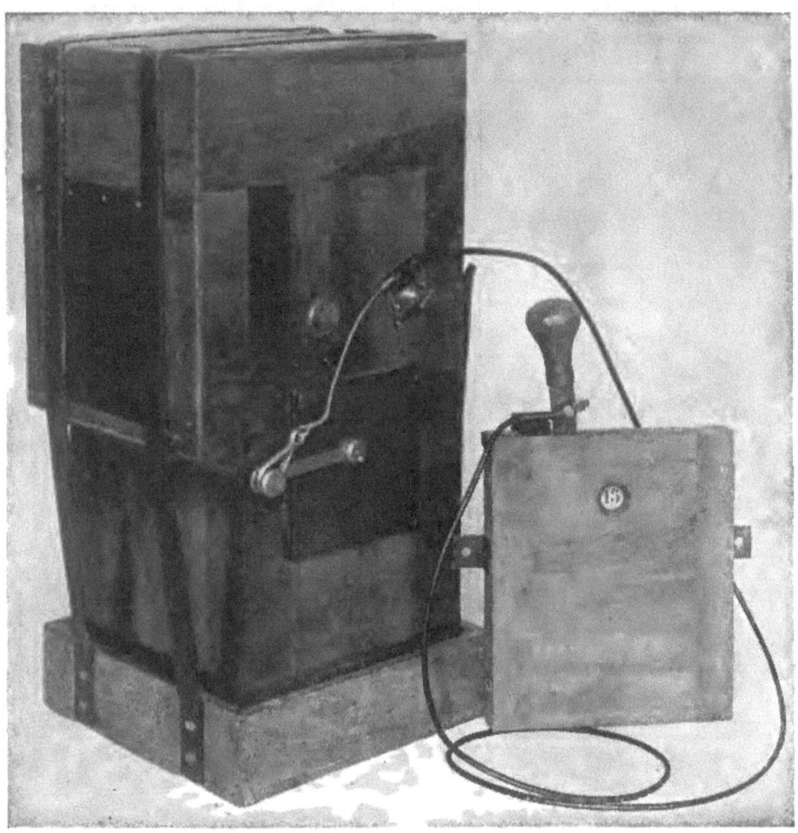

FIG. 49.—Italian (Lamperti) single-motion plate camera, on anti-vibration tray.

of carrying them in single-seaters was common. Attempts at standardization of Allied practice through the adoption of standard lens cones were, of course, out of the question with this camera. With its limitations of shutter efficiency and plate size it is doubtful whether it would have been satisfactory outside the service for which it was developed.

The limitations set by the between-the-lens shutter in this type have been overcome in an experimental camera along similar lines made by the Premo Works of the Eastman Kodak Company, and in the French Aubry model (Fig. 48). These employ focal-plane shutters which swing out of the way and are set as the exposed plate swings or drops to the receiving chamber. The dependence on gravity in this type could doubtless be avoided by positive finger mechanisms. If so, the resultant cameras, set and exposed by a single motion, would acquire a highly desirable simplicity of operation. They would have peculiar merit because of the very short interval required between exposures—a characteristic needed for making low stereo-oblique views. The cameras just mentioned have, however, departed far in form from the lines of standardized practice and have not been followed up.

CHAPTER IX

SEMI-AUTOMATIC AERIAL PLATE CAMERAS

In the hand-operated camera the limit to progress is set when the number of operations is reduced to a minimum. In cameras using the larger sizes of plates a reduction in the number of operations almost inevitably results in inflicting considerable muscular labor upon the operator. Furthermore, distance operation becomes difficult to arrange for, because the common reliance—the Bowden wire—is unfitted for heavy loads. Consequently, for setting the shutter and changing the plates we must resort to some other source of power than the observer's arm. Air-driven turbines or propellers have been used on aerial cameras, as well as clock-work, and also electric power, the latter derived either from a generator or from storage batteries. The relative merits of these sources of power form the subject of a separate chapter. Mention only is here made of the form of drive actually employed in connection with the various cameras.

The term *semi-automatic camera* is best used to designate that type in which the observer (or pilot) has merely to release the shutter, after which the mechanism performs all the operations necessary to prepare for the next exposure. There has been some difference of opinion as to whether it is ever advisable to go further than this with plate cameras. The English Service holds that completely automatic exposing, in addition to plate changing, is apt to encourage the making of many more pictures than necessary, involving carrying an excessive weight of plates. The French Service has rather generally favored entirely automatic cameras in theory, although during the war practically all the work of the French army was done by the hand-operated cameras already described.

The English L Type Camera.—The L, a modification of the earlier C and E models, differs from its predecessors chiefly in the addition of a mechanism which when connected

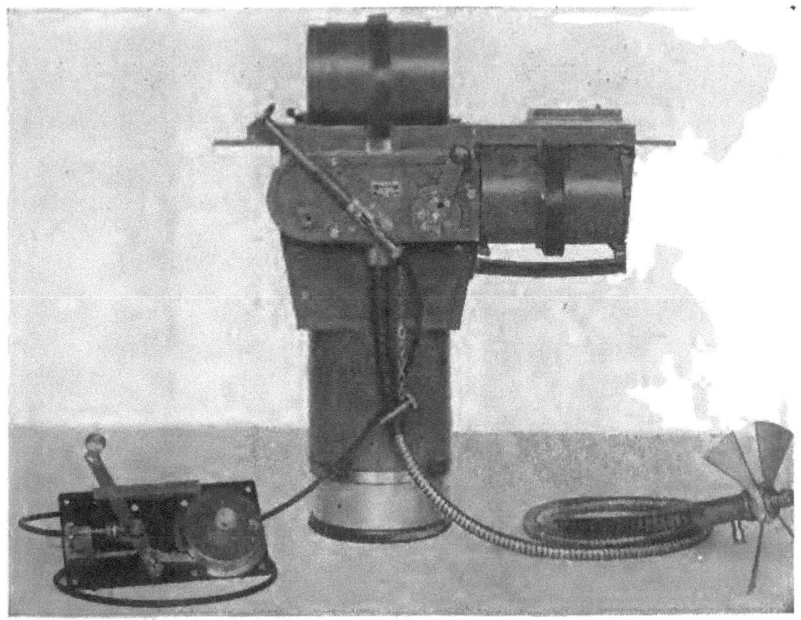

FIG. 50.—American model, English "L" type semi-automatic camera.

with a suitable source of power can be used whenever desired for changing the plates and setting the shutter. As in the C and E types, all unexposed plates are carried in a magazine above the camera, while the exposed plates are shifted in a horizontal direction to one side and fall thence to a receiving magazine.

Fig. 50 shows the American model, which is a copy, with modifications, of the original English design. Its weight with one loaded magazine is about 35 pounds. Its manner of functioning may be studied from the picture of the mechanism (Fig. 51). The part of the mechanism to the left is inoperative during hand operation, and the large toothed wheel is locked by the removable pin shown hanging on its chain in Fig. 50. To change a plate and set the shutter

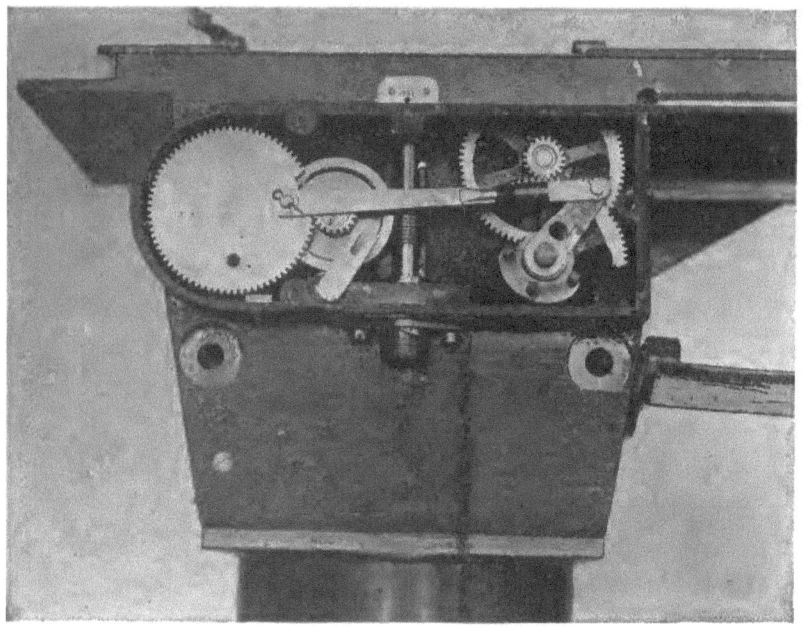

Fig. 51.—Mechanism of "L" camera.

the projecting lever (Fig. 50) is thrown over and back. This causes a sliding tray, in which the exposed plate rests, to travel to the right, over the receiving magazine, where the plate is dropped. After this the tray returns to the left exposing position. Simultaneously the shutter is wound up. Exposure is made either by pressing down upon the plunger, or better, by using a Bowden wire. Provision for both methods of exposing, one for the pilot and one for the observer, is shown in Fig. 81. The shutter is th variable-aperture type already described, provided in addition with a tension adjustment on the back of the camera. A flap behind the lens does the capping during the setting operation.

For power operation the camera is connected through a flexible shaft with a wind driven propeller (Figs. 50, 83 and 84). The locking pin is now moved over from the toothed wheel to the lever arm, so that the rotation of the worm driving the large toothed wheel forces the lever through its plate changing motion. To prevent repetition, a part of the periphery of the toothed wheel is cut out, so that it stops when its cycle is run. When the Bowden wire actuates the shutter release it forces the toothed wheel around into engagement (aided by one spring tooth) and so starts the cycle once more.

When connected with the air propeller the worm is rotated continuously. Other sources of power—an electric motor, for instance—can be attached through the same kind of flexible shaft. If an electric motor is employed it may be run continuously or it may be operated with an insulated sector introduced into the large toothed wheel so that the electric circuit is broken and the motor stops until the wheel is once more forced around by the exposing lever.

Faults of the L camera. The L camera was the mainstay of the English Air Service. In fact for the last two years of the war it was practically the only camera the English used, and they thought highly of it. It is, of course, subject to the limitation of small plate size and short focus lens. It is in many ways an inconvenient camera to handle. For instance, the upper magazine cannot be closed or removed until all the plates are passed through. Its dependence upon gravity for the plate changing operation is a fundamental weakness, responsible for its frequent tendency to jam in the air. Experience made the English observers very expert in relieving these jams. Sometimes they would turn the propeller backward (mounting it in an accessible position to provide for this contingency), sometimes they would shake or thump the camera. But while these makeshifts would serve to secure pictures—the chief object, of course, of the photographic service—they can scarcely be said to render the camera satisfactory.

Moreover, the propeller drive has not been universally approved, as it furnishes an additional mechanism to make trouble. Since it is not feasible to change from power to hand operation while in the air, the camera is put out of commission whenever the propeller or shaft is disabled. Bowden-wire controls for both plate changing lever and shutter release were common in the British service, which considered the extra operation or the extra muscular exertion unimporant when compared with the greater assurance of reliable action.

The English LB and BM Cameras.—During the closing months of the war an improved L type camera was constructed, the LB. This differs from the L in a number of detail changes, dictated by experience. The shutter is now made removable and self-capping. Pivoted lugs are provided to hold the exposed plate horizontal until the very instant it drops, in an effort to prevent jams caused by the plates piling up at an angle in the receiving magazine. The chief addition, however, is the provision of several interchangeable cones and cylinders, for carrying lenses of focal lengths from 4 to 20 inches. Fig. 95 shows the LB with 20 inch lens cylinder mounted on a bell crank support in the camera bay of an English plane.

The BM camera is but a larger edition of the LB, for 18×24 centimeter plates. It also carries several interchangeable lens cones.

The American model deRam camera.—The rotary changing box devised by Lieutenant deRam of the French army and incorporated in his entirely automatic plate camera, has been adapted by the American Air Service to a very successful semi-automatic camera. Fig. 52 shows the principle of this changing box. The pile of fifty plates, each in its sheath, is carried in a rectangular box open at top and bottom. The lower plate next the focal-plane shutter is first exposed; the pile then rotates about a horizontal axis through a complete turn. When the exposed plate arrives in a vertical position it is allowed to drop off, by the opening of cam actuated fingers, and lodges against the side of the enclosing camera box proper. Still further along in the cycle the plate is thrown off from its lodging place into a "scoop" on the top of the rotating container and laid on the top of the plate pile. Meanwhile the curtain of the focal-plane shutter winds up, at the same time that it is depressed out of the way of the revolving plate container. Although the plate changing operation depends on gravity, it nevertheless functions satisfactorily up to 30 degrees from the vertical.

The shutter in this model is the variable-aperture fixed-tension type, adjusted by pivoted idlers (Fig. 28). In the exposing position it runs within three millimeters of the plate surface, and is therefore of high efficiency for all openings. Capping during the operation of setting is performed by flaps at the bottom of the camera body. Interchangeable cones are supplied for lenses of various focal lengths.

For hand operation the changing box is turned over by means of a handle, which rotates four times for the complete cycle (Fig. 90). For semi-automatic operation an additional mechanism is provided on the side of the rectangular camera body, copied with some necessary modifications after the L camera power drive. From the observer's stand-

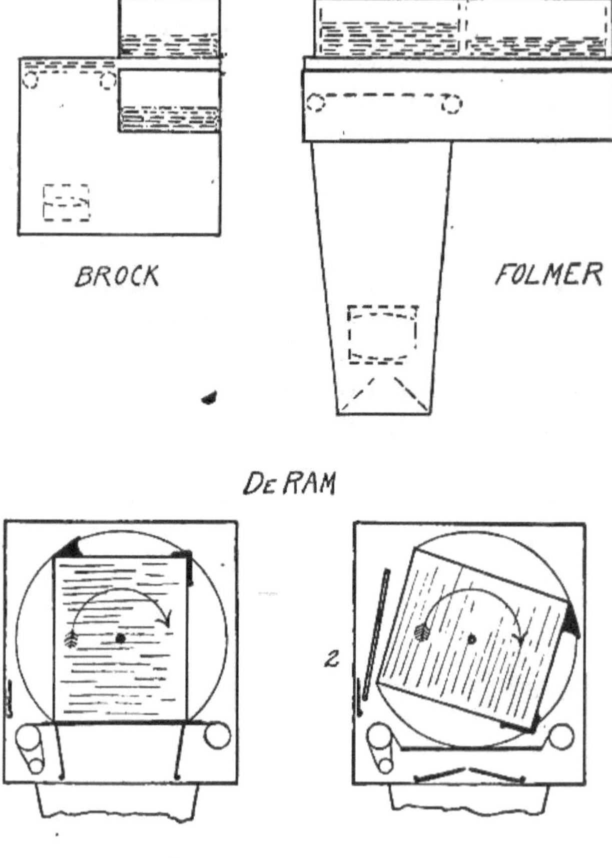

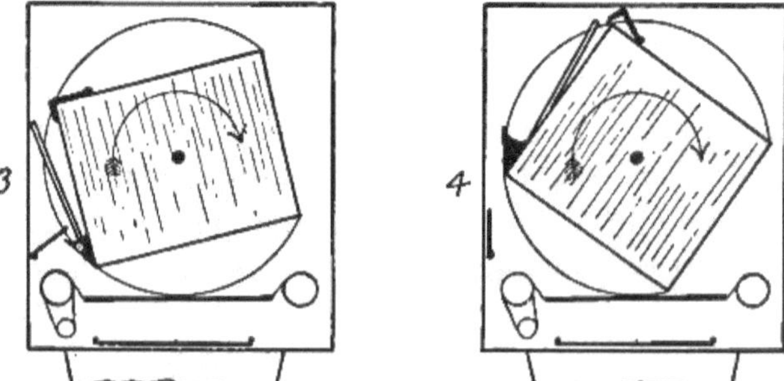

FIG. 52.—Diagram of automatic plate camera movements.

point the operation of the whole camera is the same as in the L camera, with the important exception that power operation in no way interferes with hand operation. Indeed, the hand can help out if the power flags or fails.

This camera is most satisfactorily driven by a 12 volt 1/10 HP electric motor working through a flexible shaft attached to a swivel connection at the front of the semi-automatic drive box. A change once every four to five seconds is possible, but greater speed is apt to throw the changing plate too violently for safety.

The chief practical objection to this camera is its bulk. Its great height makes it impractical for many planes. Its weight of nearly a hundred pounds is a formidable load for a plane to carry, but this is no more and probably less than that of any other camera when taken up with the same number of plates in magazines. The price paid for economizing in magazine weight is that the whole camera body, excluding the lens cone, must be carried to and from the plane for both loading and unloading.

CHAPTER X

AUTOMATIC AERIAL PLATE CAMERAS

General Characteristics.—The ideal in the automatic plate camera is to provide a mechanism which will not only change the plates and set the shutter, as does the semi-automatic, but make the exposures as well, at regular intervals under the control of the operator. Such a wholly automatic camera would leave the observer entirely free for other activities than photography and it is to meet this tactically desirable aim that the war-time striving for automatic cameras was due.

It is obvious that the one essential difference between the automatic and semi-automatic types lies in the self-contained exposing mechanism with its device for the timing of the exposures. There is no difficulty in arranging for the driving power to trip the shutter, but it is no easy matter to design apparatus which will space the exposures equally, and at the same time permit of a variation of the interval. It is indeed the crux of the problem of automatic camera design to provide for the easy and certain variation of the interval from the two or three seconds demanded for low stereoscopic views to the minute or more that high altitude wide angle mapping may permit. This problem is one intimately bound up with the question of means of power drive and its regulation, and will be treated in part in that connection. It is to be noted, however, that there are in general two modes of exposure interval regulation. One is by variation in the speed at which the whole camera mechanism is driven. The other is by the mere addition to a semi-automatic camera of a time controlled release which affects in no way the speed of the plate changing operation. In many respects the latter is the best way to make an automatic camera.

While the advantages of automatic cameras are great it must not be overlooked that a camera which can only be operated automatically is of limited usefulness. It is not suited for "spotting" at any definite instant, as, for illustration, at the moment of explosion of a bomb. It should, therefore, be the aim of the automatic camera designer to so build the apparatus that it can, at will, be used semi-automatically. In addition, to meet the contingency of any break-down in the source of power, the camera should be capable of hand operation, as in the case of the American semi-automatic deRam. In short, the automatic camera should not be a separate and different type; it should merely have an additional method of operation.

Certain desirable mechanical features of all aerial cameras have already been enumerated. Some of these may be repeated here with the addition of others peculiar to automatic cameras. As a general caution, mechanical motions depending on gravity or on springs should be avoided. Movements adversely affected by low temperatures (20 to 30 degrees below zero, Centigrade), are unsuitable. All adjustments called for in the air must be operable by distance controls whose parts are large, rugged, and not dependent on sound or delicate touch for their correct setting. The center of gravity of the camera should not change during operation (important in connection with the problem of suspension). The camera should work in the oblique as well as in the vertical position. The power required for operation must not exceed that available on the plane. Electrical apparatus, for instance, should not demand more than 100 watts.

Any devices which diminish the weight of the camera are

particularly desirable in automatic plate cameras, because of the large number of exposures which such cameras encourage. For instance, if the plates could be handled without placing them in metal sheaths we should gain a substantial reduction in weight (the sheaths weigh nearly as much as the plates) as well as in the time necessary for handling.

The Brock Automatic Plate Camera.—This camera is somewhat similar to the same designer's film camera, both in shape, in size, and in its employment of a heavy spring motor for the driving power. It uses 4×5 inch plates, and carries a 10 to 12 inch lens.

The plate-changing operation is unique. As shown diagrammatically in Fig. 52, the unexposed plates are carried in a magazine on top of the camera, the exposed ones in a magazine inserted in the body of the camera, directly below the unexposed magazine. The bottom plate of the exposed pile drops into a sliding frame and is carried along the top of the camera to the exposing position. After exposure, the plate is carried back and drops into the receiving magazine. In order for the plate to fall only the proper distance at each stage of the cycle, special plate sheaths are necessary. These are cut away to form edge patterns which clear or engage control fingers so as to ride or fall through the sliding frame as required.

The camera is entirely automatic in operation. Regulation of the exposure interval is by a special form of variable length escapement controlled through a Bowden wire, in a manner parallel to that in the Brock film camera, described elsewhere. These plate cameras were never produced in quantity.

Folmer 13×18 Centimeter Automatic Camera.—This camera, also never manufactured in quantity, is shown in Fig. 53, and a sketch of its manner of operation is included in the *ensemble* of automatic camera diagrams (Fig. 52). Its most distinctive feature is perhaps the use of a two compartment magazine. This is similar in form to the one already described in connection with the hand-held cameras, but larger, to hold eighteen 13×18 centimeter plates. The unexposed plates are placed in one compartment, and after exposure are shifted to the other. The transfer is effected

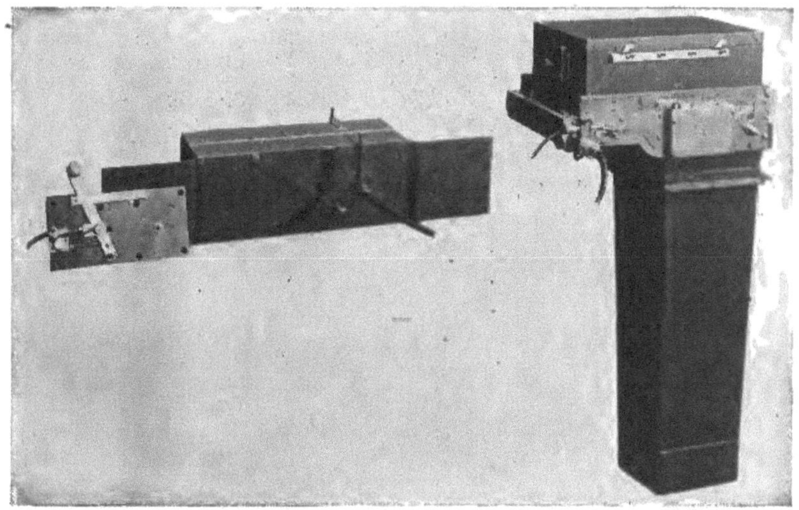

Fig. 53.—Folmer 13x18 centimeter automatic and semi-automatic plate camera.

by the motion of a rack, which is part of the magazine and which is driven by a toothed pinion, also part of the magazine, which in turn engages in a toothed wheel projecting upward from the camera body. This toothed wheel is turned first in one direction and then in the other by an arrangement of gears and levers driven by the source of power, which as shown in Fig. 53 is a wind turbine connected through a flexible shaft. Operation is either automatic or semi-automatic as desired, and the camera can be put through its cycle by hand if necessary.

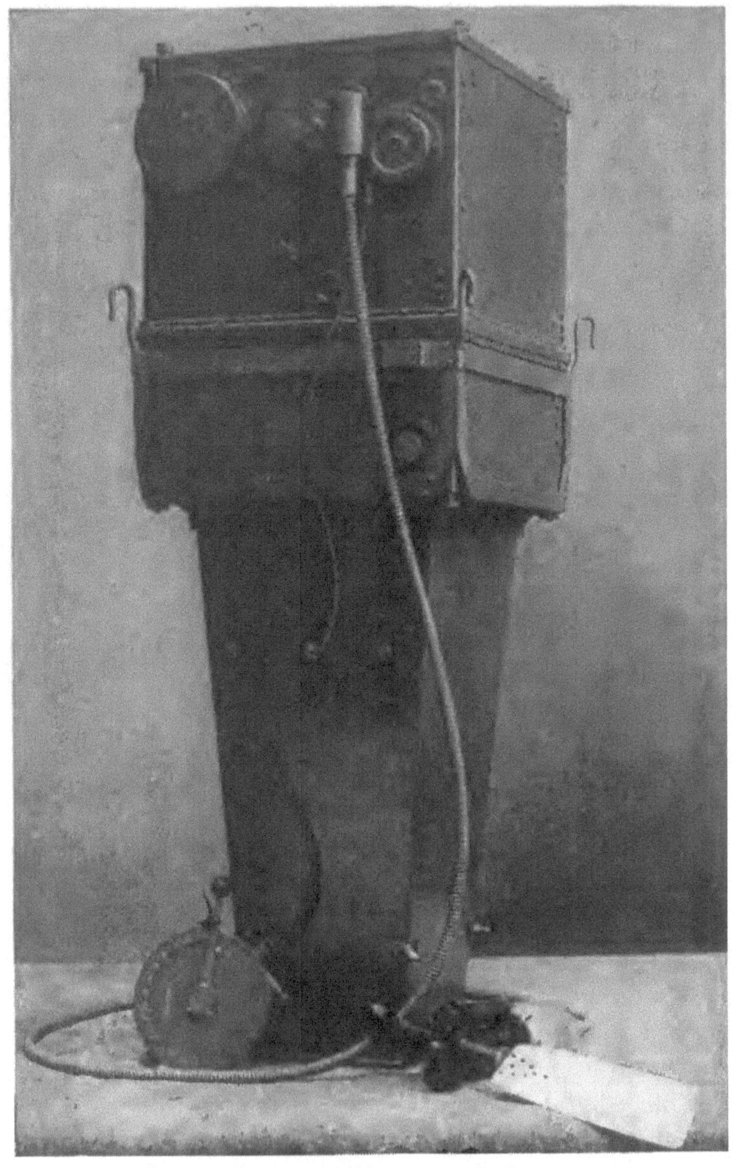

Fig. 54.—French model deRam automatic plate camera.

As with several other designs, the completion of the working model of this camera occurred after agreements had been reached by the Allies, as to plate size, standard lens cones, and other features, not easily incorporated in it, thus making manufacture inadvisable. The validity of the design for peace-time work is, of course, not affected by this fact.

The deRam Camera.—The only completely automatic plate camera actually produced commercially before the end of hostilities was the French model deRam (Fig. 54). Its plate-changing action has already been described in connection with the American semi-automatic model (Figs. 52, 90 and 91). It differs from the American model in the shutter, which is of the self-capping variety, carried on a rising and falling frame; and in the exposing mechanism. The latter embodies a clutch whose point of attachment to a uniformly rotating disc in the camera is governed through a Bowden wire, whereby the interval between the plate-changing operation and the shutter release is varied. The intervals are indicated by figures on the dial to which the observer's end of the Bowden wire is attached. The source of power for the camera is a constant speed propeller. Complete semi-automatic operation is not possible, as an interval of 1 to 2 seconds elapses between the time a single exposure is called for and its occurrence. No arrangement is provided for hand operation.

It will be noted that while this camera is a true automatic apparatus it does not meet even a majority of the requirements listed above as found desirable by experience. There exists a great opportunity for designing and developing an entirely satisfactory automatic plate camera—provided it is agreed that anything more than semi-automatic operation is ever advisable for plates.

CHAPTER XI

AERIAL FILM CAMERAS

The weight of the glass and the sheaths in the plate camera forms its most serious drawback. This weight must be reckoned at least three quarters of a pound for each 18×24 centimeter plate. Consequently, with the use of these large plates, and with the demands for ever increasing numbers of pictures to be taken on long reconnaissance flights, a serious conflict arises between the weight of the photographic equipment and the carrying capacity of the plane. Among plate cameras probably the most economical in weight is the deRam. It carries fifty 18×24 centimeter plates, and has a total weight of approximately 100 pounds. An advance to 100 or 200 plates—not feasible in the deRam construction—even if we assume the lightest possible magazines, would bring the weight of camera and plates to 150 or 200 pounds, which would be detrimental to the balance and would seriously infringe on the fuel carrying capacity and ceiling of any ordinary reconnaissance plane.

Early and persistent attention was therefore paid to the possibilities of celluloid film in rolls, as used so widely in hand cameras and in moving picture work. The two great advantages of film would be its practically negligible weight (approximately one-tenth that of plates, not including sheaths) and its small bulk, which would permit the greatest freedom in the development of entirely automatic cameras to make exposures by the hundreds instead of by the dozens. Certain disadvantages were foreseen at the outset: the difficulty of holding the film flat and immune from vibration in the larger sizes; the difficulty of quickly developing and drying large rolls; the question whether as good speed or color sensitiveness could be obtained in sensitive emulsions when flowed on a celluloid base as on glass. Early trials revealed a further problem to solve: how to eliminate the discharge of static electricity occurring at high altitudes, especially when the weather is cold.

As far as camera construction is concerned the chief problems are to hold the film flat, and to eliminate static.

Methods of Holding Film Flat.—Several means have been proposed and used for holding the film flat. Disregarding mere pressure guides at the side, which are suitable only for small area films (up to 4×5 inch), the successful means have taken three forms: *pressure of a glass plate, pressure of the shutter curtain*, and *suction*. A glass pressure plate can be used in either of two ways; the film may be in continuous contact with it or may be pressed against its surface only at the moment of exposure. The advantage of this first method lies solely in its mechanical simplicity; its disadvantage in the likelihood of scratches or pressure markings on the film. Where a glass plate is used there is always the chance of a dust or dirt film accumulating, or of the condensation of moisture, to impair the quality of the negative. There is, moreover, an inevitable loss of light (about 10%), together with some slight distortion, due to the bending of the marginal oblique rays through the thickness of the glass. In cases where a filter would normally be employed, the loss of light is minimized by using yellow glass for the plate, so that it serves for filter and film holder as well.

Pressure of the shutter curtain is utilized in the Duchatellier film camera by furnishing the edges of the curtain aperture with heavy velvet strips, whose light and gentle pressure during the passage of the shutter holds the film against a metal back. In many ways this is the simplest film-holding device; it occasions no loss of light, and needs no mechanical movements or external accessories, such as are called for in the suction devices next described. There is always danger of markings on the film, if the velvet is not of the right thickness and softness, and the operation and speed control of the shutter are necessarily complicated by the additional frictional load.

Suction of the film against a perforated back plate has been found a very successful means of securing flatness. Suction at the moment of exposure may be produced by the action of a bellows, which has been compressed beforehand by the camera-driving mechanism. Continuous suction can be produced either by a continuously driven pump, or by a Venturi tube placed outside the fuselage. The Venturi tube (Fig. 55) consists of a pipe built up of two cones, placed vertex to vertex, to form a constriction. When air is forced through this at high velocity suction is produced in a small diameter tube taken off at the constriction. A suction of two centimeters of mercury, acting through holes about one centimeter equidistant from each other in the back plate, has been found adequate to hold flat a film 18×24 centimeters.

One merit of suction applied only at the moment of exposure is that the film-driving mechanism does not have to work against the drag of the suction. Continuous suction, on the other hand, gives a longer opportunity for flattening out kinks in the celluloid, and easily permits movement of the film during the exposure, either for the purpose of permitting a longer exposure or for the purpose of preventing distortion due to the focal-plane shutter. A disadvantage of continuous suction is the production of minute scratches on the celluloid surface as it drags over the suction plate. These are ordinarily too small to cause trouble, but may show up when printing is done in an enlarging camera.

Static discharges are produced by the friction of the celluloid against the pressure back or other surfaces with which it comes into contact. They show in the developed film as branching tree-like streaks (Fig. 56) and in cold dry weather may be numerous enough to ruin a picture. The dis-

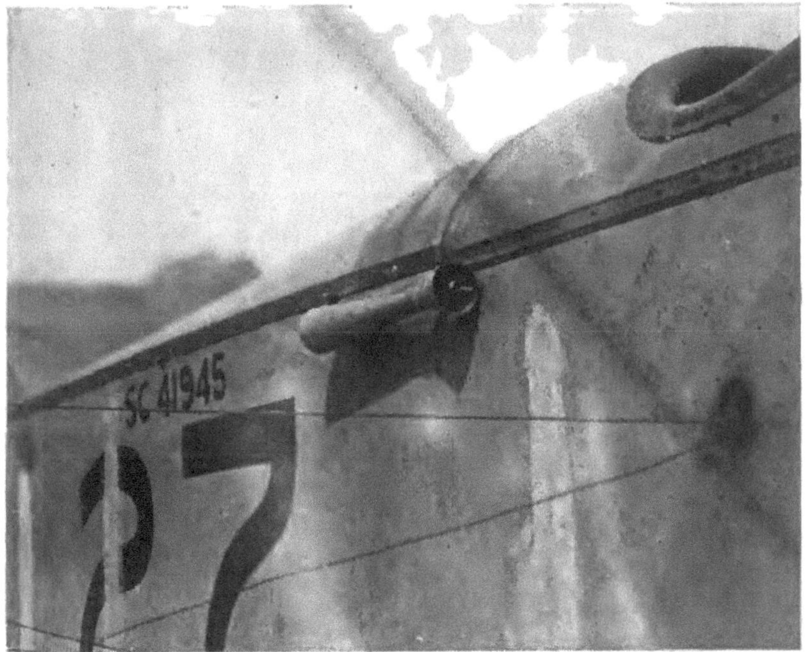

Fig. 55.—Venturi tube on side of plane.

charges are noticeably less frequent with film coated on the back with gelatine ("N.C."), but the extra gelatine surface is extremely undesirable. When handled by developing machines, as large rolls must be, this back gelatine surface becomes scratched and bruised in a serious manner. Plain unbacked film is much to be preferred if the static can be obviated.

To avoid static, it is necessary to provide for the immediate dissipation of all acquired electrical charges. Experiments made by the United States Air Service have shown that nothing is so good as rather rough cloth, thoroughly impregnated with graphite, held in close contact with the celluloid during as great a portion of its travel as possible. In the United States Air Service film camera which uses

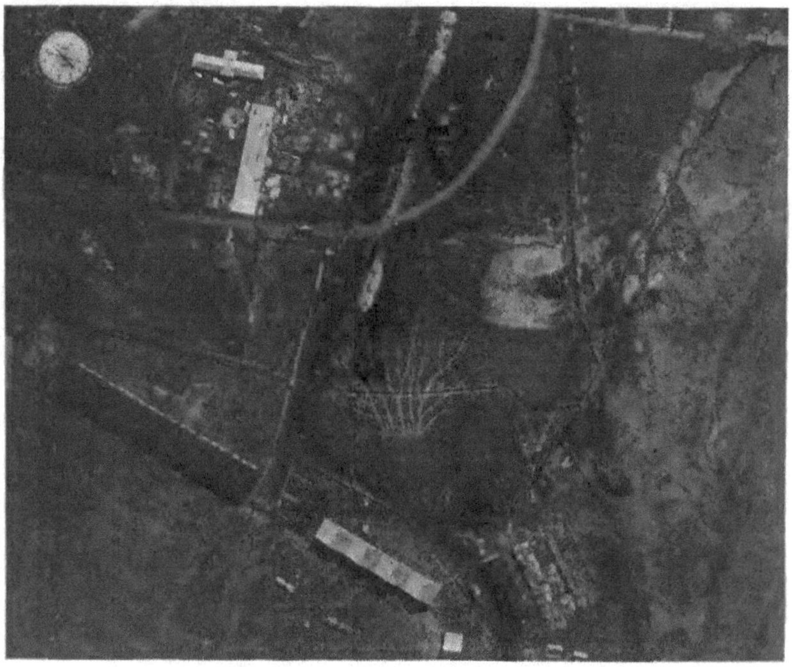

Fig. 56.—Print from film camera negative, showing static discharge, and (upper left-hand corner) record of altitude and compass direction made by Williamson film camera auxiliary lens (Fig. 58).

suction through a perforated back plate, the plate has been covered with thin graphited cloth, and similar cloth sheets are pressed against the film rolls by sheets of spring metal (Fig. 65). In cameras with this equipment no trouble has been experienced with static.

Representative Film Cameras.—*The English F type* (Williamson). This is one of the earliest cameras designed for film, as is indicated by the nature of the power drive, which presupposes that the camera is to be carried on the outside of the fuselage. Its essential features are shown in Figs. 57 and 58. It consists of a rectangular box with a cone at the front on which is mounted a propeller, intended to be rotated by the wind made by the motion of the plane. This drives, through a governor controlled friction clutch, a train of gears which draws the (5×4 inch) film across the focal plane, sets and exposes the shutter at regular intervals.

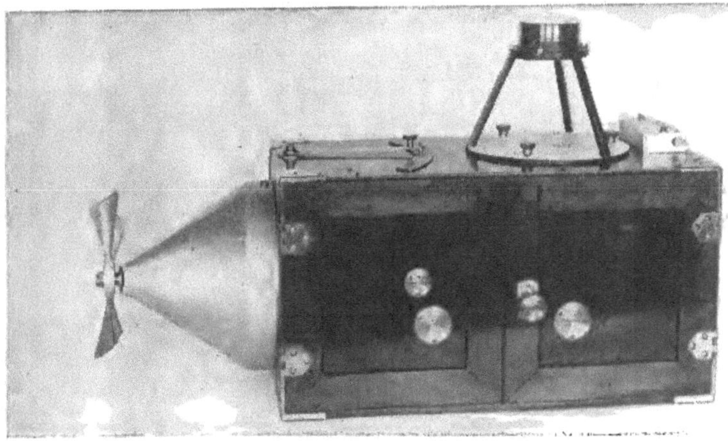

Fig. 57.—English type "F" (Williamson) automatic film camera.

Above the camera, supported on a tripod, are a compass and altimeter, both recording on a single dial, illuminated from below by the light reflected from a circular white disc painted on top of the camera. An image of the dial is thrown on a corner of the film by a lens, whose shutter is actuated in synchronism with the main focal-plane shutter. No special means are provided for holding the film flat. Special film with perforated edges is used.

The camera was designed for mapping work on the Mesopotamian and other fronts where no maps at all existed.

The Duchatellier camera is essentially a film magazine to fit on the standard French deMaria camera bodies, of the 18×24 centimeter size. In its simplest form it embodies a shutter (the regular focal-plane shutter of the camera being removed) and a film-moving mechanism, both actuated by a single motion of the hand. Automatic and semi-automatic operation are accomplished by an auxiliary mechanism to

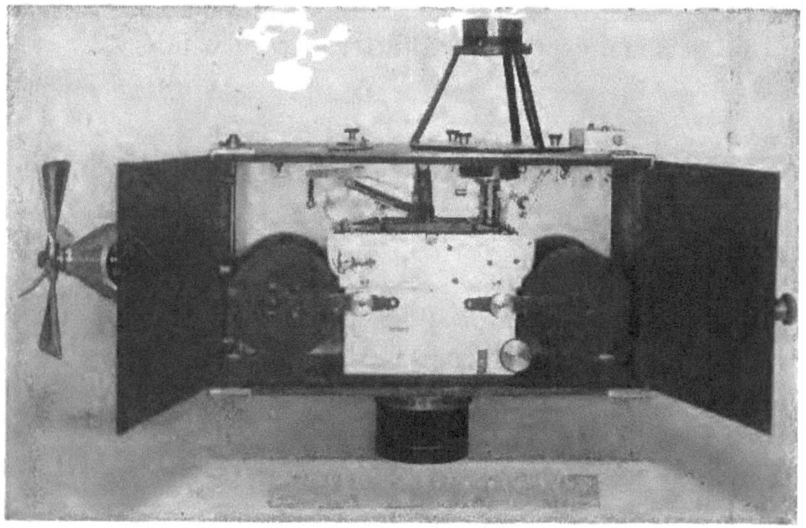

Fig. 58.—Interior of type "F" camera, showing lens for photographing compass and altitude readings.

which Bowden wires from the hand lever are attached. The motive power is an air propeller. Variation of speed is obtained by changing the point of contact of a roller on a friction disc, the disc being directly connected to the propeller shaft, the roller to the camera drive shaft.

The most distinctive features of the Duchatellier camera is its use of the focal-plane shutter to hold the film flat during the exposure. As already explained, this is accomplished by pressure, velvet strips on the shutter edges keeping the film close against the back plate. The return of the shutter curtain to the "set" position is accomplished by locking it to the film by perforating points, so that it is pulled across as the film is wound. This introduces between each pair of pictures a strip of tremendous over exposure, as wide as the

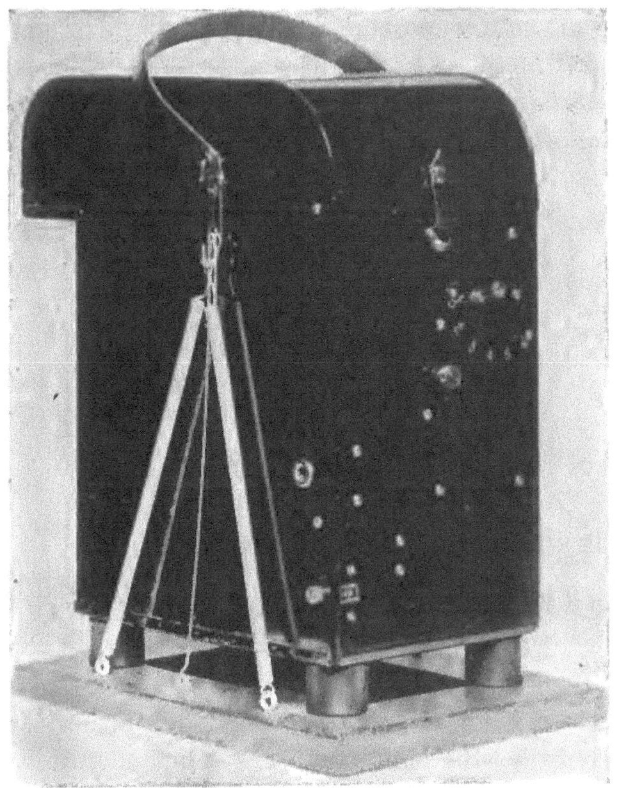

FIG. 59.—G. E. M. automatic film camera.

curtain opening. A fixed-aperture variable-tension shutter is used. The magazine carries a roll of film long enough for 200 exposures, feeding the long way of the picture. When film needs to be changed in the air, this is done by changing the entire magazine, including its shutter.

The G. E. M. camera (Fig. 59) is a very light self-contained clockwork-drive camera taking 36 pictures six inches square. The film is unrolled from a small-diameter feeding roller on to a large-diameter receiving roller to which the driving mechanism is attached. By this means approximately equal spacing of pictures on the film is assured. The film is held flat by continuous contact with a glass plate, which is made of yellow glass, so that it serves at the same time as a color filter.

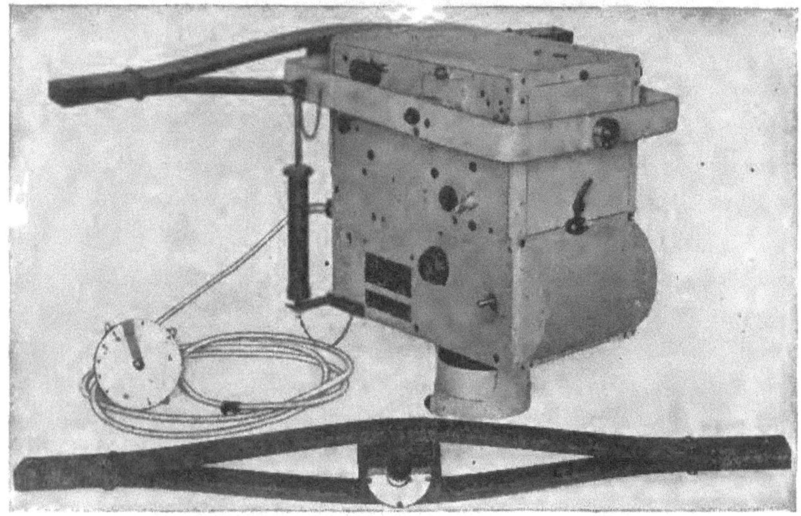

FIG. 60.—Brock automatic film camera.

The lens—of 8 to 12 inch focus—is equipped with a single speed between-the-lens shutter. The operation of the camera is entirely automatic. The interval between pictures is controlled by varying the clockwork speed, through a lever on the outside of the camera box. Protection of the camera from vibration is sought by supporting it on four spring cushions mounted on a solid frame, to which the camera is held by spiral springs attached to its sides.

The Brock Film camera (Fig. 60) is an entirely automatic,

very compact self-contained camera, taking one hundred 4×5 inch pictures. The motive power is clockwork, regulated in speed by an escapement controlled by a flexible shaft carried to a dial which may be fastened to the instrument board or to some other convenient part of the plane. The lens is 6, 12, or 18 inch focus. The shutter is of the fixed-aperture variable-tension type, of long travel, and with a flap behind the lens for covering during the setting period. None of the special means above described for holding the film flat are provided. A metal plate resting on the back, and a flat metal frame in front with a 4×5 inch aperture, are considered sufficient check on the excursions of the small-sized film. A ball bearing double pivoted frame serves to support the camera in a pendulous manner, permitting it to assume a vertical position after tilting. Damping of oscillations and vibration is arranged for by two pneumatic dash pots.

The *German film mapping camera,* shown in Fig. 61, is distinguished by a number of special features. The size of the pictures, 6×24 centimeters, is unusual. It has its advantages, however. Since the short dimension is in the line of flight, the maximum width of field covered by the lens is utilized (Fig. 17). This of course necessitates a larger number of exposures to complete a strip, which is perhaps an added advantage, since the narrower the individual pictures the better the junctions will be, especially if large overlaps are made. This proved to be the case with captured German mosaics. Difficulty is experienced in making overlaps on a turn (Fig. 62), but this is not a vital objection. The shutter has a fixed aperture, narrower at the center than at the ends, to compensate for the falling off in illumination away from the center of the lens. No safety flap is needed because the curtain moves in opposite directions on successive exposures, thereby also compensating for shutter distortion, as has already been discussed. Shutter speed is controlled by varying the tension of the actuating spring.

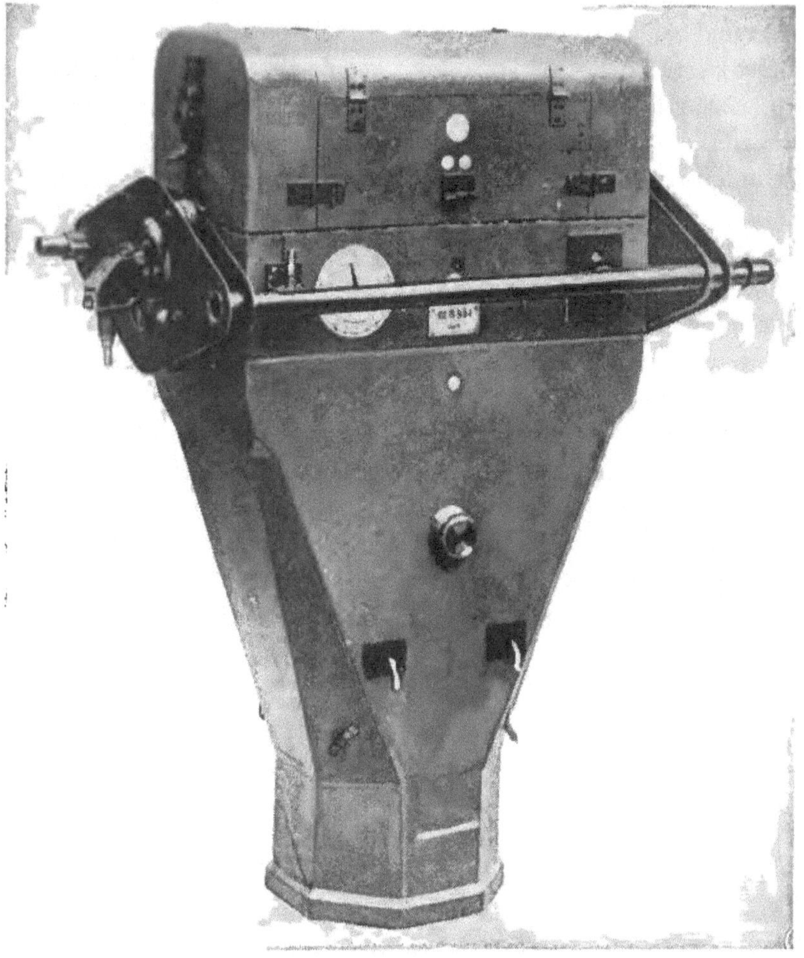

FIG. 61.—German automatic film camera.

The camera is driven by an electric motor, connected to a set of gears, whose shifting provides for speed variation. The film is moved by rubber rollers which are cut away for

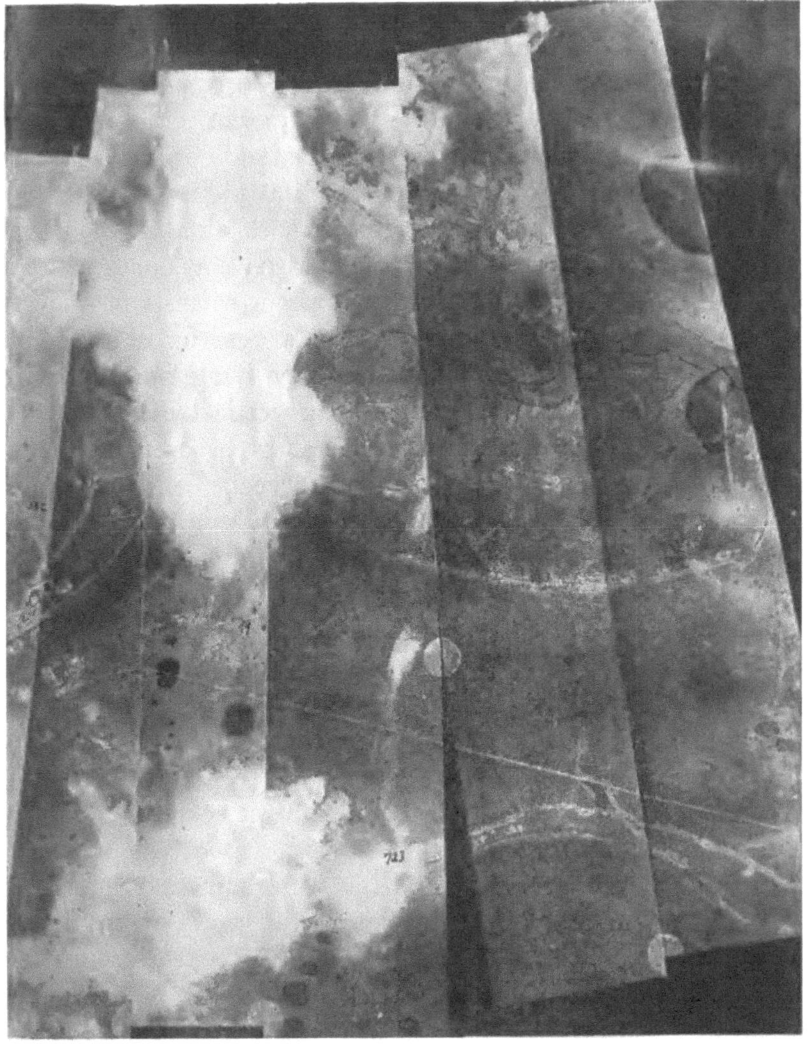

FIG. 62.—Method of joining and printing film from German camera.

part of the circumference, allowing the film to stand still until they bite again. A yellow glass pressure plate holds the film during the exposure and serves as color filter also

(Fig. 63). An electric heater is provided near the shutter, as in all the later German cameras.

United States Air Service automatic film camera—Type K (Figs. 64, 65, 92, 93, 98, 99). This is an entirely automatic camera, manufactured by the Folmer and Schwing Division of the Eastman Kodak Co., taking 100 pictures of 18×24 centimeter size at one loading. As with all the American cameras of this size, it uses the standard lens cones of any desired focal length. The camera proper consists of a compact chamber in which the film rollers are carried at each end forward of the focal plane, the shutter lying between. In consequence of this arrangment the vertical depth of the camera is the absolute minimum—short of decreasing the length of the optical path by mirror arrangements—making it possible to suspend the camera diagonally in the American and British planes, for taking oblique pictures.

Flatness of the film is secured by a suction plate covered with graphited cloth and connected with a Venturi tube. The top cover is removed for re-loading. The shutters on the first cameras of this type are of the variable-tension fixed-aperture design, though later ones have the variable-aperture curtain controlled by an idler, as used in the American deRam. An auxiliary curtain shutter serves to cap the true shutter during setting.

The operation of the film driving mechanism is comparatively simple. It consists of a train of gears, driven by a flexible revolving shaft attached to some separate source of power capable of speed variation. The action of the gears is to move the film, set the shutter and then expose it; in the earlier cameras with the film continuously moving. In the first cameras constructed the space between the pictures varies as the film rolls up, due to the increasing diameter of the roll. In later cameras the film roller is disengaged from

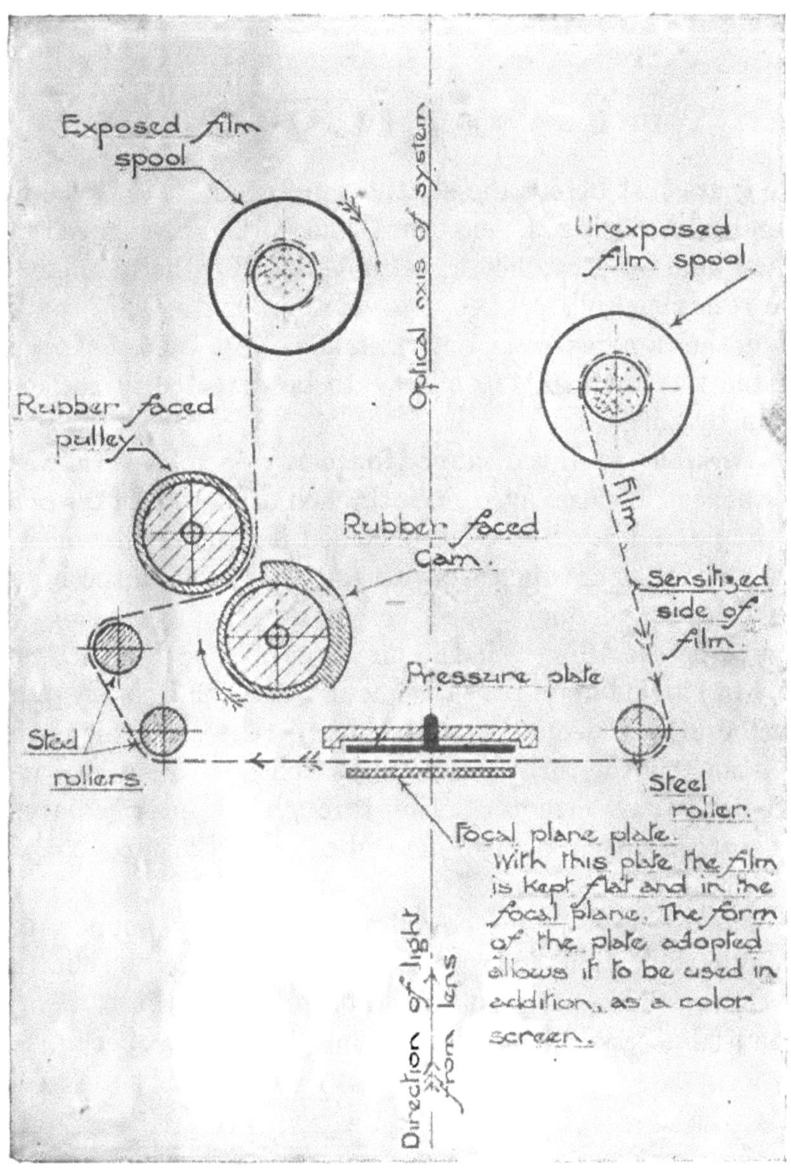

FIG. 63.—Film winding and exposing mechanism in German film camera.

the gears just before the shutter is tripped, so that the film stands still during the exposure, and is then re-engaged at a new point on a ratchet wheel governed by the diameter of the receiving roll, whereby the pictures are equally spaced. In all the cameras, punch marks made at the time of exposure enable the limits of the picture to be detected in the dark room by touch.

Variable speed is arranged for in any one of several ways. For peace-time uses a turbine attached to the side of the plane is simple and positive, and, provided it is made of sufficient size—which is not the case with the one shown in the Figure—will give adequate speed regulation upon varying the aperture through which the air enters. The Venturi tube may be carried upon the same mount, or a small rotary pump can be attached on the same shaft. Where the high wind resistance of the turbine is an objection the camera is driven electrically, by a motor acting through the intermediary of a variable speed control described in the next chapter (Fig. 68).

The camera weighs complete about forty pounds, and the film rolls about four pounds. The latter can be changed in the air without great difficulty provided the camera is mounted accessibly and so that the top may be opened.

CHAPTER XII

MOTIVE POWER FOR AERIAL CAMERAS

As long as circumstances permit, hand operation still remains the most reliable and satisfactory method of driving a camera. It is always available, can be applied to just the amount desired, and at the time and place needed. For instance, in a magazine of the Gaumont type (Fig. 40), what is needed is the periodic application of a very considerable force rather quickly, and while this can be done quite simply by hand, no mechanism has even been attempted to go through this same operation automatically. Instead, the fundamental design of automatic magazines has been made along other lines calculated to utilize smaller forces more steadily applied.

It must be granted, however, that for war planes, and particularly for single seaters, cameras should be available which are capable of operating semi-automatically or automatically. This necessarily means the employment of artificial power, whose generation, transmission to the camera and control as to speed present a mechanical problem of no small difficulty.

Available Sources of Power.—The sources from which power may be drawn on the plane are four, although the various combinations of these present a large number of alternative approaches to the problem. These sources are:

1. The engine of the plane.
2. Wind motors.
3. Spring motors.
4. Electric motors.

These may first be considered largely from the descriptive standpoint, leaving questions of performance and efficiency for separate treatment.

Power may be derived directly from the engine through a flexible shaft, similar to that used for the revolution counter. This source of power is inherently the most direct and efficient, since the engine is the seat of all the lifting and driving energy of the plane. There is no loss through transformation into other forms of energy, such as electrical; or by the use of more or less inefficient intermediary apparatus, such as wind propellers. Against the direct drive of the camera from the engine may, however, be urged that the usual distance between engine and camera is too great for reliable mechanical connection, as by flexible shafting. Objections also arise from the standpoint of speed. This cannot be controlled by the camera operator; and varies over too wide a range, as the engine changes from idling to full speed, to fit it for automatic camera operation. The first objection may be met by that combination of methods of power drive which consists in transmitting the power electrically; that is, by letting the engine operate a generator from which cables run to a motor close to the camera. This method, of course, sacrifices efficiency, and it breaks down when the engine speed drops below the speed necessary to generate the requisite voltage. This defect may in turn be met by floating in storage batteries, which brings up the whole question of electrical drive, to be treated presently. While use of the engine for direct drive or for generating electric current has not been adopted in the American service, it is known that some German planes were supplied with electric current in this way.

Coming next to the wind motors, these possess one very great merit: they utilize a motive power that is always present as long as the plane is in motion through the air.

On the other hand, the process of using the main propeller of the plane to pull another smaller propeller through the air appears a roundabout way to utilize the driving power of the airplane engine. Yet on the whole it is probable that some form of propeller or wind turbine is the simplest and most convenient device we have for the operation of airplane auxiliaries. As long as the amount of power required is small, such inefficiency as is inherent in its use is offset by its convenience and reliability. An advantage of the propeller is that its speed is almost directly proportional to that of the plane through the air, a desirable feature in automatic cameras provided the proportionality is under control. Yet it is just in this matter of varying the speed at will that the propeller presents difficulties, to be met only by additional mechanisms for gearing down or governing. Propellers have the practical disadvantage that they present an easily bent or broken projection to the body of the plane (Figs. 83 and 84). The strength of small propellers for operating auxiliaries is never so much in question with reference to their resistance to whirling and thrust of air as it is to their ability to withstand the inevitable knocks and careless handling that will fall to their lot. The propeller bracket is just what the pilot is looking for to scrape the mud off his boots before climbing in.

The wind turbine has the advantage over the propeller that its speed can be varied rather simply by exposing more or less of its face to the wind. A turbine fitted with an adjustable aperture for admitting the wind is shown in Fig. 64, in connection with the type K automatic film camera. The turbine has the advantage of being compact and lying close against the body of the plane. In the form figured, altogether too much head resistance is offered—just as much for low as for high speeds—but with proper design this need

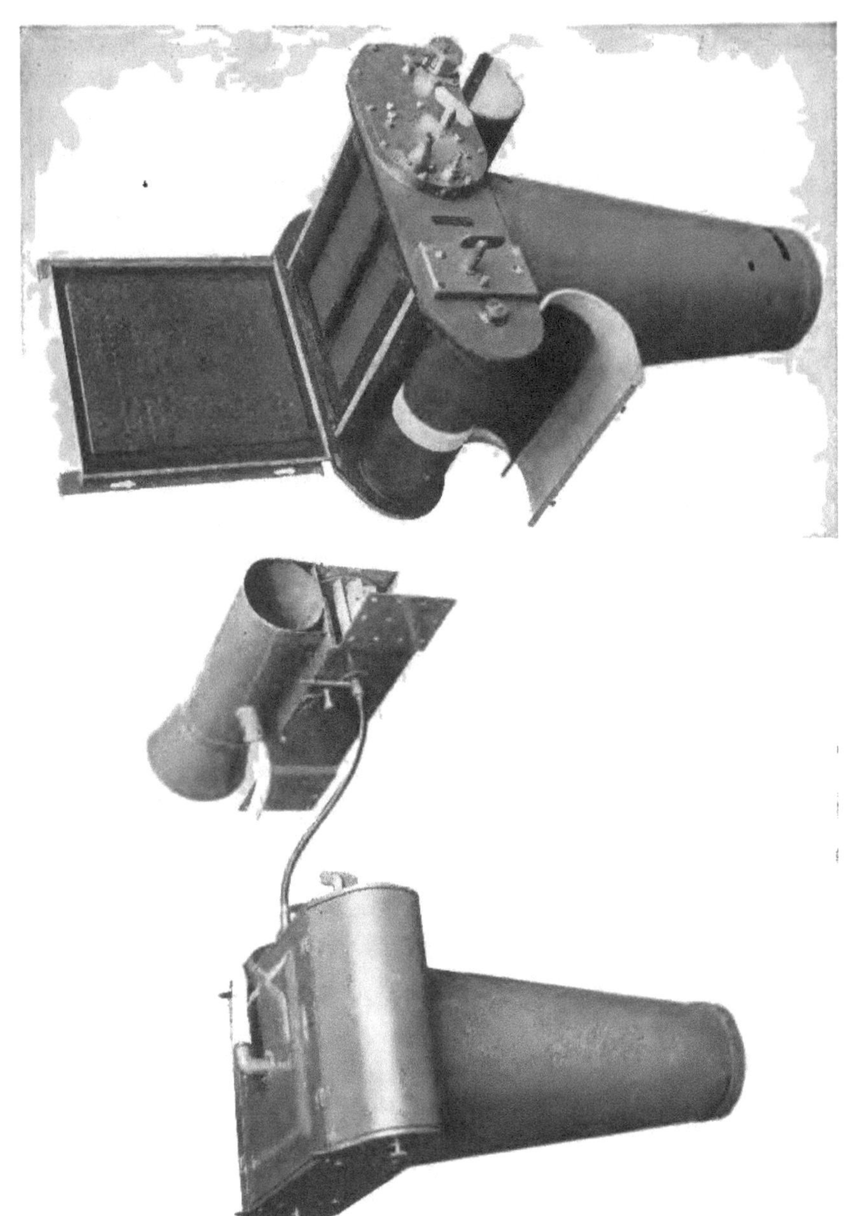

not be the case. It is, moreover, quite too small to give the needed speed regulation, as it only begins to operate near its full opening.

Spring motors have the very real advantage that by their use the camera can be made entirely self contained. The simplest application of the spring motor would be to the semi-automatic camera, where no close regulation of speed is required. In such a camera the operation of exposing the shutter would release the spring, which would then change the plate or film and re-set the shutter, repeating this operation as long as the spring retained sufficient tension. Small film hand-cameras of this type, using self-setting between-the-lens shutters, have been designed, though not for aerial work. The possibilities of using springs as motive power in semi-automatic cameras have not apparently been seriously considered.

When a spring motor is used for automatic camera operation it at once becomes necessary to add to the motor an elaborate clock mechanism for controlling and regulating its speed of action. Springs are much better fitted for giving power by quick release of their tension than by slow release, and the necessary clock mechanisms for their regulation become very heavy, as well as complicated and delicate, when they are made large enough to do any real work. For their repair they require the services of clock makers rather than the usual more available kind of mechanic.

Coming next to electric motors, we meet with a source of power of very great flexibility both in its derivation and in its application. If a source of electric current is already provided for heating and lighting as it is on the fully equipped military plane, and if it has sufficient capacity to handle the camera, its use is rather clearly indicated, irrespective of how efficiently or by what method it is produced. Especially is this the case, from the standpoint of economy and simplicity, if a propeller-driven generator is the source of current, and the alternative power drive is an additional propeller for the camera. If, on the other hand, the camera must have its own source of electric power, the advantages and disadvantages must be closely scrutinized. In this case either a generator must be provided, or resort be made to storage batteries, or a combination of the two.

Ruling out a special propeller-driven generator, we are left with either the generator driven from the engine or the storage battery. Inasmuch as storage batteries are practically indispensable with generators, in order to maintain the voltage constant at all speeds, it is on the whole advisable to rely upon batteries alone. An advantage of their use is that the power plant is entirely within the plane: All projections such as propellers are avoided. Another merit is that the power is drawn upon only as needed. Against storage batteries is their weight, the need for frequent charging, and their loss of efficiency at low temperatures—a loss so serious with those of the Edison form as to preclude their use.

When once the source of electrical energy is decided upon, its method of application needs to be considered. Here we meet at once the peculiar merit of electrical energy, namely, the ease and convenience with which it may be transmitted. All we need is a pair of wires, led to any part of the plane by any convenient route and connected by simple binding posts. It may with equal ease be turned on or off by merely making or breaking a contact with a switch. For operating semi-automatic cameras this feature may be utilized in the interest of economy, if the power is automatically turned off as soon as the plate-changing operation is finished. Exceptionally reliable make and break contacts are necessary to insure the success of this latter scheme.

Two methods of transforming the supply of electrical energy into mechanical motion are available. The first is by the use of a solenoid and plunger. This is a device practically restricted to semi-automatic cameras, in which the operation consists of a straight to-and-fro motion, initiated at the will of the operator. It has been used little if at all. The second motion is the continuous rotary one secured by the use of an electric motor. This motion is the most practical one for the continuous operation of any mechanism, but on the other hand requires that the imposed load be reasonably uniform at all times through the cycle of operations. Assuming that the camera mechanism is of this character, the motor may be attached directly to the camera, or if it must be so large as to cause danger by vibration, it may be connected through a flexible shaft. This use of an electric motor is very practical for semi-automatic cameras such as the "L" or the American deRam, in planes supplied with a suitable source of current.

When it comes to entirely automatic cameras, where uniform and regulatable speed is required, as in making overlapping pictures for mapping, the electrical drive is not so conveneint. The shunt-wound motor runs at nearly constant speed, while the series-wound motor in which the speed can be regulated by the interposition of resistance, has nothing like a sufficient range of variation for the purpose (at least five to one is imperative) before it fails to carry the load. Hence we must either incorporate in the camera some mechanism for varying the interval between exposures while the speed of the motor remains constant, or introduce an auxiliary device to effect the required transformation in speed. If we do use an auxiliary device the train of apparatus, consisting of battery (or generator), motor, speed control and camera, is altogether too long; it is apt to cause annoying delays in connecting up in an emergency, and it offers an excessive number of chances for break-down.

Performance and Efficiency Data.—The first step in deciding upon methods of power drive, and indeed in deciding whether power drive is feasible at all, is to assemble definite data as to the power required to drive representative cameras. Approximate figures for some of the cameras described in previous chapters are:

L camera,	26 watts,
deRam,	60 watts,
"K" film,	30 watts.

These requirements—not exceeding 1/10 horse power—are insignificant in comparison with the total of 100 to 400 horse power available for all purposes from the plane's engine.

Propeller characteristics. Data on the performance of small propellers are somewhat meagre. However, the results of the rather extensive researches on large ones, suitable for driving planes, may be applied, with proper reservations, to give a fair guide to the study of the application of small propellers for driving plane auxiliaries.

The first factor to be considered is the thrust or *head resistance* offered by a propeller to motion through the air. This varies as the *square of the velocity*, as the *density of the medium*, and as the *area of the body* projected normally to the wind, the formula being

$$T = cdaV^2$$

where T = thrust, d = density, a = area, V = velocity. Data on the L camera propeller are shown in Fig. 66, where its thrust both when free and when loaded with the camera is given, as well as that of a solid disc of the same diameter as the propeller. For this propeller, which is double-bladed, and six inches in diameter, $cda = .000275$ with the load on.

The total thrust amounts to only about three pounds when the plane velocity is 100 miles per hour. The head resistance of the whole plane is a matter of hundreds of pounds, so that the propeller resistance is quite negligible.

The next factor is the speed of revolution of the propeller, expressed in revolutions per minute. This varies with the

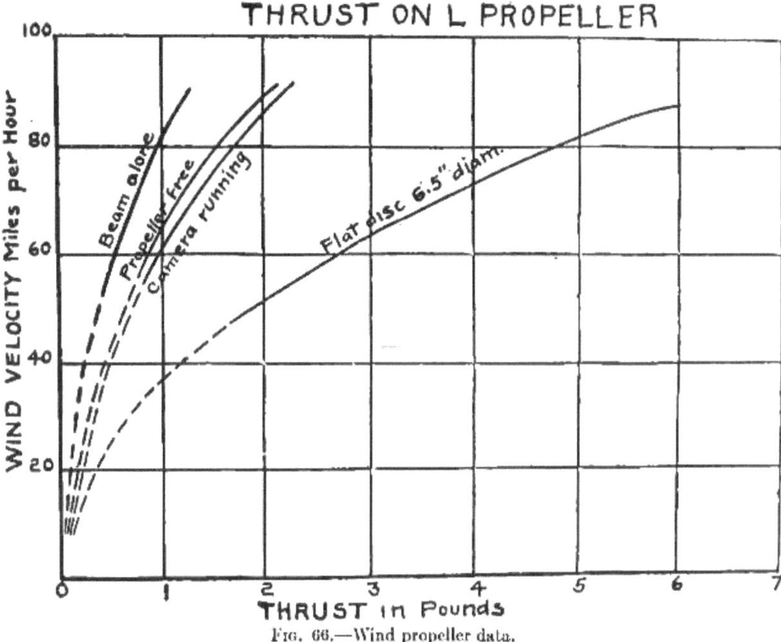

FIG. 66.—Wind propeller data.

design—the number of blades, their area, and pitch. For a given design the speed of revolution is *directly proportional to the speed of motion through the air*, and to *the density of the air*. Representative data for the L camera propeller are shown in Fig. 67. It will be noted that the speed goes up to 8000 for 120 miles per hour air speed. This illustrates the necessity for great strength to withstand centrifugal force. Propellers should be constructed of tough material, and subjected to whirling tests up to speeds considerably in excess of any the plane will attain in any maneuver. At low speeds the linear relationship fails, as a critical velocity is reached—about 3500 r. p. m. for this propeller—where it refuses to turn.

The fact that the speed of the propeller depends on the density of the air has an interesting corollary, which is that a propeller adequate at low altitudes will fail at high ones. The density of the air varies with altitude according to the following figures:

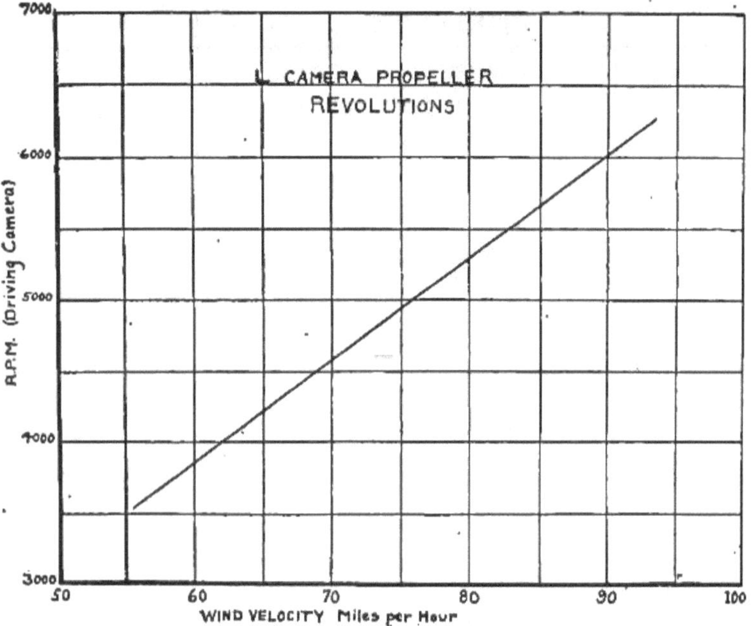

Fig. 67.—Relation between air speed and propeller revolutions.

At 3000 meters,	72 per cent. of sea level
5000 meters,	59 per cent. of sea level
6000 meters,	52 per cent. of sea level

If we take the r. p. m. at 90 miles per hour at sea level as 6000, then at the above altitudes the speeds will be 4300,

3500, and 3000, respectively. The last figure is below that for which this size of propeller stalls with its normal load, as noted in the last paragraph. Consequently, if flying is to be done at these altitudes a larger propeller must be carried, which will still deliver enough power at the lower density.

The next factor to be considered is the *power furnished by the propeller.* As a representative figure may be quoted the performance of the L propeller. This gives 27 watts at 3600 revolutions per minute (56 miles per hour). From this figure the performance of other propellers may be deduced from the basic laws, which are: that the *power varies as the density of the medium* and as the *cube of the velocity* (assuming constant efficiency). Since the power delivered by the six inch diameter L propeller is already adequate at 60 miles per hour, the necessary dimension to function satisfactorily at 100 miles per hour would need to be only a little more than three inches, except for the desirability of a safety factor for high altitudes and low air densities.

The *efficiency* of the propeller is defined by the relation—

$$\text{Efficiency} = \frac{\text{power delivered by the propeller}}{\text{power supplied to the propeller}}$$

The denominator of this fraction is the thrust times the velocity, for which the curves of Fig. 66 supply us data for the L propeller. Using the figures 3600 r. p. m., 56 miles per hour, and 27 watts, we find the efficiency to be about 50 per cent. This increases with the velocity, with a possible upper limit of 70 to 80 per cent. Since the main propeller of the plane is not over 80 per cent. efficient we have at most an efficiency of 64 per cent. in using a propeller drive, as compared with taking the power directly off the engine.

In considering the use of *spring and clock-work motors* we meet at once with the problem of comparing the effect on the performance of a plane of a carried weight, as against a head resistance. The efficiency of a spring motor is measured in terms of its weight, that of a propeller in terms of its head resistance. The general answer to this question is given by the relation that *a pound of dead weight is equivalent to* ⅕ *pound head resistance.*

In order to apply this relation to the study of spring motors for driving cameras, data are necessary on the power delivery per pound weight of such mechanisms. Such data are not easily accessible, largely because clock-work has not generally been seriously considered as a motive power for large apparatus. To arrive at an approximate figure we may take the fact that in an 8×10 inch film camera designed by one of the manufacturers who have utilized clock-work, the motor weighed 30 pounds. This is equivalent to six pounds head resistance. Now the type K, 18×24 centimeter film camera is operated, even with the addition of a friction drive speed control, by means of the L camera propeller. As shown in Fig. 66, at 100 miles per hour the head resistance of this propeller is still less than three pounds. Consequently, it appears that from the efficiency standpoint the clock mechanism is quite outclassed by the wind propeller.

Coming next to the *electric motors,* the L camera and the K are both operated satisfactorily with a 1/20 horse power motor, weighing 6 pounds. For the deRam a 1/10 horse power motor has been adopted.

Taking up efficiency considerations, we have, if the current is supplied by a generator from the engine, a transformation factor of 70 to 80 per cent. from mechanical to electrical energy and a similar factor in using a motor for the camera. When batteries are employed the matter of weight *versus* head resistance again arises. The batteries found most satisfactory for operating the K and deRam cameras are of the six-cell 12 volt lead type. Their capacity is 40 ampère hours at three ampères or 36 at five ampères—more than is necessary for a single reconnaissance, but a practical figure when economy of charging and replacement are considered. The weight of this unit is 27 pounds. To this must be added the weight of the motor—6 lbs.—making a total of 33 pounds, equivalent to a head resistance of nearly 7 pounds. This is more than twice the propeller head resistance invoked to do the same work.

These considerations of efficiency have been gone into because they are usual in studying any engineering problem and because of the insistent demand from the plane designer that every ounce of weight and head resistance be saved. Actually, as already stated, the load imposed by any method of power drive is trivial in comparison with the whole load of the plane. There is, however, an important reservation to be made, which applies against clock-work and batteries: This is, that while the equivalent head resistance of any camera motive power carried as dead weight is small, its effect on *balance* may not be so. While the use of a propeller need not disturb the plane's balance, the weight of the camera alone, without any driving apparatus, is already seriously objected to on this score. The merely mechanical superiority of the propeller as a source of motive power is on the whole rather marked.

Control of Camera Speed.—In the semi-automatic camera the only control required on the speed of the operating motor is at the upper and lower limits. It must not go so fast as to anticipate the completion of any steps in the cycle of camera operation, such as the fall of plates or pawls into position, which would jam the camera. On the other hand, it must not be so slow that pictures cannot be obtained with the requisite overlap for maps or stereoscopic views. In the American deRam camera the cycle of operations cannot safely be put through in less than four seconds, a short enough interval for most purposes. It is also highly desirable in the semi-automatic camera to have the motive power capable of stopping completely. This saves wear and tear on both motor and camera mechanism.

In the automatic camera an extreme range of speed is called for by the several problems of mapping, oblique photography, and the making of stereoscopic views. For mapping alone, the shortest likely interval may be taken as that required for work at approximately 1000 meters altitude, for a plane speed of 150 kilometers per hour, which demands an interval of six seconds with a ten inch lens on a 4×5 inch plate. For vertical stereos at the same altitude and speed this interval is divided by three, and low oblique stereos need even quicker operation. Hence a range of from 1 to 30 pictures per minute should be provided for. This requirement is difficult to meet with any simple mechanism.

From the standpoint of simplicity in speed regulation the wind turbine of adequate vane surface has much to recommend it. It is only necessary to present more or less of its vane area to the wind in order to secure a considerable range of speed. The method of doing this by a shutter interposed in front is uneconomical, but it is probable that the design can be so altered that more or less of the turbine is exposed beyond the side of the plane, possibly by varying the angle, to secure the same result without introducing useless head resistance. A serious practical objection to the turbine lies in the large vane surface necessary to give adequate power combined with proper speed variation. In the automatic film camera (Type K) this area should be as much as 40 to 50 square inches.

The wind propeller does not lend itself at all well to speed variation. It cannot be partially covered from the air stream, as can the turbine, because of the resulting strain on its mount. A possible form of variable speed propeller, one which, however, has not yet been practically developed, is a propeller with controllable variable pitch. If this could be made mechanically sound it would be well-suited for camera operation. That such a propeller could be worked out is indicated by the good performance of a constant speed propeller developed for radio generators and used on the French deRam camera (Fig. 54). Parenthetically, it may be questioned whether a constant speed propeller is really desirable with an airplane camera. What is required is not exposures at a definite time interval—although most of the data are in that form—but exposures at definite intervals with respect to the motion of the plane, which practically means with reference to its air speed. Rather than build a camera calculated to give exposures at intervals of so many seconds when it is attached to a constant speed propeller, we would do better to use a propeller which responds to the speed of the plane, in conjunction with some form of tachometer to show the rate at which exposures are being made. This in turn should be coördinated with the indications of a proper camera-field indicating sight.

One solution of the problem of speed control with a propeller of practically fixed speed, is to use a governor and slip clutch as in the English Type F film camera (Fig. 57). Here the propeller shaft and the camera driving axle are connected by two friction discs. That on the camera mechanism is forced against the other by a spiral spring, whose tension is controlled by a ball governor. If the camera speed becomes too high the governor reduces the tension on the spiral spring and the discs slip over each other. The point where this slipping occurs is determined by the position of the governor as a whole, and this is controlled by a lever on top of the camera.

Another speed control device, perhaps more positive but certainly more complicated and wasteful of power, consists of a large flat disc, driven by the propeller or electric motor, and from which the camera is driven by a shaft from a smaller friction disc which may be pressed against any point from

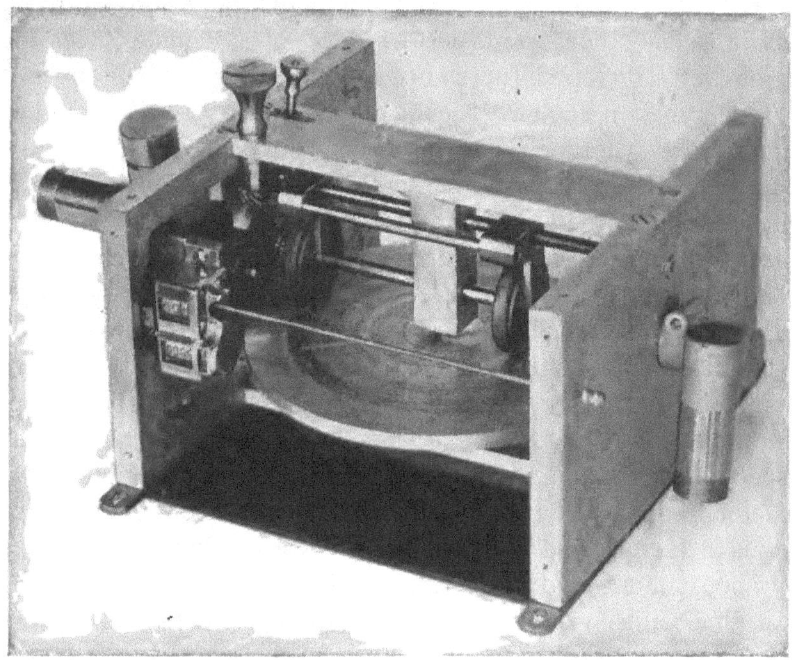

FIG. 68.—Friction disc speed control.

the center to the periphery of the larger disc. The speed range attainable in this way is limited only by the size of the large disc. An application of this idea is shown in the speed control (Fig. 68), designed for the American Type K camera when operated on an electric motor or on a simple propeller. The same idea is utilized in the Duchatellier film camera, in connection with the constant speed propeller already described.

On the whole it is eminently desirable from the standpoint of power operation that the automatic camera should embody its own means for altering the interval between exposures, so that all the external attachment needed is a single connection to a source of power either of constant speed, as an electric motor, or of speed proportional to that of the plane, as with a simple wind propeller. This makes the camera largely independent of the nature of the power supply, whereas a camera designed for a special variable speed device is of little use on a plane where this is not available.

Transmission of Power to the Camera—It has already been pointed out that the ease of transmission of electrical energy makes it particularly convenient for use in a plane. All other sources of power, except clock-work incorporated in the camera, require flexible shafting, so that the question of bearings and connections becomes a serious one, especially when the shaft runs continuously for long periods at very high speeds.

The shafting found most suitable is the spirally wound form commonly known as dental shafting. This must be encased in a smoothly fitting sheath, flexible enough to permit of easy bends. The ends of the shaft should be equipped with square or rectangular pins to fit into corresponding slots in the motor and camera shafts. The ends of the shaft casing may be fitted either to attach by bayonet joints or by smoothly fitting screw collars. At the point of attachment to the camera it is desirable to have some form of junction adjustable as to the direction from which the shaft may be connected, so that it need be bent as little as possible. A right angle bevel gear offers one means of doing this. Bearings, such as those of the propeller, should be of the ball variety, while heavy lubrication, such as vaseline, should be freely used, both in the bearings and in the shaft casing.

An important feature of any power drive system should be a safety device, so that the power will race in case of any jam or stoppage in the camera. This will often prevent serious damage through the breakage of some relatively weak part of the camera mechanism on which the whole force of the driving apparatus is suddenly thrown. The "L" camera propeller is fitted with a spring friction clutch with the idea that if the camera refuses to operate the propeller will slip instead of wrenching the shaft to pieces.

CHAPTER XIII

CAMERA AUXILIARIES

Distance Controls and Indicators.—All operations connected with the exposing and changing of plates (except the changing of whole magazines) should be arranged for accomplishment at a distance. Other operations, such as changing the shutter speed or the interval between exposures in an automatic camera, which are usually done on the ground, may sometimes be satisfactorily left for performance at the camera. Conditions of extreme inaccessibility may, however, make it necessary to carry even these controls to a distance. Indicators of the number of exposures already made, and of the readiness of the camera for the next exposure, may be attached to the camera, but often are more profitably placed at a distance. Distance control and indication are especially necessary if the pilot makes the exposures—a common English practice in two seaters, and the only recourse in single seaters.

When electric power is available, electrical distance control devices are perhaps the simplest kind, as they transmit motive power without displacing or jarring the camera. Solenoids suffice for the simple pressing of releases or for counting mechanisms, while small service motors may be utilized for operations involving more work. A standing practical objection to electrical control lies in the necessity for using contacts, which are apt to be uncertain under conditions that involve vibration.

The Bowden wire—a wire cable carried inside a heavy nonextensible but flexible sheath—constitutes the most satisfactory mechanical means for transmitting straight pulls.

By means of "the Bowden" a pull may be transmitted so as to be made entirely relative to two parts of the same body, calling forth no tendency of the body as a whole to move. Thus in the L camera shutter release (Fig. 50), the releasing lever with its attached counter is several feet distant from the camera. If the plate bearing the lever and sheath end is rigidly fastened down, the pressure exerted on moving the lever acts between the lever and the end of the sheath. This pressure passes immediately to the other end of the sheath, while the pull on the wire is transmitted to its farther end on the camera. In this way the conditions at the lever are reproduced, but with the advantage that, due to the flexible cable and sheath, any vibration of the lever support is damped out.

Due to its stretching, there is a pretty definite limitation to the feasible length of the Bowden wire. This length is about four feet. Where according to English practice the pilot makes the exposure, a considerably longer wire and sheath are called for. In this case the effective length of the release is increased by giving the pilot a second releasing lever, connected to the first by a rigid rod (Fig. 69). The releasing lever, wire, and all mechanical parts of the Bowden release should be made much stronger than would be indicated by bench tests of the camera. In the air it is impossible to decide either by sound or by delicacy of touch whether the mechanism has acted, so that the observer is apt to pull much harder than necessary and to strain or break the release if it is weak.

The Bowden wire is used in the American service only for shutter release. In the English service it has been used for plate changing with the L camera.

Sights.—In airplane photography the need for a finder or sight is fully as great as in everyday work. A new condi-

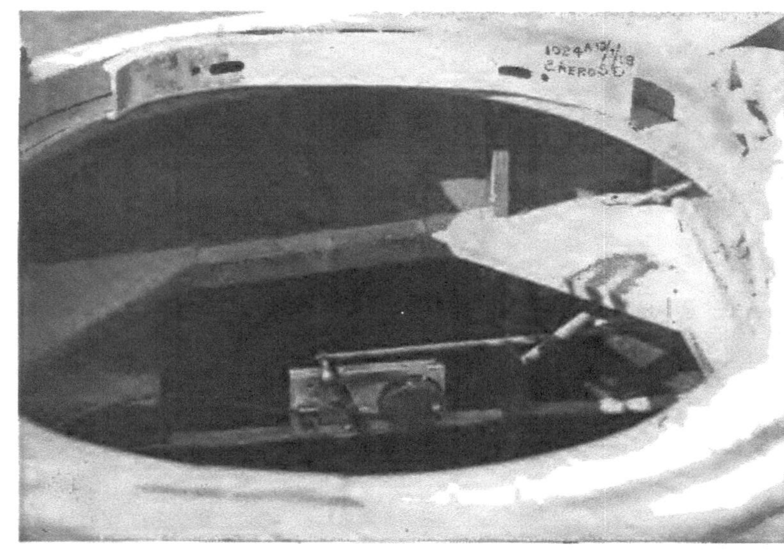

Fig. 69.—Bowden wire release in rear cockpit, with rod connected to similar release for the pilot.

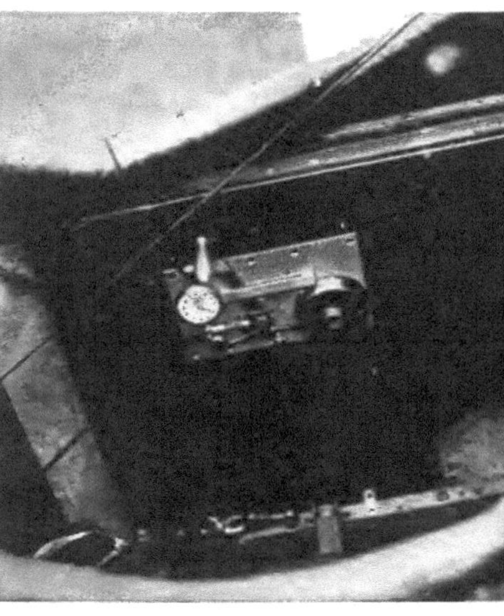

Fig. 70.—Bowden wire release with stop watch attached, for use in timing for overlaps.

tion, however, prevails, for except with hand-held cameras, and even to some extent with them, the operation of pointing the camera involves pointing the whole vehicle that carries the camera. The pointing of airplane cameras is therefore akin to the sighting of great guns. While the observer may perform the actual operation of taking the picture, the responsibility for covering the objective rests with the pilot. Teamwork counts equally with tools. Airplane camera sights may accordingly be divided into two classes: sights attached to the camera, for use principally with hand-held apparatus, and sights attached to the plane, for the use of pilot, of observer, or of both.

Sights for Hand-held Cameras.—The simplest form of sight attached directly to the camera is modeled on the *gun sight*, consisting of a forward point or bead and a rear V. This sight of course serves merely to place the objective in the center of the plate and gives no indication of the size of field covered. Another simple sight of rather better type is the *tube sight*—a metal tube of approximately one inch diameter and three inches length, carrying at each end pairs of wires crossed at right angles. The camera is in alignment when the front and back cross wires both exactly match on the object to be photographed. The best way to mount the cross-wires is with one pair turned through 45 degrees with respect to the other, so that it is at once apparent which is the front and which the rear pair (Figs. 31 and 39).

Sights to indicate the size of the field are usually less needed on hand cameras than on fixed vertical cameras. Yet certain circumstances make them most desirable, for instance in naval work where a complete convoy must be included on the plate. A sight of this kind can be made up of two wire or stamped metal rectangles, a large one in front and a smaller one behind, of such relative sizes and separations that the true camera field is outlined when the eye is placed in position to see the two rectangles just cover each other. The dimensions should be so chosen that the correct position of the eye is approximately its natural location with respect to the camera when this is held in the hands in the plane. It is usual to provide the rectangular sights with cross-wires to indicate the center of the field. Alternative rear sights are simple beads or peep-holes—the use of the bead assuming that the camera is held at about the right distance from the eye for the rectangle to indicate the field. The peep-sight is not a desirable form, as it is hard to hold the camera as near the face as is necessary. These various types of rectangle sights are well illustrated in the cameras shown in Figs. 38, 40 and 186. They are all made so as to fold down flat on the camera and to snap quickly open when needed. The springs to support the sights must be fairly strong, and the surface presented to the wind as small as possible. Wire frames give very little from the pressure of the wind, but flat metal frames are apt to be bent back.

The position of the sight on the camera is important. If the observer can stand, or if he sits up well above the edge of the cockpit, the conventional position of the sight on a pistol, namely, on top, is unobjectionable. But if the observer sits very low, as he usually does, then the sight should be on the bottom of the camera, thereby avoiding any need for the observer to raise his head unduly into the slip stream. Similarly, if the camera is used over the side for verticals, as it is in flying boats, a sight on the top is impractical, since it requires the observer to lean out dangerously far (Fig. 185).

Sights Attached to the Plane.—Any of the sights just described can be attached to cameras fixed in the plane, but they would be useless in the positions ordinarily occupied by the camera. It has therefore become common practice to attach the camera sight to some accessible part of the plane. The most primitive method of sighting is merely to look downward over the side—a method in general use to the very end of the Great War. One step in advance of this is to mark a large inverted V on the side, with its vertex at a point where the observer can place his eye and so see the fore and aft extension of the field of view covered by the camera. This kind of sight was common on the French "photo" planes. On some of the English planes the tube sight was carried on the outside of the cockpit. Any of the sights described can be carried on the inside of the fuselage, provided a hole is cut in the floor. For satisfactory sighting a hole in the floor is really necessary, as it enables the terrain on both sides of the vertical to be seen. One drawback to the simple hole, however, is that it cannot be made large enough to show the whole field from the ordinary height of the observer's eye, thus forcing him to bring his head down near the floor. This difficulty is gotton over in a very beautiful way by the use of the *negative lens sight* shown diagrammatically in Fig. 71.

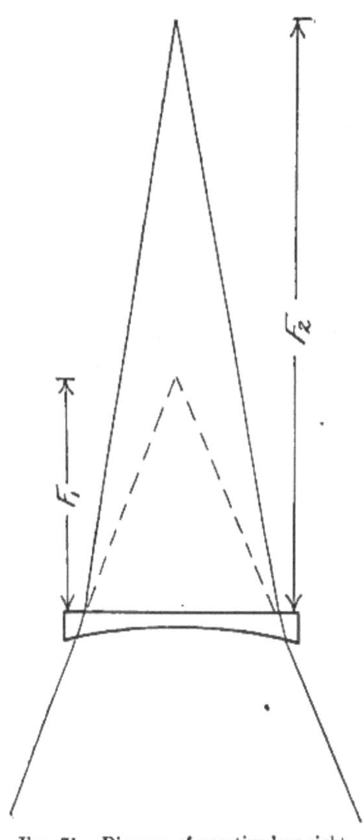

Fig. 71.—Diagram of negative lens sight.

Let F_1 be the distance at which the edge of the hole (or a rectangle marked on the lens) appears the size of the camera field (if the hole is the size of the plate, F_1 is the focal length of the camera lens). Let F_2 be the distance from the floor to the observer's eye. What is desired is a concave lens which will diverge the rays from their normal meeting point at F_1 to a new meeting point, F_2. The focal length of lens required is given at once by the simple lens formula—

$$\frac{1}{F_1} - \frac{1}{F_2} = \frac{1}{F}$$

Thus if F_1 is 12 inches, and F_2 is 36 inches, F will be 18 inches. The lens is to be marked with a rectangle showing the shape and size of the camera field, and a central mark such as a cross. An upper rectangle, or a bead, or a pair of cross wires a few inches below the lens, may be used for the other sight. For precision work the sight above or below the lens should be adjustable in position, especially where the camera suspension permits the camera to be adjusted for the angle of incidence of the plane.

A negative lens sight should be placed in the observer's cockpit, if he takes the pictures, and also in the forward cockpit, so that the pilot may be accurately guided in his part of the task. In addition, it is advisable to place a negative lens well forward in the pilot's cockpit, to enable him to see the country some distance ahead. The lenses should be plano-concave with the flat side upward; otherwise, all the loose dirt in the airplane settles in the middle of the concave depression. A negative lens sight in a metal frame forming a completely self-contained unit ready for mounting in the plane is shown in Figs. 72 and 73.

Devices for Recording Data on Plates.—*Numbering devices.* The number of the camera is impressed on negatives taken with the American L camera through the agency of a small transparent corner of celluloid. It would be entirely possible to incorporate a rotating disc which should turn by the operation of plate changing and carry a series of numbers, so that each exposure could be numbered serially. Numbering of individual plates is more commonly done by holes, notches, or even numerals, in the turned over portion of the sheaths, which are then recorded photographically when a picture is taken (Fig. 75). The chief objection to this method

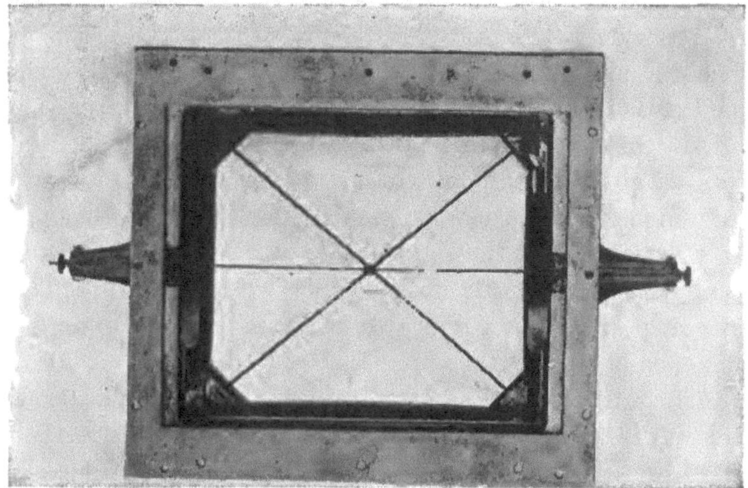

Fig. 72.—Negative lens and mount, viewed from above.

is the difficulty of keeping the sheaths together in sets, especially as individual ones become damaged or lost. In practice there is also danger of the sheaths being carelessly loaded in wrong order.

The more ambitious idea of recording on the plate all the information given by the instrument board of the plane occurs independently and spontaneously to all aerial photographic map makers. These ideas vary from attempts to photograph the actual instrument board on every plate—a difficult task indeed with the instruments and camera placed as they are in the ordinary plane—to the incorporation of compass, altimeter, and inclinometer in the camera itself.

Figure 58 shows the plan adopted in the English F type film mapping camera already described, for photographing a compass and an altimeter on the film. Here the combined compass and altimeter dial is above the camera, and is

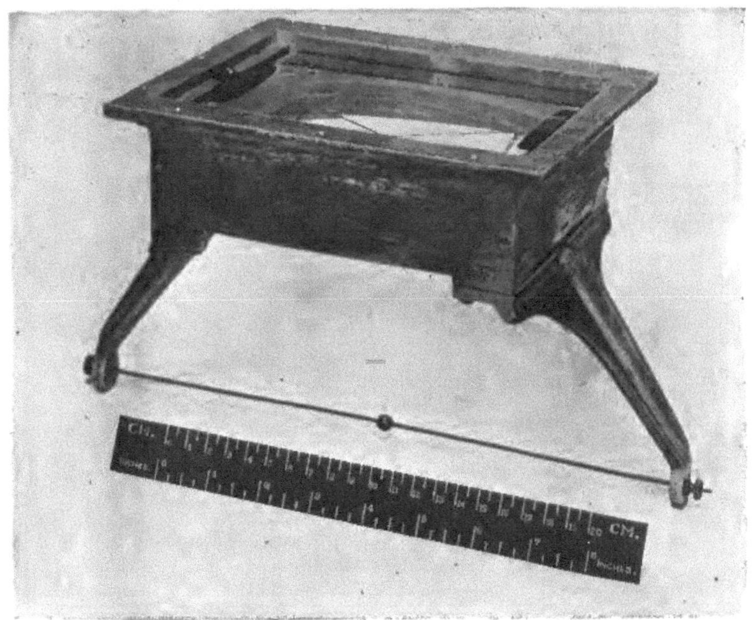

FIG. 73.—Negative lens and mount, side view.

mounted in a cell with a glass bottom. Below it is a lens focussing the needles and compass points on the plane of the film. The light for photography is furnished by a diffusely reflecting white surface on top of the camera, illuminated by the sky. (The camera was carried outboard.) In Fig. 56 is shown a picture with the compass image impressed upon it.

Figure 74 shows a type of inclination indicator found in some captured German cameras. It consists essentially of

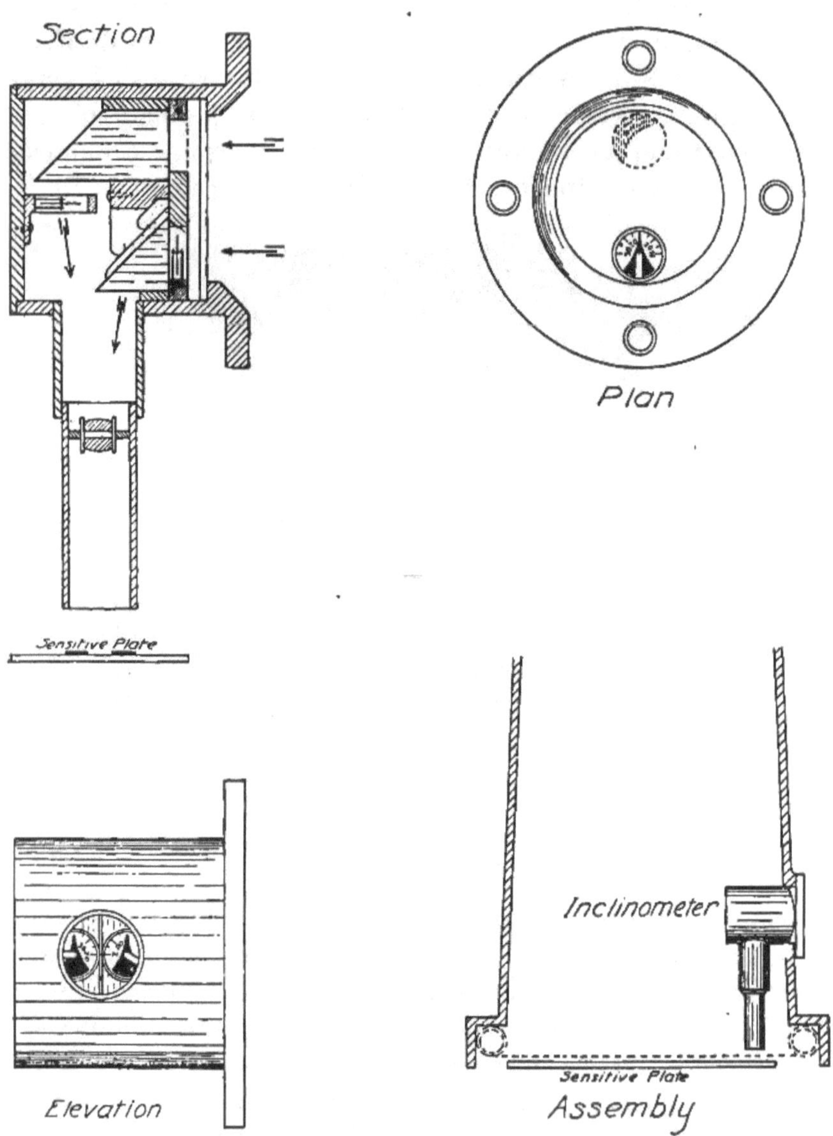

Fig. 74.—Diagram of inclinometer used in some German cameras.

two small pendulums or plumb-bobs; one to indicate lateral, the other longitudinal inclination, arranged to be photographed in silhouette on the plate, as shown in the lower part of the diagram and in the print from a captured negative (Fig. 75).

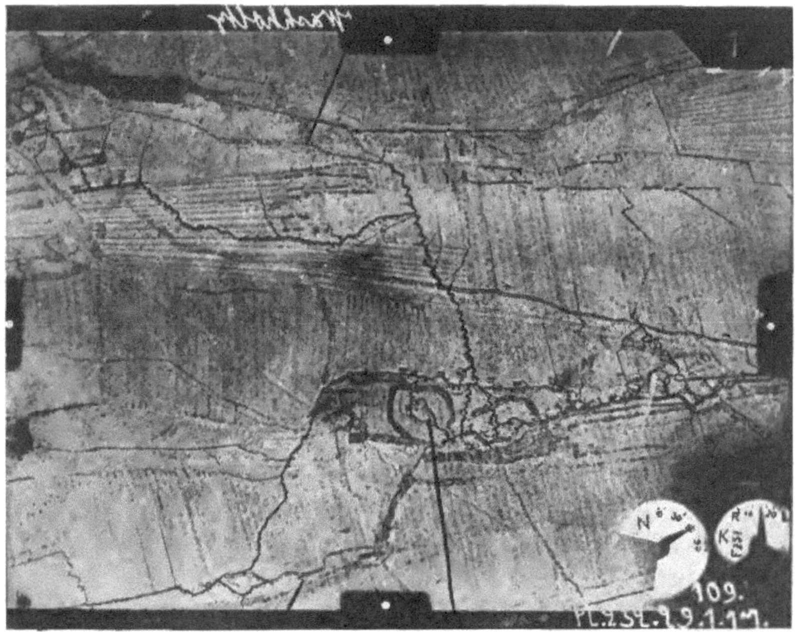

Fig. 75.—Photograph made with German camera, showing inclinometer record, four points for locating diameters and center of plate, and (upper right-hand corner) number of the plate sheath.

Both these devices suffer from the deficiencies of the instruments they photograph. The compass and the inclinometer, as already mentioned in the discussion of airplane instruments, only behave normally in straight-away flying, failing to indicate correctly when the plane is subject to accelerations in any direction. In general all attempts to record directional data in the camera are of little promise, unless either the instruments or the camera are automatically held level by some gyroscopic device. If the instruments are so controlled, rather elaborate means for photographing them are necessary. If the camera is stabilized, the inclinometers are unnecessary, and the compass behaves rationally.

Another scheme for indicating inclinations, which is not subject to the above objections, is to photograph the horizon either on a separate film or on the same sensitive surface, simultaneously with the principal exposure. The difficulty here is the practical one that it is only feasible in localities of great atmospheric clearness. Ordinarily, especially anywhere near the sea-coast, the horizon is too rarely seen to be a reliable mark (Fig. 4). It is possible, however, that this objection could be overcome by the use of specially red sensitive plates and suitable color filters, as discussed in the chapter on "Filters." The method would in any case be useless in mountainous country.

The difficulties discussed with reference to direction indicating instruments of course do not hold with the altimeter. Ordinarily, though, the altitude changes slowly enough to permit of sufficiently accurate records being made by pencil and pad. For high precision map making a photographic record of altimeter readings has a legitimate claim. As we have seen, a small altimeter is incorporated in the English F camera, but the bulk which a really precision altimeter would assume would be a bar to its use in this way. A time or serial number record on the plate or film, synchronized with a similar record on the film of an auxiliary camera which photographs the altimeter and other instruments, may be the simplest way to preserve the majority of the desired data.

Devices for Heating the Camera.—Parts of the camera mechanism which depend on the uniformity of action of springs or upon adequate lubrication are susceptible to change with variation of temperature. At high altitudes low temperatures are met which may freeze ordinary machine oils or may cause springs to seriously alter their tension, even to break. To meet this difficulty, and probably also to dispel the occasional condensation of moisture on the optical parts, the German cameras are equipped with an electrical heating coil placed just below the shutter, and arranged to connect with the general heating and lighting current of the plane. Two contacts are ordinarily provided, for offsetting the effects of temperatures of −15 and −30 degrees centigrade. An additional function of this heating coil is perhaps to maintain the sensitiveness of the plates or film.

III

THE SUSPENSION AND INSTALLATION OF AIRPLANE CAMERAS

CHAPTER XIV

THEORY AND EXPERIMENTAL STUDY OF METHODS OF CAMERA SUSPENSION

General Theory.—In addition to the limitation of exposure set by the ground speed of the plane another limitation is set by the *vibration* of the camera. This may be caused either by the motor, or by the elastic reactions of the plane members to the strains of flight. Unlike the movement of the image due to the simple motion of the plane, movements due to vibration may be eliminated by proper anti-vibrational mounting of the camera.

The effect of vibration may show as an indistinctness of the whole image—this is its only effect with a between-the-lens shutter—or as a band or bands of indistinctness parallel to the curtain opening (Fig. 76). These are due to shocks or short period vibrations during the passage of the focal-plane shutter.

The obvious remedy for vibration troubles is to mount the camera on some elastic, heavily damping support, like sponge rubber or metal springs. Such a mounting should, however, be designed on sound principles derived from a proper analysis of the nature and effect of the possible motions of the camera. Otherwise, the vibrational disturbances may be increased rather than diminished by the camera mount. Such an analysis, based merely on general mechanical principles, shows that all motions of the camera are resolvable into *six*. These are three *translational* motions, namely, two at right angles in one plane such as the horizontal, and one in the plane at right angles to this (vertical);

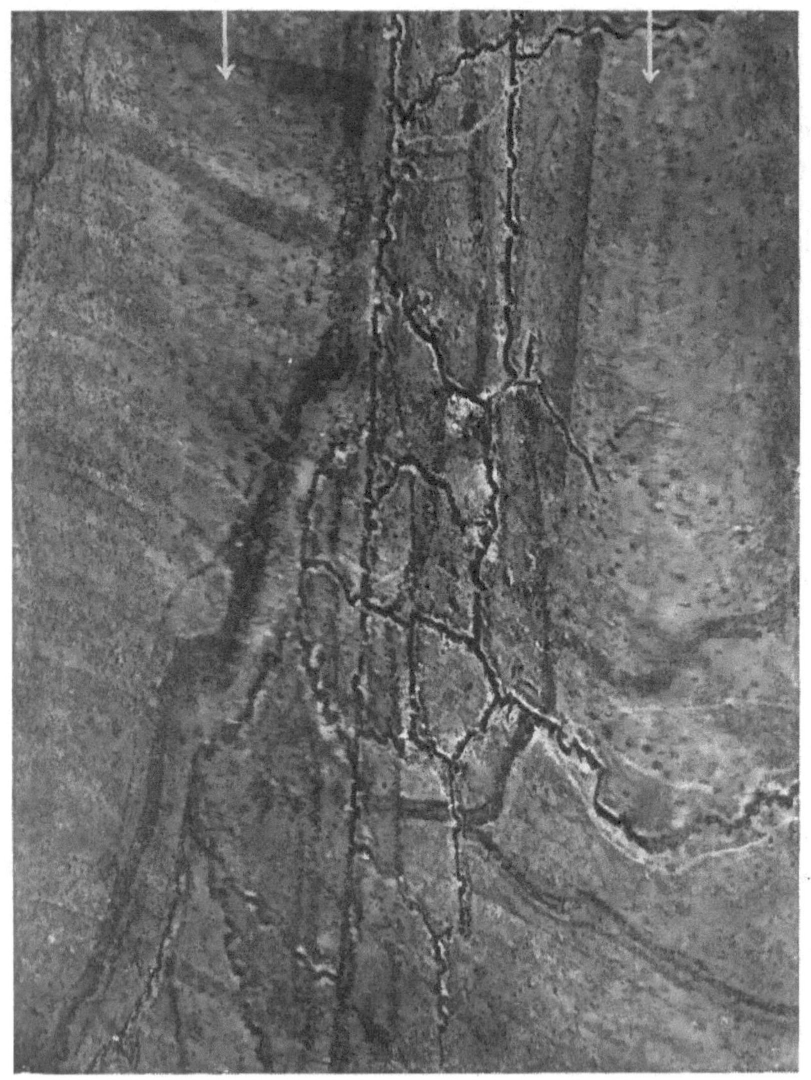

Fig. 76.—Captured German photograph, showing zones of poor definition due to vibration during passage of focal plane shutter aperture.

and three *rotational* motions, one about each of the above directions.of translational motion as an axis (Fig. 77).

Brief consideration will show that only the latter—the rotational motions—are of any importance when the small displacements due to vibration are in question. To illustrate the negligible effect of vibrations which merely move the camera parallel to itself in any direction it is only necessary to imagine the camera moved parallel to the ground through a large distance, such as 10 centimeters. Now 10 centimeters motion of the camera at 3000 meters elevation means, with a 25 centimeter camera lens,

$$\frac{.25}{3000} \times 10 = \frac{1}{1200} \text{ centimeter}$$

motion on the plate, which would be only a tenth the distance separable by a good lens. If we reduce this motion to the small fraction of a centimeter which vibration would actually produce, it is evident that such vibration is of absolutely no importance. Similarly, if we imagine the camera, under the same conditions, moved vertically with reference to the ground by ten centimeters, the scale of the picture would merely be changed by $\frac{1}{12000}$ or by $\frac{1}{1000}$ centimeter on a 12 centimeter plate, again quite negligible.

When we consider motions of rotation, however, the case is quite different. If the camera is mounted so that the effect of any vibration is to rotate it around a horizontal axis, this is exactly equivalent to rotating the beam of light from the lens so that it sweeps across the plate. Thus a millimeter displacement of the lens of the camera with the plate remaining fixed gives approximately a millimeter motion of the image. Consequently, a rotation producing only a fraction of a millimeter's relative motion of lens and plate during the period the curtain aperture is over a given point would cause fatal blurring—and the visible vibration of plane longerons and cross members is easily of half millimeter amplitude or more. Reduced to angular units it is easily shown that a rotation of one degree per second—which is quite slow as plane oscillations go—is beyond the limits of toleration. Translational motions of large amplitude may be allowed, but the mounting of the camera must not permit these translations to be at all different for different parts of the camera.

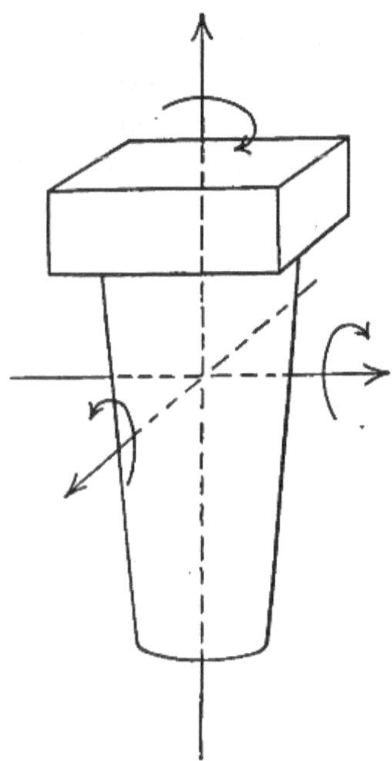

Fig. 77.—Diagram showing possible motions of the airplane camera: three of translation and three of rotation, and their combinations.

The proper way to eliminate vibrational effects is to devise a mounting that will transmit only the translational shocks or that will transform the rotational ones into translations. Platforms pivoted and cross-linked so as to be free to move only parallel to themselves (described in the next chapter) represent one attempt to reach this result. Quite the simplest and most scientific form of mounting to achieve this end is to support the camera solely *in the plane of the center of gravity.* The principle here involved is easily grasped if we note that when we jar a camera supported above or below its center of gravity, the effect is to start the camera vibrating with the center of gravity oscillating pendulum-like about the point of support. The closer the center of gravity to the center of support, the smaller the moment of the body about the latter point.

Experimental Study of Methods of Camera Support.—Conclusive evidence as to the merits of any system of camera mounting can be obtained only under conditions that eliminate the effect of other variables which may be equally efficacious in diminishing the effects of vibration, but which have only limited application. Very brief exposures—$\frac{1}{500}$ second and less—will, for instance, result in good pictures with almost any condition of vibration. Hence a sharp picture offers no proof of the merits of a camera mounting unless it is known that the exposure was no shorter than the limit set by the ground speed of the plane. In fact it may be said that the chief object of studying methods of camera suspension is to increase the allowable exposure to a maximum, thus lengthening the working hours and multiplying the useful working days for aerial photography.

The most satisfactory method of test yet developed is to fly over a light or a group of lights on the ground with the camera shutter open. In the first use of this method, which originated in the English Service, such flights were made at night, but later it was found that good results could be got by placing the lights in a forest and making the tests when the sun was fairly low. One of the group of lights must be periodically interrupted, at a known rate, to furnish the time intervals.

Some characteristic "trails" obtained by this method of test are shown in Fig. 78. The first trail is that given by a camera rigidly fastened to the fuselage. The second and third show hand camera trails, made by an inexperienced and by an experienced observer, respectively. They show by comparison with the other figures that the human body is an excellent block to vibration, but in unskilled hands a poor check to rapid erratic (probably rotational) motions of the camera. The fourth is the trail given by a camera

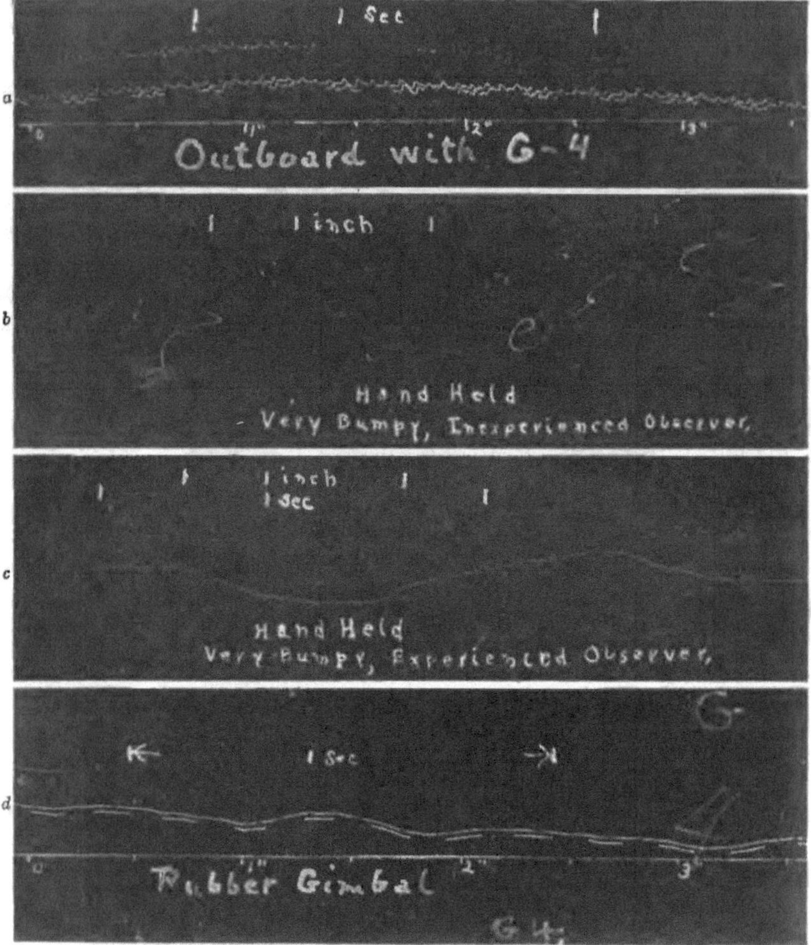

Fig. 78.—Tests of camera mounting, made by flying over a bright light against a dark background. (a) Rigid fastening on side of plane; (b) held in the hand, inexperienced observer; (c) held in the hand, experienced observer; (d) camera mounted at center of gravity on gimbals bedded in sponge rubber.

supported by gimbals bedded in sponge rubber accurately in the plane of the camera's center of gravity. Other trails are shown in the next chapter in connection with the description of practical camera mountings. Clearly the best suspension is that giving the smallest amplitude of displacement during the interval of time covered by an average exposure. It is, in fact, possible to determine from these trails the permissible exposure for any assumed permissible blurring of the image. The rigid mounting trail indicates very bad conditions, calling for literally instantaneous exposures. The center of gravity trail, at the other extreme, shows practically no limitation of exposure in so far as vibration is concerned, thus bearing out the theoretical conditions above discussed. An interesting conclusion from these experiments is that a rapidly running motor gives less harmful vibration than a slow one, although in the war it was a common practice to throttle the motor before exposing. As might be expected, the greater the number of cylinders, the shorter the period and the smaller the amplitude of the vibration.

Pendular Camera Supports.—The design of the camera support may be approached from a different standpoint, namely, with the aim of carrying the camera so that it will tend to hang always vertical. In mapping this is of fundamental importance. It is, indeed, a question whether aerial mapping will ever be worthy of ranking as a precision method unless the camera can be mounted so that its pictures are taken in the horizontal, undistorted position.

The simplest way to hold the camera vertical is to mount it on gimbals, with its center of gravity below the point of support. When so mounted the camera swings as a pendulum. Delicacy of response to variation of level is obtained by leaving a considerable distance between the center of gravity and the center of support. Oscillation about the vertical position is to be prevented by some system of dash pots or other damping. A suspension of this kind is furnished with the Brock film camera (Fig. 60).

It will be seen at once that the relation of center of gravity to center of support called for here is in direct contradiction to the requirements for eliminating vibration. Either one requirement or the other must be sacrificed, or else a compromise made in which neither delicate response to inclination of the plane nor fully satisfactory freedom from vibration is attained. This is a very serious objection to the pendular support. But the really vital objection to the pendular support is that it performs its function only very partially. It is entirely satisfactory only under conditions of steady flying, as in a uniform climb or glide, with the plane tail or nose heavy, or in flying with one wing down. As soon as we introduce any acceleration, as in making a turn, the camera follows the plane and not the earth.

It is true that mapping photography is done from a plane flying as level as possible, and that except under bad air conditions it holds its course with very little turning, if handled by a skilled pilot. Nevertheless, a surprisingly small deviation from straight flying causes quite serious variations from the vertical. It is of interest to calculate how large may be the horizontal accelerations that accompany swervings from a straight course which one might think insignificant. For instance, consider the horizontal acceleration due to a turn having a radius of a kilometer when the plane is moving at 100 kilometers per hour. If a is the acceleration, v the velocity of the plane, and r the radius, we have from elementary dynamics that

$$a = \frac{v^2}{r}$$

Substituting the values chosen, we have—

$$a = \frac{100{,}000^2}{3600^2 \times 1000} = .77 \frac{\text{meters}}{\text{sec}^2}$$

The acceleration of gravity being $9.80 \frac{\text{meters}}{\text{sec}^2}$ we have that the ratio of the horizontal acceleration to the vertical is

$$\frac{.77}{9.80} = .078$$

This is the tangent of the angle of deviation from the vertical, from which the angle turns out to be about $4\frac{1}{2}$ degrees, a very considerable error, rapidly multiplied as the speed of the plane is increased. It is, indeed, open to question whether the average deviations from the vertical are not apt to be less with the camera rigidly fixed to the plane, if guided by a skilled pilot who will hold the ship level at the expense of "skidding" the slight turns he must make to hold his direction.

Gyroscopic Mountings.—The ideal support for the aerial camera will undoubtedly be one embodying gyroscopic control of the camera's direction. By proper utilization of the principles of the gyroscope it is to be expected that not only can the camera be maintained vertical, but it may be supported anti-vibrationally as well. At the present time the problem of gyroscopic control is in the experimental stage, so that only the elements of the problem and the possible modes of solution can be laid out.

The gyroscope consists essentially of a heavy ring or disc rotating at a high speed on an axis free to point in any direction (Fig. 79). If mounted so that the axes of the supporting gimbals pass through the center of gravity of the rotating disc, the result is a *neutral* gyroscope. Its characteristic is that its axis maintains its *direction fixed*, but this fixity is with respect to space and not with respect to the gravitational vertical. Consequently, as the earth revolves the inclination of the gyroscopic axis changes with respect to the earth. In latitude 45° this change is approximately a degree in five minutes. Furthermore, the action of friction in the supports, which can never be entirely eliminated, also acts to slowly alter the direction of the gyroscopic axis. Therefore, unless some *erector* is applied even the gyroscope will not perform the task required of it.

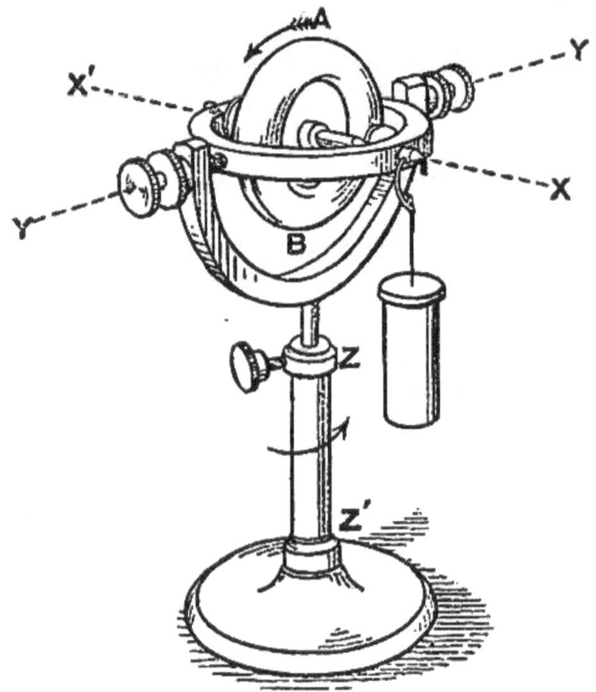

Fig. 79.—Diagram of simple gyroscope.

Before discussing possible forms of erectors it may be noted in general, first, that these must depend upon gravity; second, that such being the case, they must respond to the resultant of gravity and any acceleration, that is, to the *apparent or pseudo-gravity*. As already seen, this pseudo-gravity, during a turn, is exactly what limits the usefulness of the pendular support, and necessitates recourse to the gyroscope. The problem thus becomes one of making an erector-gyroscope combination which will respond to true gravity and not to pseudo-gravity.

In general this problem would be insoluble, since there is no difference in the nature of the acceleration of gravity and that due to centrifugal force. A way out is offered, however, by the fact that true gravity acts continuously and at a small angle to the axis of the gyro, while the components which cause the pseudo-gravity are of short duration, liable to rapid changes of direction, and, on a turn, act at a large angle. What we require, therefore, is an erector which will respond slowly but surely to the *average* acceleration, which is downward, but too sluggishly to be affected by the shorter period accelerations due to turns or rolls. Slowness of response is a matter of the erecting forces being small and of the mass and angular velocity of the gyro disc being large. The success of the compromise called for depends on the relative times taken for the gyroscope to tilt seriously from the true vertical, due to the causes above mentioned, and for the average turn or roll. Fortunately the former is a matter of minutes, the latter of seconds or at the worst of fractions of a minute. More than this, since the roll or turn is apt to be of much greater angle than any normal deviation of the gyroscopic axis from the vertical in the same time, we are offered the possibility of some device for filtering out the deviations which alone are to effect the erector. Forinstance, by shunting the restoring force whenever it is called upon to act through more than a predetermined small angle.

As to the method of erecting the gyroscope, its characteristic property must be kept in mind. This is that the axis does not tilt under an applied force in the direction it would if the gyro were not rotating, but around an axis at right angles to that of the applied couple. Thus in Fig. 79, if a weight is attached as shown, the disc does not incline downward toward the weight, around the axis Y, Y', but *precesses* about the vertical axis Z, Z'. Some means is therefore needed to translate the pull which any gravitational control, such as a freely swinging pendulum, would give, into a pull with *at least a component* at a finite angle to this.

In the Gray stabilizer several metal balls are slowly rotated in a tray above the center of gravity of the gyroscope. Specially shaped grooves or compartments limit the freedom of motion of these balls so that when the gyro is inclined the balls travel at different distances from the center on the ascending and descending sides. By this scheme a couple is produced about the axis through the center and the low point of the disc, which tilts the apparatus to the gravitational vertical. In an alternative form the balls are carried past the low point by their momentum and are prevented from returning by the walls of the containing compartment, which have meanwhile been advanced by the rotation of the erector as a whole. The net result is to shift the center of gravity of the system of balls in the proper direction to erect the gyro. The rectifying action is purposely made quite slow so that the displacements of the balls due to pseudo-gravity will be averaged out.

In a design due to Lucian, small pendulums work through electric contacts to actuate solenoids which in turn move small weights in the appropriate directions to give the desired tilt. Response is made fairly quick and delicate, and pseudo-gravity, due to turns and rolls, is rendered inoperative by the contacts breaking whenever the pendulums swing more than three or four degrees. This can only happen if they move too quickly for the erecting forces to act, reliance being here placed on the characteristic differences of action in respect to time of real and pseudo-gravitational forces.

Besides the neutral gyroscope as just considered there is the pendular or top type, in which the center of gravity is not in the plane of the supports. In general this type depends on a couple resulting from the gravitational pull and the inevitable friction of the supports to slowly tilt the axis to the gravitational vertical. This type is slower to respond

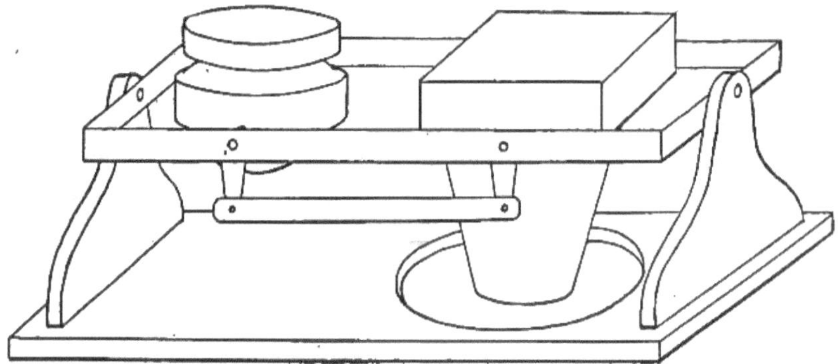

Fig. 80.—Diagram of camera linked to gyroscopic stabilizer.

than the designs in which a definite couple in the proper direction is provided and it reaches the true vertical only through a circuitous path.

Three methods of controlling a camera by a gyroscope are suggested. One is to fasten the gyroscope rigidly to the camera and mount the whole system on gimbals. A second is to mount both camera and gyro side by side on gimbals, linking the two so that the camera is moved parallel to the gyro (Fig. 80). A third method is to utilize the gyro to make electric contacts to operate motors which in turn move the camera.

Considerable weight and space are required for a gyroscope capable of stabilizing a camera. The rotating disc should be about half the weight of the camera, and with its mounting may be expected to double the room required for the camera alone. Motive power for maintainng the gyro in continuous rotation may be supplied by an air blast, or the gyro may be made up as an induction motor—the latter necessitating an alternating current supply.

In view of the space and weight limitations in a plane it is a question still to be decided whether it is more economical to stabilize the camera or to stabilize an inclinometer and photograph its indications simultaneously with the release of the shutter which takes the aerial picture.

CHAPTER XV

PRACTICAL CAMERA MOUNTINGS

General Considerations.—Camera mountings as used during the war were far from being developed on the basis of scientific study or test. At first the need for special supporting apparatus was not realized, and the suspensions later in use were largely field-made affairs, often dependent on adjustments made according to individual taste. Through lack of accurate methods of test and of conclusive evidence on the subject, it was quite common to find extremists who, on the one hand, denied the efficacy of suspensions in general, and on the other ardently supported crazily conceived supporting arrangements which accurate comparative test show to be even worse than useless.

In the French service, despite numerous types of suspension available, the very general practice was to lift the camera from its support and hold it between the knees. Or else the hand was pressed on the top of the camera during exposure, more reliance being placed on the damping qualities of the body than on any of the rubber or spring mechanisms.

As is clearly shown by the experimental data described in the last chapter, a correctly designed supporting device, carrying the camera accurately in the plane of its center of gravity, accomplishes practically perfect elimination of vibrational troubles. So important is the use of a mount and so important is it that the mount should be correctly dimensioned and adjusted for the camera, that an entirely different attitude should be adopted from the prevalent one which focuses attention on the camera and regards the mounting as a mere auxiliary to be left more or less to chance. *The mounting should be considered an integral part of the camera.* The man in the field should receive camera and mount together, leaving as his only problem the attachment of the complete camera—and—mount unit to the plane. This may be arranged, by proper designing, to be a simple matter

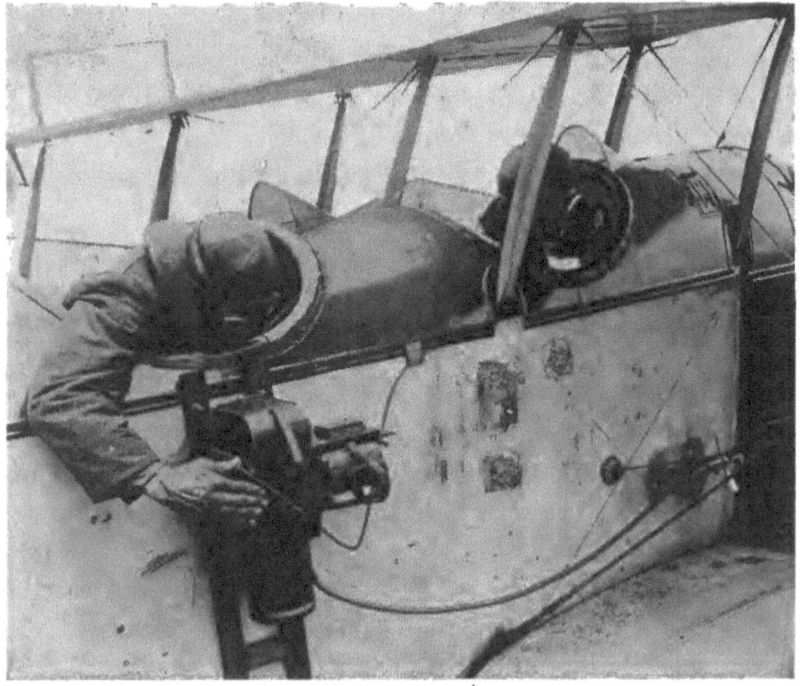

Fig. 81.—"L" camera mounted outside the fuselage. Observer using exposure plunger, pilot using Bowden wire release.

of rigid bolting or strapping, requiring ingenuity perhaps but not the scientific knowledge which is required for mounting design.

Outboard Mountings.—In the English service the camera was first attached to the plane outside the fuselage by a rigid frame, to which the camera was strapped or bolted (Fig. 81). Obvious objections exist to placing the camera in this position, such as the resistance of the wind and the difficulty of changing magazines. However, in the earlier English planes with their fuselages of small cross section no other accessible place for the camera was to be found. Vibrational disturbances with the rigid outboard mounting are quite serious, as is so clearly indicated by the trace shown in Fig. 78. Extremely short exposures are alone possible, and a very large proportion of the pictures are apt to be indistinct.

Floor Mountings.—A step in advance of the outboard mounting is to support the camera snout in a padded conical frame on the floor of the plane (Fig. 82). This mounting avoids the objection on the ground of wind resistance that holds with the outboard, and has possibilities of being worked out as an entirely satisfactory support. Yet to be satisfactory, the point of support must lie in the plane of the center of gravity of the camera, and the camera must be of a type that preserves its center of gravity unchanged in position as the plates are exposed. Unless these conditions are fully met the floor mounting gives results little better than does the outboard.

Cradles or Trays.—Floor space in the cockpit being unavailable in the battle-plane, due to duplicate controls, bomb sights, etc., the English service was driven to the practice of carrying the camera in the compartment or bay behind the observer. Here it was attached either to the structural uprights or longerons, or to special uprights and cross-pieces built into the plane to serve photographic ends. As an intermediary between the camera and the supporting cross-pieces there was introduced the camera *tray* or *cradle*. This is essentially a frame carrying sponge rubber pads into which the camera is more or less deeply bedded. Figs. 83 and

84 show an American L camera cradle based on the design of the English L camera tray. Thick sponge rubber pads support the two ends of the camera top plate, and additional pads are provided to hold the nose of the camera. Careful

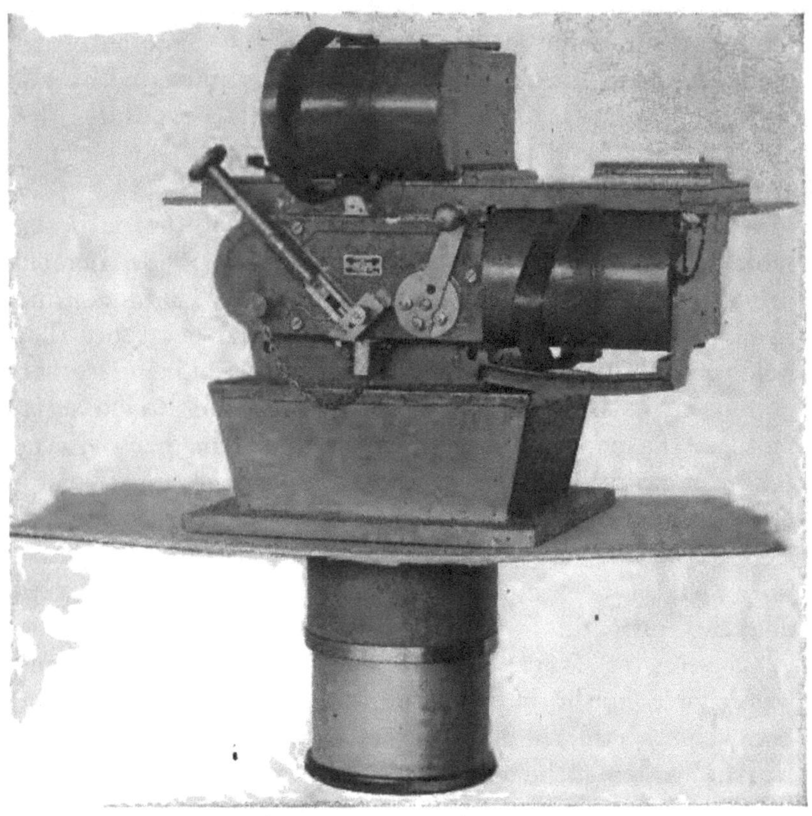

FIG. 82.—"L" camera in floor mounting.

tests show this cradle to be superior to the outboard mounting, but still leave much to be desired. Its performance is better with the nose of the camera left free.

Tennis-ball Mounting.—A very simple mount used by the French consists of a frame enclosing the nose of the camera, and carrying four tennis balls, on which the whole weight rests (Fig. 40). If the center of support is in the plane of the center of gravity and if the four balls are of uniform

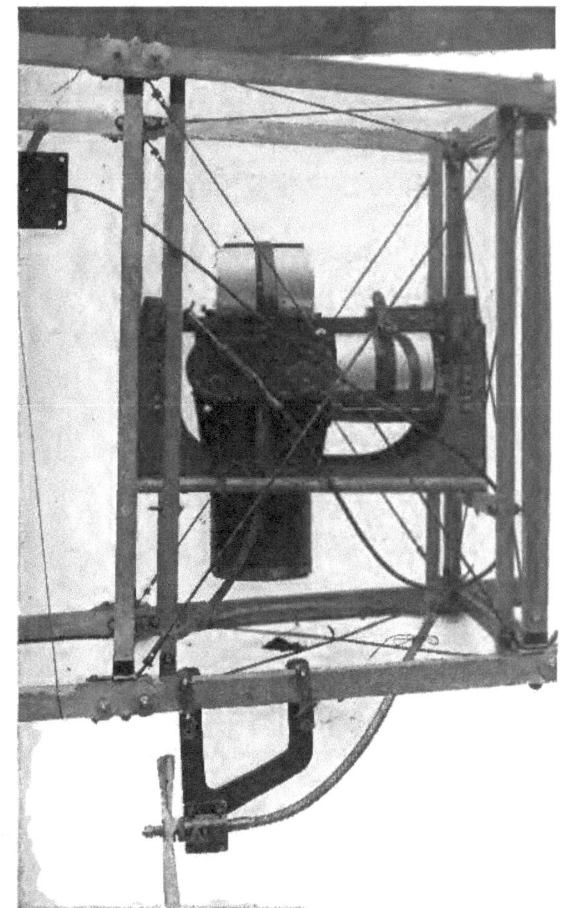

FIG. 83.—"L" camera and cradle mount in skeleton DeHaviland 4 fuselage, side view.

age and elasticity, this form of support is good. As provided by the camera manufacturer, the tennis ball frame fits much too far down on the camera. Another application of the tennis ball idea was frequently made in the French service, in which the balls were close up under the shutter housing (Fig. 85). Additional support was, however, given to the camera nose by flexible rubber bands, the success of the whole being largely a matter of the adjustment of the tension on the bands.

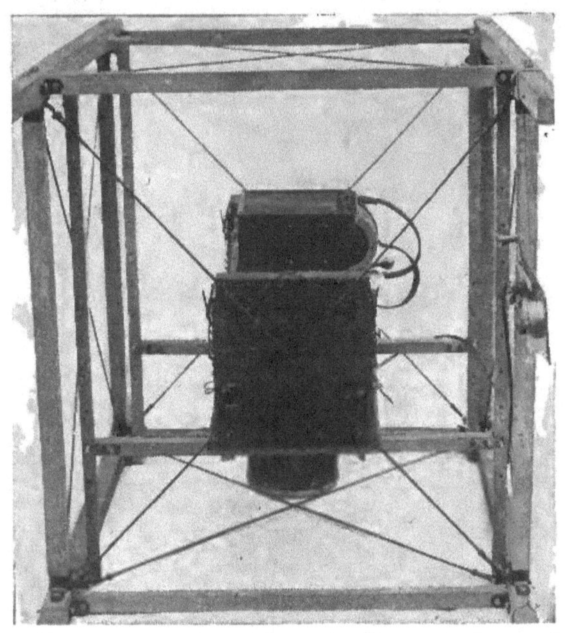

FIG. 84.—"L" camera and cradle mountin skeleton deHaviland 4 fuselage, front view.

Parallel Motion Devices.—A form of suspension favored by the French consists of parallel bell cranks, rigidly linked together and held up by springs. Mountings of this sort are illustrated in Figs. 86, 87, 88 and 96. The guiding principle is that any sort of shock will be transformed into a straight up-and-down or side-wise motion of the camera, which is harmless. The mounting as adapted by the English surrounds the camera body, making the plane of support somewhere near the center of gravity. In certain of the French suspensions employing this principle the whole camera is hung below the bell cranks (Fig. 86), and then the

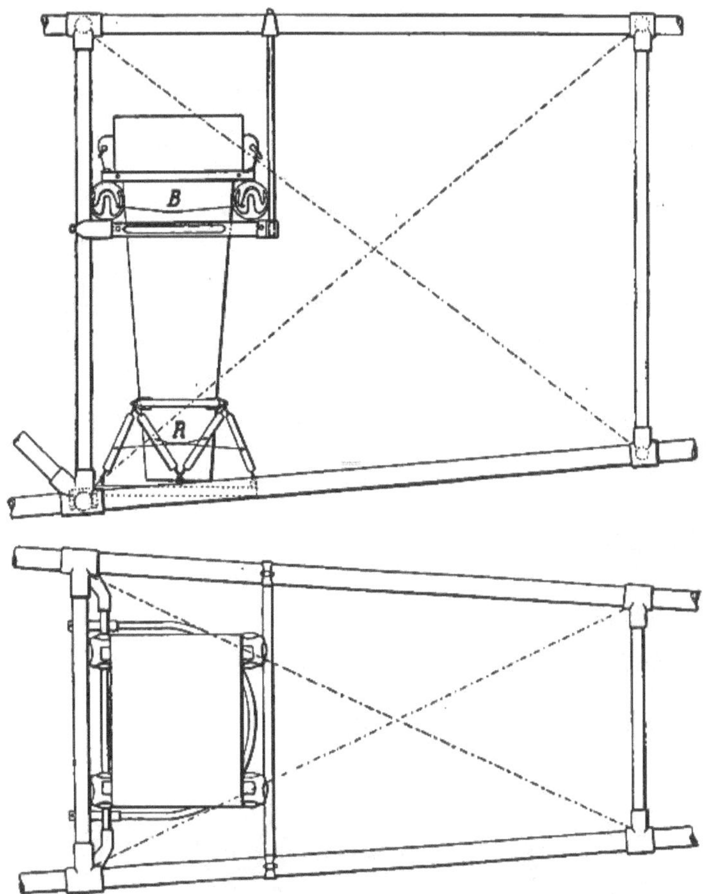

FIG. 85.—Tennis ball suspension, assisted by elastic bands attached to nose of camera.

nose is restrained by heavy rubber bands. The net result is largely a matter of adjustment.

Tests on the English design made in the United States

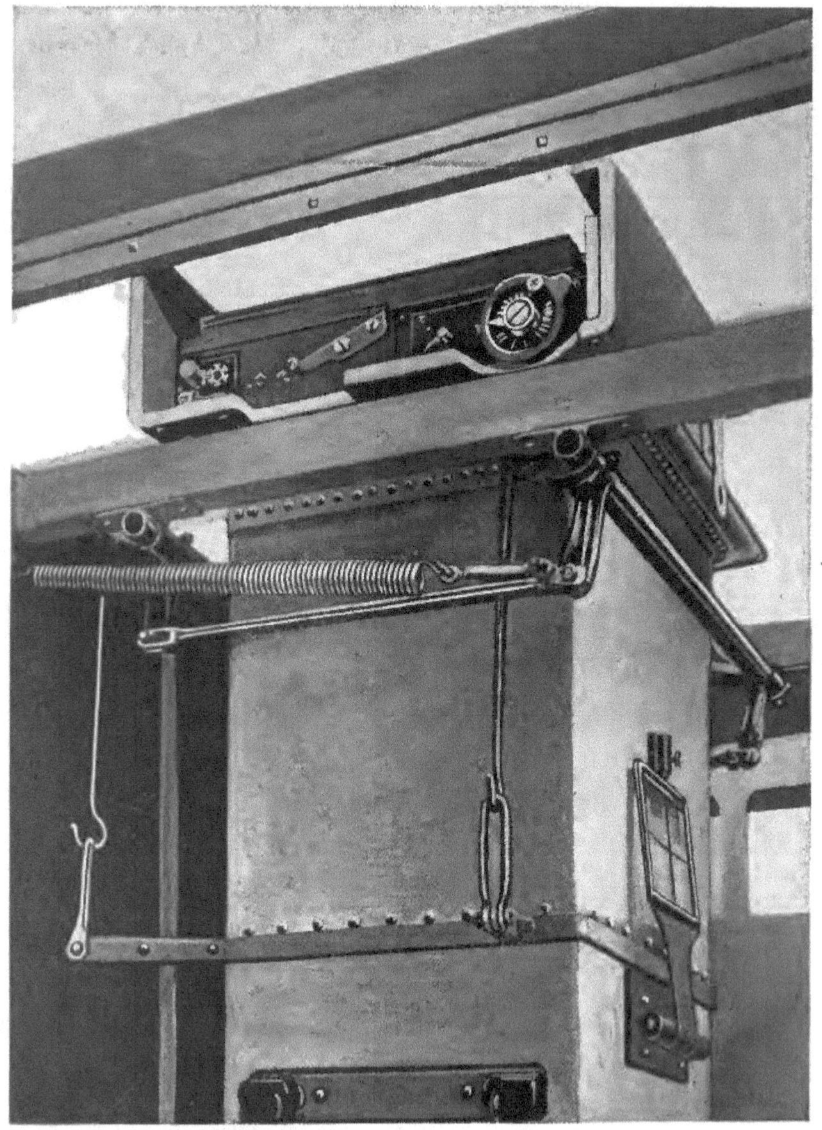

FIG. 86.—French spring and bell crank suspension.

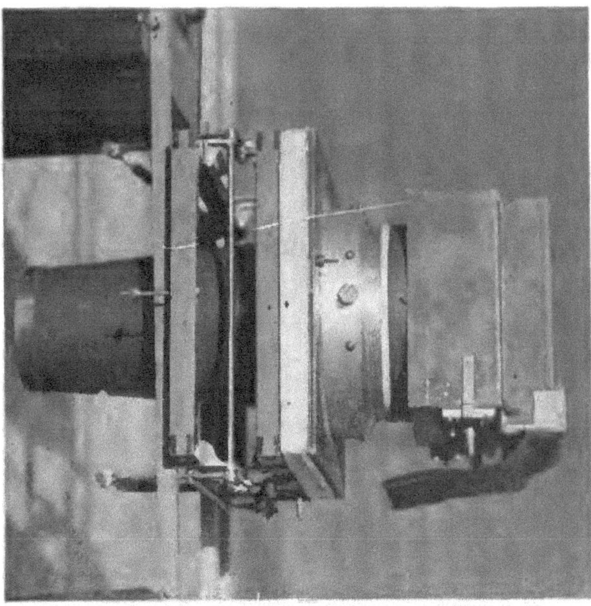

Fig. 87.—U. S. hand-operated 18x24 centimeter plate camera on bell crank mount with rotating turret.

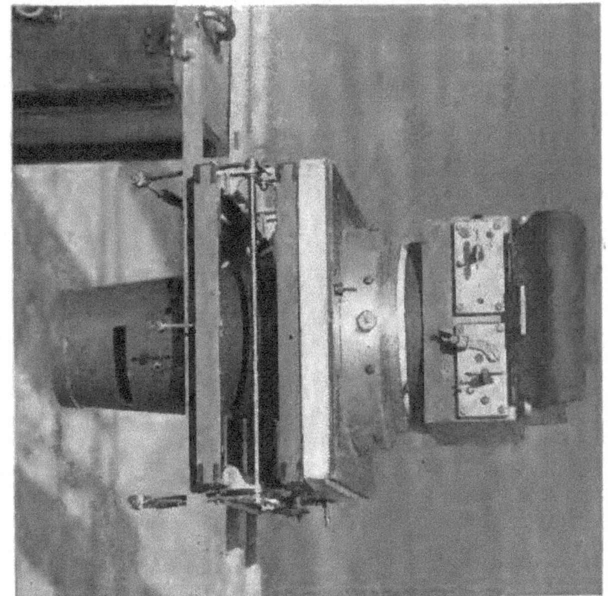

Fig. 88.—Same camera in plate changing position.

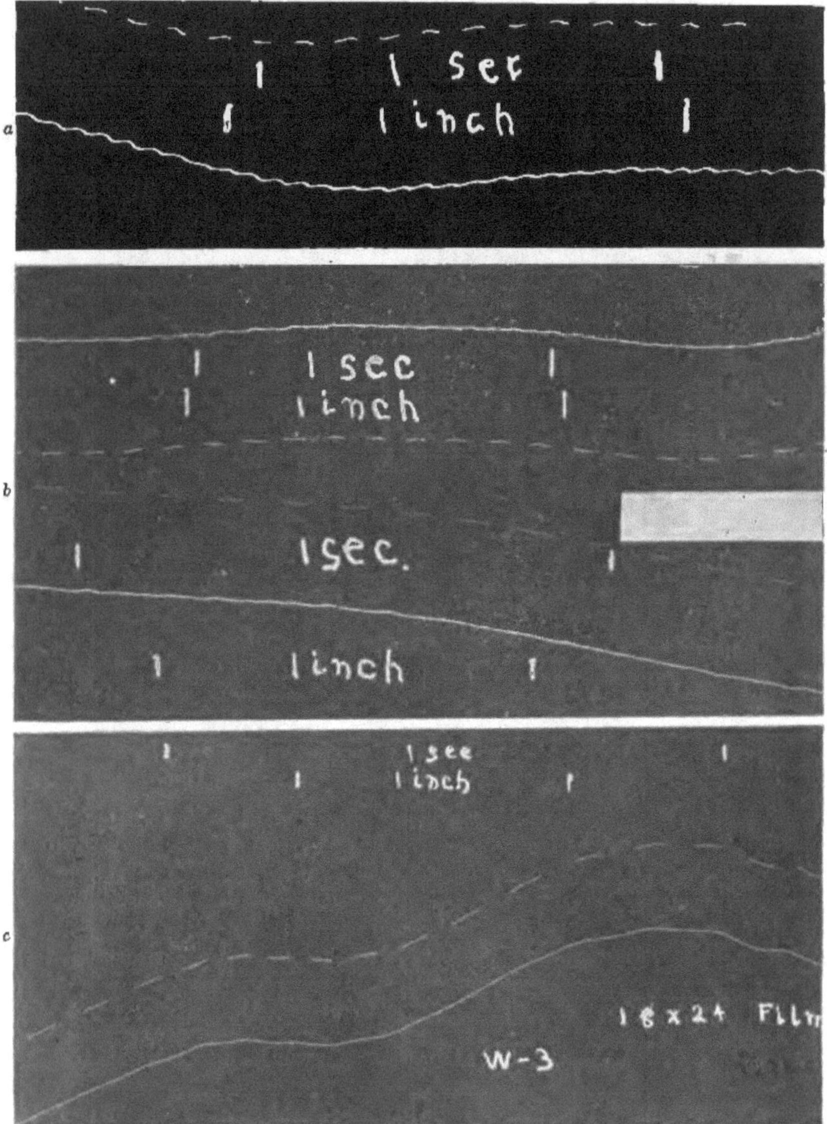

Fig. 89.—Tests of camera mountings: (a) deRam camera on bell-crank-and-spring mount, below the center of gravity; (b) same, at center of gravity; (c) type "K" film camera on universal mounting (Fig. 88).

Air Service appear to show that the chief virtue of the mounting lies in the approximation of the point of support to the center of gravity in the English cameras. A deRam camera supported by its cone, so that its center of gravity was considerably above the center of support gave rather poor results (Fig. 89*a*), but when the bell cranks were attached near the center of gravity, highly successful results were obtained (Fig. 89*b*). The French deRam camera as ordered for the American Expeditionary Force was fitted with a bell crank supported in this position.

Figures 90 and 91 show a bell crank mounting furnished with a rotating turret. This was designed to facilitate the changing of magazines in the English B M camera, which is swung around through 90 degrees from the exposing position to bring the magazine near the observer. The camera shown in the mounting is the American hand-operated model (type M), in which there is the same necessity for turning in order to manipulate the bag magazine easily. The camera is shown in both exposing and plate changing positions. An important detail of these mounts is a *safety catch*, which must be fastened before the plane lands, in order to prevent the shocks of landing from producing oscillations sufficient to throw the camera out of the mount.

Center of Gravity Rubber Pad Supports.—Given a camera whose center of gravity does not change during operation, a simple and entirely adequate anti-vibration support is furnished by a ring of sponge rubber in the plane of the center of gravity. But if provision has to be made for oblique views or for adjusting the camera to the vertical, something more elaborate is necessary.

Mountings for the American deRam and for the Air Service film camera, embodying the results of complete study of the anti-vibration problem, are shown in Figs. 90,

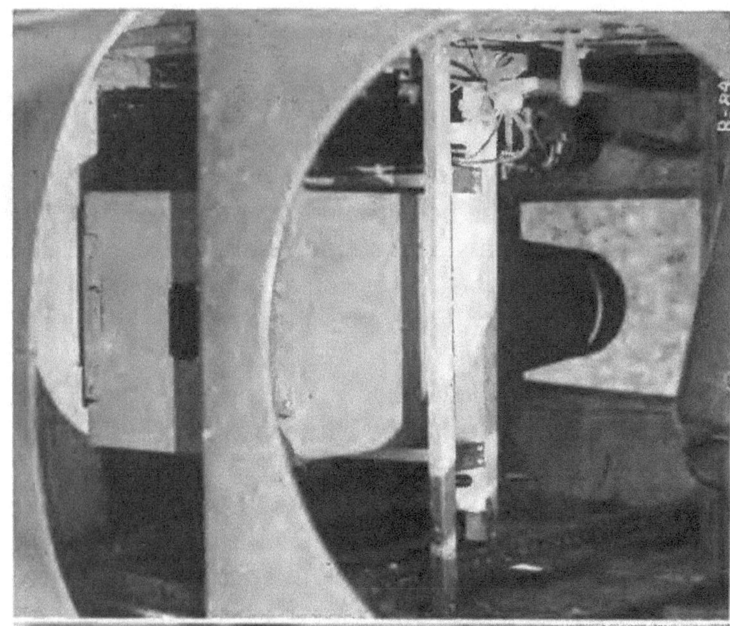

FIG. 90.—U. S. model deRam camera on anti-vibration mounting adjustable for the angle of incidence of the plane.

FIG. 91.—U. S. deRam camera and mount installed in photographic deHaviland 4 (Fig. 100). Viewed from above the observer's cockpit.

92 and 93. Trusses carrying the cameras on pivots rest on four pads of sponge rubber which are mounted on frames

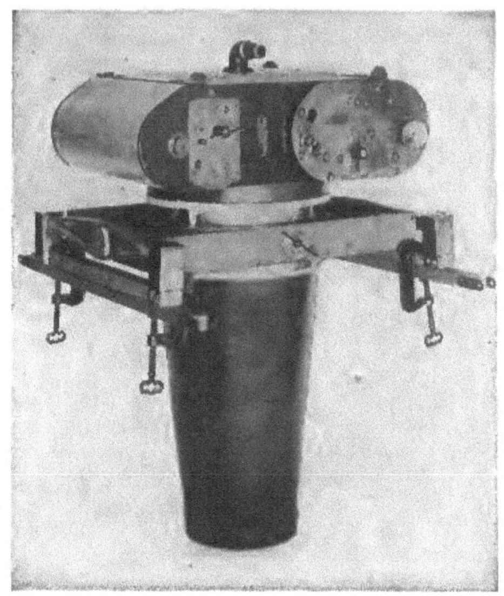

FIG. 92.—U. S. type "K" film camera on universal mounting, vertical position.

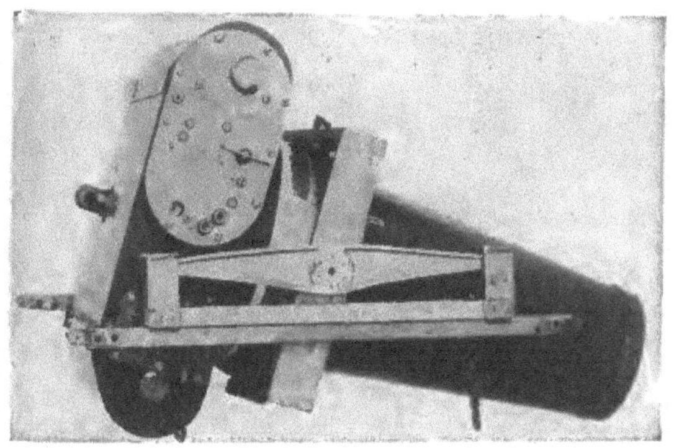

FIG. 93.—U. S. type "K" film camera on universal mounting, oblique position.

correctly spaced ready for attachment to the cross-pieces of the airplane camera supports. In the deRam (Fig. 90) the pivots, attached to the camera body, permit it to be leveled fore and aft, to compensate for the inclined position of the fuselage assumed at high altitudes or in some condi-

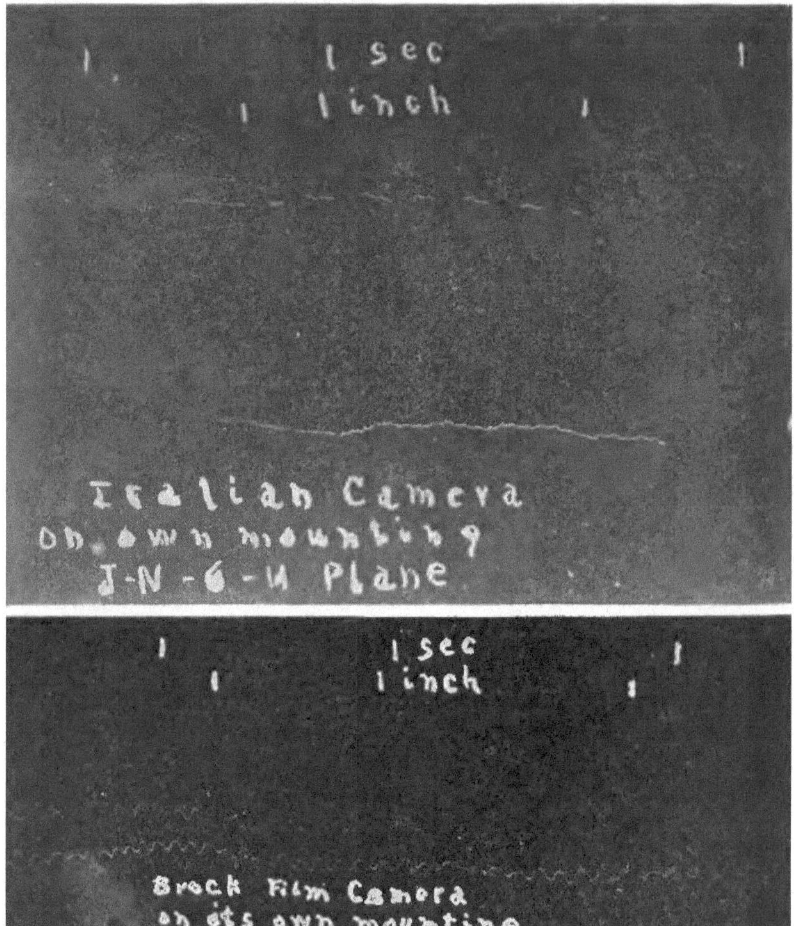

FIG. 94.—Tests on two types of camera mount: (a) Support at bottom of camera; (b) support above center of gravity.

tions of loading. This will sometimes amount to as much as 11 or 12 degrees, which is very serious, since one degree causes (with an angular field of 20 degrees) about one per cent. difference of scale at the two sides of the plate. The film camera mounting carries the camera in a conical ring, and is pivoted not only for vertical adjustment, but for the taking of obliques as well (Fig. 93). These mounts transmit practically no vibration.

A caution must be noted with regard to center of gravity mountings. Any change in the camera, in particular the substitution of a short for a long lens cone, must be made so as to cause no alteration of the relative positions of the center of support and the center of gravity. Either the short cone must be weighted, or additional supporting pivots must be provided in the plane of the new center of gravity.

The Italian and G. E. M. Mountings.—These mounts (Figs. 49 and 59) are similar in that the protection from vibration is furnished by an elastic support at the bottom of the camera. Tests show that these two cameras give very similar results, of the unsatisfactory sort to be expected from this kind of mounting in view of the lessons of the last chapter on the proper point of support. Fig. 94, *a*, shows a trace given by the Italian mount. The permissible exposure, on the criterion adopted, is very short with either mount, about $\frac{1}{200}$ second.

The Brock Suspension.—This consists of a pair of frames into which the camera is fitted by ball bearing pivots, so that it is free to move in any direction (Fig. 60). In order to permit gravity to control the direction of the camera, the point of support is made considerably (ten inches) above the center of gravity. Air dash pots are provided for damping the swings. As already explained, the pendular method of support is in basic contradiction to the requirements for vibration elimination. Tests of the Brock suspension, shown in Fig. 94, *b*, indicate it to be of low efficiency in damping out the short period vibrations which are responsible for poor definition.

CHAPTER XVI

THE INSTALLATION OF CAMERAS AND MOUNTINGS IN PLANES

Conditions to Be Met.—The characteristic difficulty in installing the airplane camera is that there is no place for it. After the gasoline supply, the armament, the wireless, the oxygen tank, the bombs, and other necessities are taken care of there is neither space available nor weight allowable. Where space may be found it will be inaccessible, or accessible only through a maze of tension and control wires; or it will be in a position where any weight will endanger the balance of the plane. Plane design has in fact been more or less of a conflict between the aeronautical engineer, who is designing the airplane primarily as a machine to fly, and the armament and instrument men, who look upon it as a platform for their apparatus. Lack of appreciation of the extreme importance of aerial photography resulted, during a large part of the war, in the camera installation being neglected until the plane was supposedly entirely designed, and even in production. At that stage the installation could be but a makeshift. Only in the later stages of the war, when plane design became a matter of coöperation between all concerned, were fairly convenient and satisfactory arrangements made for the camera. Always, however, the rapid succession of new plane designs, with various shapes of fuselage and details of structure, made camera installation in the war plane a matter calling for the greatest ingenuity.

The problem was met in part by constructing both cameras and mountings in sections, to be laboriously wormed in through inadequate apertures, in part by later structural changes in the planes, such as the substitution of veneer rings or frames for the tension wires. In certain cases the rear cockpit controls were omitted, thereby freeing accessible and often adequate space for the larger cameras. Rear controls were never used in the German planes, so that their standard practice was to carry the camera forward of the observer. This, together with the general restriction to the 13×18 centimeter size plate, made the installation problem less difficult in the German aircraft than in the Allied.

Practical Solutions.—An important feature of camera installation has already been mentioned, but may well be repeated for emphasis. The camera and its anti-vibration mounting should always be considered as a unit, and should be so designed that simple bolts or straps will suffice to fasten it in its place in the plane. Even should the spacing of the structural parts of the plane not correspond to that anticipated by the mounting design, the ingenuity of the man in the field may be depended upon to make the necessary alterations or additions to the plane. The design of the camera suspension itself cannot, however, be left to uneducated ingenuity.

Assuming the camera and mounting supplied, the next step—a very difficult one—is to insure uniformity in the structures to be built into the planes for the purpose of supporting the camera mountings. With this uniformity must, however, be combined the greatest possible flexibility to provide for various designs of cameras.

In the English service the standard camera installation consists of wooden uprights with cross bars athwart the plane, adjustable as to height (Fig. 95). A distance between the cross bars of 13¼ inches has been standardized, and all camera cradles and mountings are notched or otherwise spaced to fit this dimension. The installation adopted in

14

the American planes is similar, but with a distance of 16 inches between cross bars. These uprights and cross bars are ordinarily situated in the bay behind the observer, but can be placed in any available space. Fig. 83 shows, in a model bay, the arrangement of uprights and cross bars in

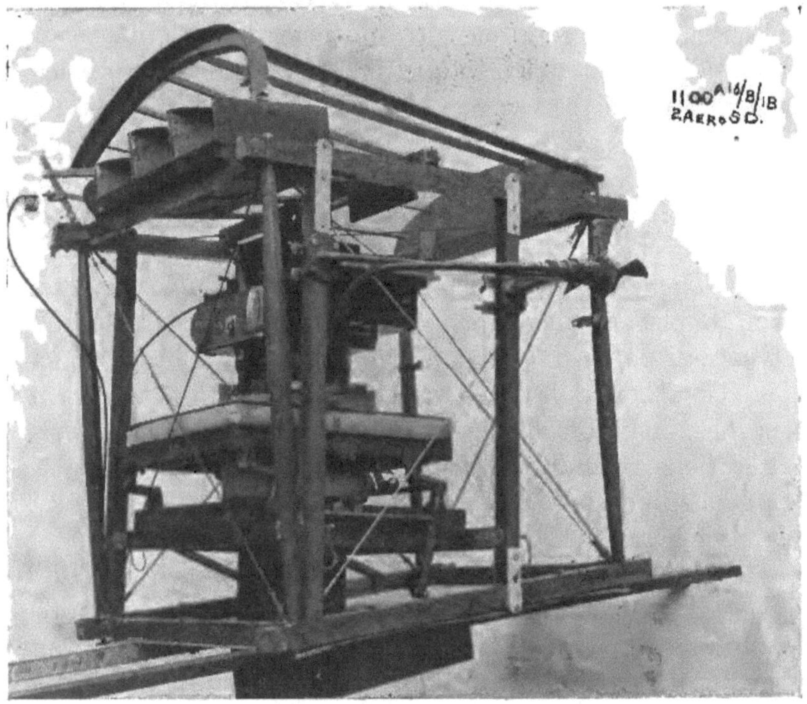

FIG. 95.—"L-B" camera with 20-inch lens, mounted on bell-crank suspension in skeleton fuselage. Stream-lined hood below to cover projecting end of lens cylinder. Propeller and Bowden release in place.

the American DH 4, with the L camera in place in its cradle. It is just possible to introduce camera and cradle separately from the observer's cockpit through the tension wires, and, by uncomfortable reaching, magazines may be changed.

A step in advance is made when the top tension wires and superstructure are replaced by a rigid frame with an opening large enough to admit the entire camera and mounting. When this is done considerably larger cameras may be accommodated in the same sized bay, as shown in Fig. 96. A further advance, from the standpoint of accessibility and convenience of installation, follows when the tension wires between observer's and camera bay are replaced by a plywood ring, as shown in Fig. 97. Here the only serious limitations are those due to the vertical height of the camera, and of course its weight.

Openings for the lens to point through are simply provided in the fabric covered aircraft, by cutting through the canvas and stiffening the edge of the hole by wire. Tension wires are often in the way. They may either be disregarded, since they merely cut off a little light, or replaced in part by metal rings, as shown in Fig. 96. In veneer covered fuselages the hole must of course go through the wood. This may be undesirable, since the veneer is depended on to furnish structural strength, a point which further emphasizes the importance of the photographic requirements being thoroughly considered while the plane is being designed.

Single seater or scout planes do not lend themselves to the insertion of such standardized uprights and cross-pieces, because of their small size and the common utilization of all space inside the fuselage for gasoline tanks and control wires. Some French scouts, whose fuselages are very wide, due to the rotary engines, have been fitted with compartments for contemplated automatic film cameras. The most commonly used camera in the single seater was, however, the Italian 24-plate single-motion apparatus (Fig. 49). This camera and its carrying tray occupy very little lateral space and have in actual practice been carried beneath the seat or pushed up through an opening in the bottom of the fuselage under the gasoline tank. Whatever criticism may be

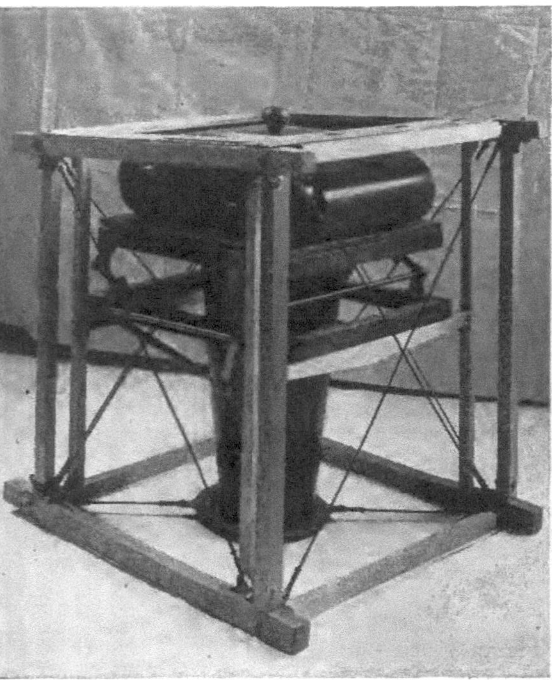

FIG. 96.—U. S. type "K" film camera on bell-crank mount, in camera bay of deHaviland 4. Veneer frame at top of bay in place of usual cross-wires.

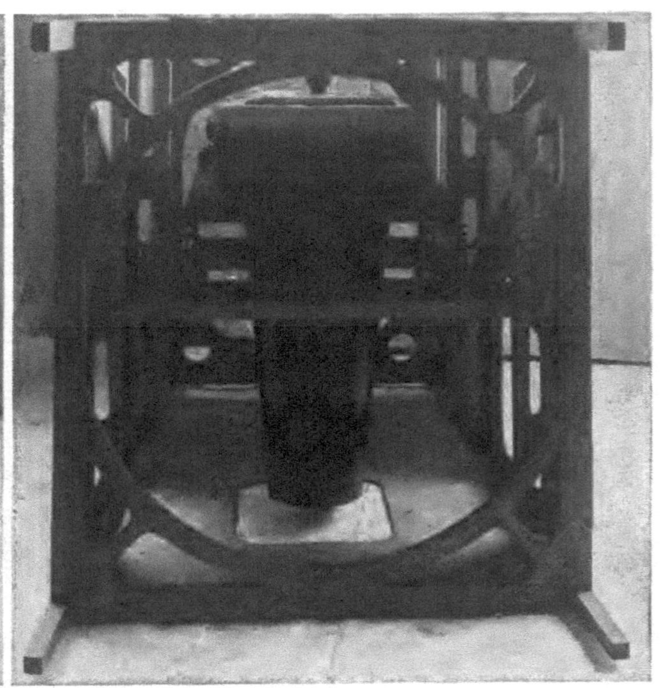

FIG. 97.—Section of fuselage of veneer construction affording superior accessibility to camera.

made of the adequacy of the mounting, it must be said that the camera, as used, is perhaps the most eminently practical of all developed in the war, as its use on scouts testifies.

Special Photographic Planes.—As cameras grew in size, the difficulty of installing them in planes built without regard to photographic requirements greatly increased. Few planes could carry even the 50 centimeter focus camera obliquely without the necessity of poking its nose through the side where it would catch wind and oil; while the 120 centimeter camera could be carried obliquely only in the fore and aft position. Even vertical installation of the latter camera was really feasible in but few planes; sometimes the camera was carried to the exclusion of the observer—and, in fact, this size was never used by the English, whose fuselages were small in cross-section.

This situation led, late in the war, to steps toward producing planes designed primarily for photographic reconnaissance. In these the camera would be entirely accessible, and cameras of any size could be carried in any desired position. One scheme which properly belongs under this heading was the provision of a special removable photographic cockpit, for the front or nose of a twin-motored three seater. Other noses, for bombing and heavy machine guns, were also planned, all to be interchangeable. Since the regular photographic bay with uprights and cross-pieces was also provided to the rear, this special photographic ship could on occasion do two classes of work, such as long focus spotting and short focus mapping.

The most completely worked out photographic plane was probably the model designated P1 by the United States Air Service. This is a modified de Haviland 4 in which the rear controls have been removed and the cowling raised and at the same time made squarer in cross-section. The space formerly occupied by the rear controls provides ample room for all types of camera. These are carried on uprights at the standard distance apart, 16 inches, with cross-pieces adjustable as to height. The camera space is accessible not only from the observer's cockpit, but from above, upon folding back the metal cover. Doors at the bottom and at each side permit not only vertical but oblique exposures. The latter are not interfered with by the wings, as they would be in some designs of plane if the camera occupied the same position relative to the cockpits. Fig. 91 shows the deRam camera in place, as seen from the rear. Figs. 98 and 99 show the 18×24 centimeter film camera, set both for vertical and oblique views.

Negative lenses are provided for both pilot and observer, the one for the pilot permitting him to see from a point far ahead to directly underneath, while the observer's is furnished with cross wires below and etched rectangles of the camera field sizes on the upper surface. Windows of non-breakable glass assist in sighting obliques. The accompanying picture (Fig. 100) of the plane showing an oblique camera in position gives an excellent idea of its appearance. Its special features are worthy of copying in peace-time photographic aircraft.

Installation of Auxiliaries.—It is quite necessary that the camera lens be protected from splashing mud and often from oil spray due to the motor. For this purpose an easily opened and closed door is essential, unless the camera is carried well up in the plane. An alternative, possessing certain advantages, is to incorporate into the camera protecting flaps operating in front of the lens, which open only when the exposure is made. If the camera projects beyond the fuselage, *stream lined hoods* (Fig. 95) must be provided to protect the camera nose with the minimum of air resistance.

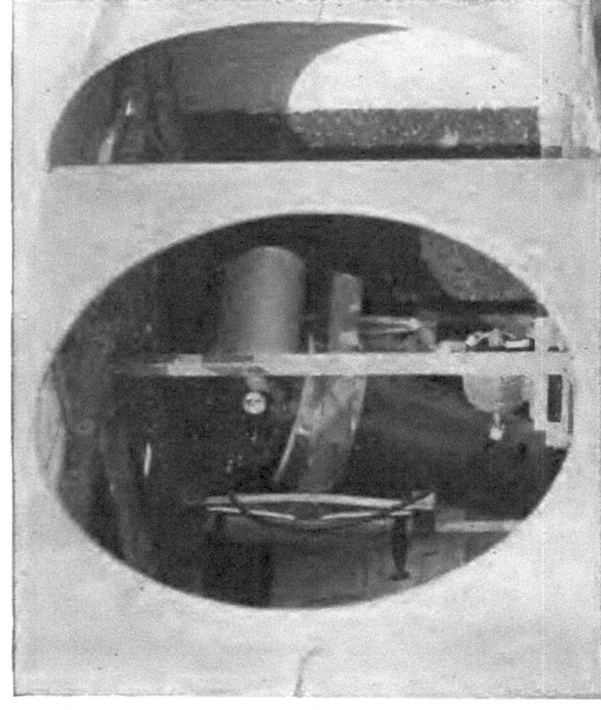

Fig. 98.—20-inch focus automatic film camera mounted obliquely in photographic DH-4.

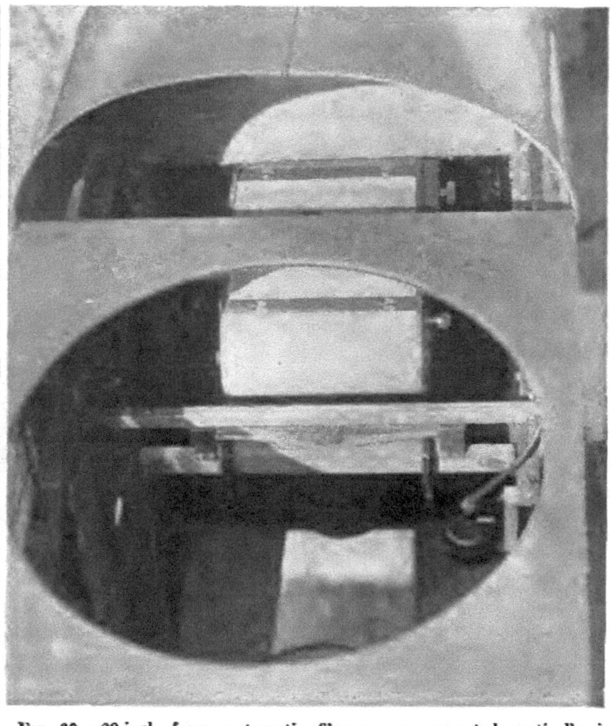

Fig. 99.—20-inch focus automatic film camera mounted vertically in photographic DH-4.

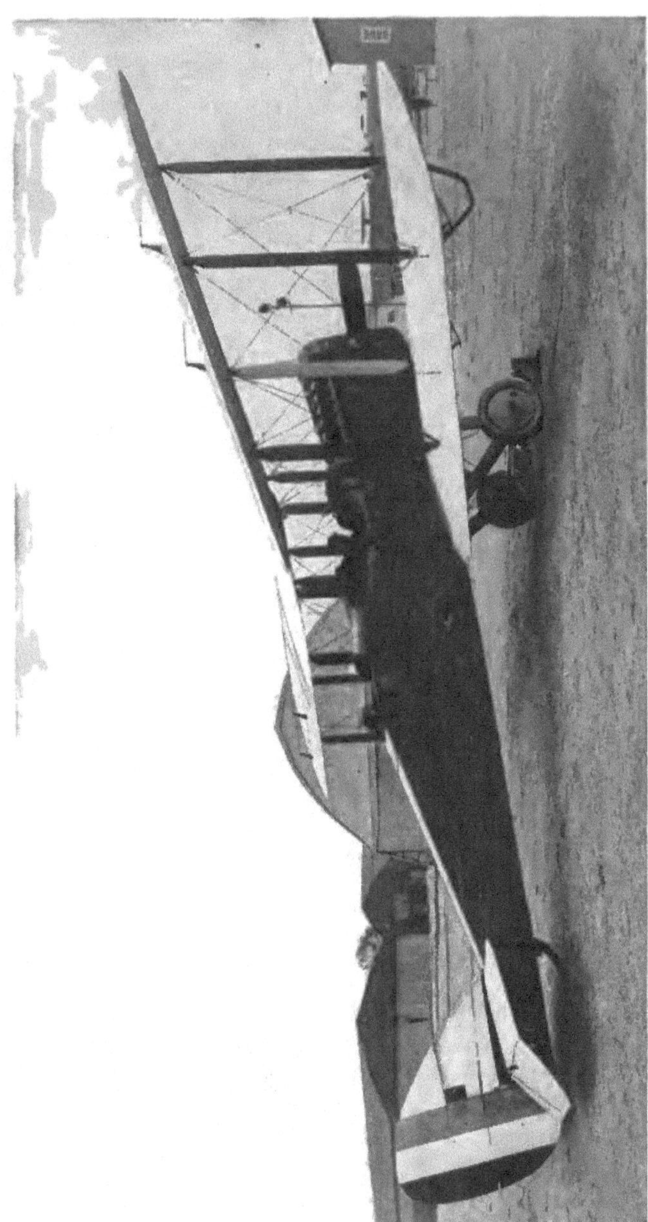

Fig. 100.—DeHaviland 4 re-constructed as a special photographic plane, showing 20-inch camera mounted for oblique photography.

The mounting of the regular camera auxiliaries—releases, sights, propellers, speed controls, motors—is usually a great bother, due to lack of space and to the severe restrictions on methods of fastening. Screws in longerons or uprights are taboo. Metal straps to go around structural parts are the approved device, but with variations in the size of these members, the holes, straps, bolts and nuts provided are very apt not to fit. Changes of construction, such as that from skeletons covered with fabric to veneer bodies, also interfere with any standard means of attachment, and leave this, like many other problems in war-time aerial photography, to the resourcefulness of the man in the field.

Magazine racks must be tucked away in any available space. Under the seat is a position frequently utilized. Especially with plates is it desirable to carry the extra magazines in a position to interfere as little as possible with the balance of the plane. In the DH 4 this means that they should be carried if possible forward of the observer, even though he must turn completely around to get and insert each magazine.

IV
SENSITIZED MATERIALS AND CHEMICALS

CHAPTER XVII

THE DISTRIBUTION OF LIGHT, SHADE AND COLOR IN THE AERIAL VIEW

The general appearance of the earth as viewed from above has already been described and illustrated (Figs. 10 and 11). It remains to deal with the earth's appearance in a more analytic and quantitative manner, in order to decide upon the characteristics to be sought in our photographic sensitive materials.

Range of Brightness.—The absence of great contrasts so apparent in the view of the earth from a plane is confirmed by photometric observations. These show that the average landscape, as seen from the air, rarely presents a range of brightness of more than seven to one, even when seen without the presence of veiling haze. It is to be remembered that shadows constitute no important part of the aerial landscape. Vertical walls in shadow, which form a substantial part of the surfaces seen by an observer on the ground, are invisible or greatly foreshortened from the air. Moreover, they are never contrasted against the sky, which is photographically often the brightest part of the ordinary picture. To the aviator's eye shadows on the ground are only of any length at early and late daylight hours. Even at these times they cover but a small area, since the number of high vertically projecting objects in a representative landscape is small. Lacking shadows, the brightness range is only that between various kinds of earth, water, and vegetation. Chalk (from freshly dug trenches), reflected sunlight from water, or marble buildings, furnish almost the only extensions to the brightness scale as above given.

Diurnal and seasonal changes. During the winter months on the Western Front photography from the air was only possible for two or three hours around noon, on clear days. This calls attention to another factor of prime importance, namely, the large variation in the intensity of

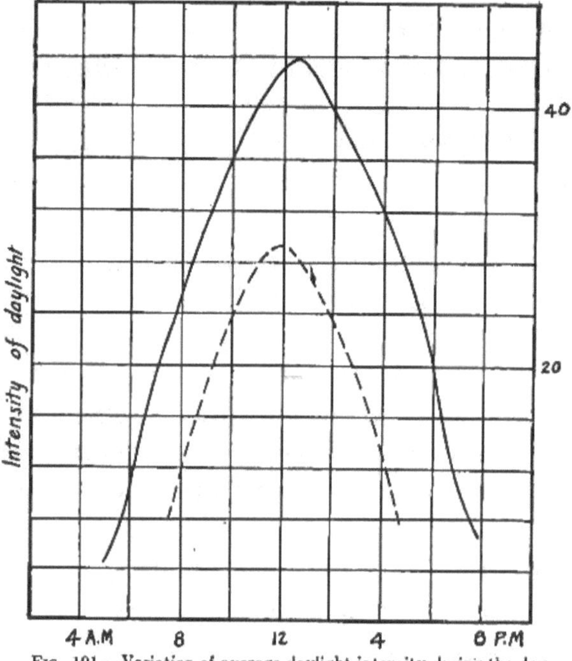

FIG. 101.—Variation of average daylight intensity during the day.

daylight during the course of the day and during the course of the year.

Measurements showing typical variations from morning to night are exhibited in Fig. 101, from which it appears that there is an increase in illumination of four to five times from 8 o'clock—when it would be considered full daylight for purely visual observation—until noon, while there is a corresponding decrease by four o'clock. Fig. 102 shows sets of measurements by two different authorities which give the average intensity of daylight for each month throughout the year. From December to July there is an increase of approximately ten times. From both sets of data it therefore appears that—neglecting the frequent occurrence of clouds which reduce the illumination to a half or a quarter or even less—a variation in illumination of forty or fifty

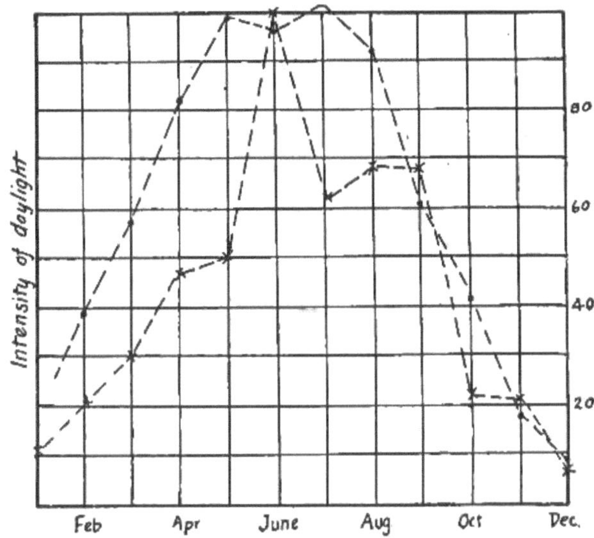

Fig. 102.—Variation of intensity of daylight through the year; two different sets of measurements.

times occurs between mid-day in summer and morning in winter. In the photography of stationary objects on the ground this range of intensities is easily taken care of by selection of lens stop and shutter speed. On the airplane it is quite otherwise, because the shutter speeds called for at the lower illuminations are much slower than the motion of the plane will allow.

Haze.—At low altitudes the brightness range is substantially that which would be obtained by photometric measurements of soil and vegetation made at the earth's surface. At higher altitudes, especially above 2000 meters, this brightness range is materially decreased by atmospheric haze. The significance of this lies in the fact that for safety from anti-aircraft guns, war-time aerial photography must be carried out at very great elevations. Toward the end of the Great War photographic missions traveling at from 5000 to 7000 meters were the rule. At these heights, even in very clear weather, a veil of bluish-white haze reduces the already small contrasts still more. Some means for overcoming the effect of this haze becomes imperative, therefore, in order to secure in the picture even the normal contrast of the object.

Haze is to be sharply distinguished from clouds or fog. Clouds and fog consist of globules of water vapor of large size, opaque to light. Haze, on the contrary, is more opaque to some colors than to others, or is *selective* in its veiling effect. Its scattering action on light is greatest in the violet and blue of the spectrum, decreasing rapidly through the green, yellow, and red, the exact relation being that the scattering is inversely as the fourth power of the wave-length. It is, consequently, possible to pierce or cut haze by using yellow, orange, or red color screens. It is this possibility which has led to the extensive use of yellow or orange goggles for shooting and for naval lookout work. In aerial photography the equivalent is to be found in *color filters*, used with color sensitive (orthochromatic or panchromatic) plates, which have been found essential for all high altitude work.

Color.—Visual observation from the airplane is aided in no inconsiderable degree by the differences of *color* that exist between various objects of nearly the same brightness. This means of distinguishing differences of character fails in the photographic plate, which is color-blind; that is, it reproduces all objects as grays of varying brightness. It is color-blind in another sense as well, in that it evaluates colors as to brightness differently from the way the eye does, overrating blues and violets and underrating yellows and reds. This first kind of color-blindness is a positive disadvantage, for it leaves available for differentiating objects only their brightness differences. The second kind of color-blindness may on occasion actually be an advantage. For it may happen, by accident, or by design (through the skilful use of color filters), that objects appearing nearly the same to the eye appear different in the plate. More will be said about this in connection with the use of filters for the detection of camouflage.

The range of hues seen in the aerial landscape is not large. Greens (grass and foliage) predominate, followed by browns (earth), neither color being bright or saturated. Over towns or cities we find that grays (roads) and redder browns (brick) are conspicuous. Blues are practically never seen, although it is to be noted that a fair share of the illumination of the ground is by blue sky light and that the haze itself is bluish. Consequently, the general tone of a landscape is much bluer than one would be apt to imagine it from consideration of the general green and brown character of the constituent objects. A color photograph from the air would greatly resemble a pastel in its low range of tones and the absence of bright colors.

The Photographic Requirements Dictated by Brightness and Color Considerations.—Considering only the demands made by the character of the view presented to the airplane camera, and leaving out of account other limitations to photographic operations in the plane, certain requirements as to sensitized materials may be outlined. First of all, the photographic process must not reduce, but should rather be capable of exaggerating, the range of brightness of the object.

Preferably the seven-to-one range of the object photographed should be lengthened out to the full range of the printing paper, which may be two to three times this. With such an increase of range, those minute differences of brightness are accentuated, on which the detection of many objects depends.

Next, the plate or film must be sensitive to the portion of the spectrum transmitted by a yellow or orange filter which will cut out the effect of haze. This calls for orthochromatic or panchromatic plates, depending on the depth of filter required. Next, if the objects to be photographed differ little in brightness but are different in color composition, we may have to rely on color filters of peculiar transmissions, capable of translating these color differences into brightness differences. These will, in general, call for fully color sensitive, or panchromatic plates.

In conclusion it may be pointed out that the endeavor in ordinary orthochromatic photography—to reproduce the visual brightness of colors in the photographic print—has no real justification in aerial work. Neither in respect to color values nor in respect to brightness range is it the object of aerial photography, especially for war purposes, to present a truthful tone reproduction. Its aim is rather the adequate differentiation of detail, by whatever means necessary.

CHAPTER XVIII

CHARACTERISTICS OF PHOTOGRAPHIC EMULSIONS

The purely photographic problem in aerial photography, as distinct from the instrumental one, is the selection of photo sensitive materials which will yield useful results under the conditions peculiar to exposure from the air. After such materials have been found by extensive field tests, it is pre-eminently desirable to determine their characteristics in such terms that the kind of plate or film may thereafter be specified and selected on the basis of purely laboratory tests. Specification must be made in terms of the ordinary sensitometric constants of the photographic emulsion—its speed, contrast, fog, development factor, its color sensitiveness, its ability to render fine detail, and its grosser physical properties such as hardness and shrinkage.

Sensitometry.—The most generally used system of sensitometry is that of Hurter and Driffield, commonly referred to as the "H & D." By this system, in order to determine the characteristics of a given photographic plate, it is necessary to take a series of graduated exposures, a standard illumination of the plate being varied in known amount by a rapidly rotating disc cut to a series of different openings, or by some other suitable means. The negative thus obtained is developed in a standard developer for a definite time, at a fixed temperature, and is then measured for transmission on a photometer. The following terms are defined and used in plotting the results:

$$\text{Transparency} = T = \frac{\text{intensity of light transmitted}}{\text{intensity of incident light}} = \frac{I}{I_o}$$

$$\text{Opacity} = O = \frac{\text{intensity of incident light}}{\text{intensity of transmitted light}} = \frac{I_0}{I} = \frac{1}{T}$$

$$\text{Density} = D = -\log_{10} T = \log_{10} O$$

Hurter and Driffield pointed out that a negative would give a true representation of the differences in the light and shade of the object if it reproduced these differences by equivalent differences in opacity. This is equivalent to stating that if the densities are plotted against the logarithms of the corresponding exposures, a straight line should be obtained at 45 degrees to the axis of exposure times. If the line is at another angle the opacities of the negative will be *proportional* to the brightness of the object photographed, but the *contrast* will be different.

A typical H & D plot is shown in Fig. 103. It will be noted that two curves are shown. These are obtained with different developments, and illustrate the fact that the contrast or proportionality between exposure differences and opacity differences is a matter of time of development. Each of these curves exhibits certain characteristics which are common to all made in this way. There is primarily a *straight line portion*, where opacities are proportional to illumination. This is commonly called the region of *correct exposure.* The slope of this straight line portion—the ratio of $\frac{\text{density}}{\text{log exposure}}$ — is the *development factor,* commonly denoted by "γ," a gamma of unity denoting exact tone rendering. Below the region of correct exposure is a "toe," or region of smaller contrast, called the region of *under exposure.* Above the correct exposure region is another where the opacity approaches constancy (afterwards decreasing or "reversing"), called the region of *over exposure.*

The *speed* of a plate on the H & D scale is given by the intersection of the straight line portion of the characteristic curve when produced, with the exposure axis. This intersection point, called the *inertia*, is the same irrespective of the time of development, as is shown in Fig. 103. The numerical value of the speed is obtained by dividing 34 by the inertia, when the exposure is plotted in candle-meter-seconds.

If a plate is developed until no more density and contrast can be obtained, its development factor is then γ_∞, (gamma

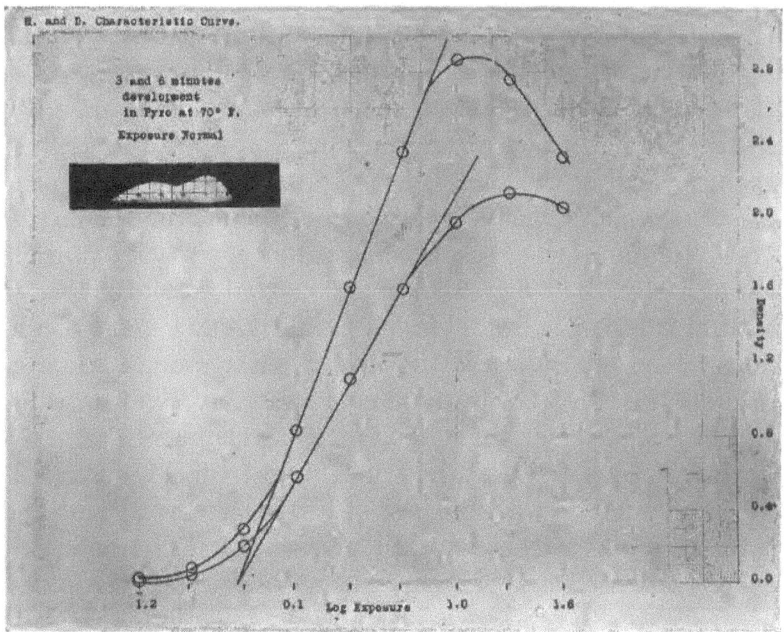

FIG. 103.—Typical characteristic curves of photographic plate.

infinity), and the larger this is the more a plate can be forced in development. If the plate fogs in its unexposed portions this fog is measured and recorded in density units along with the other constants. The speed of development is represented by the *velocity constant*, commonly symbolized by κ.

The length of the straight line portion determines the *latitude* of the plate, or the range of permissible exposures to secure a "perfect negative." Thus if we assume that an object has a range of brightness of 1 to 30, then a plate with a straight line characteristic extending over a range of 1 to 120 would have a latitude of $\frac{120}{30}$ or 4. That is, the exposure could be as much as four times the necessary one, and still give the same result on a sufficiently exposed print. If the latitude of the plate is too small, the shadows will fall in the under exposure region, the high-lights in the over exposure portion of the characteristic curve, with consequent poor rendering of contrasts.

Criteria of Speed.—In airplane photography *speed* is of paramount importance, but great care must be exercised to insure that all the factors are considered which can contribute toward yielding the desirable pictorial quality in the brief exposure which alone is possible from the moving plane. A "fast" plate on the H & D scale is not necessarily suitable for aerial work, when we remember that accentuation of natural contrast is desirable, particularly under hazy conditions. For, as is shown in Fig. 104, it is a common characteristic of "fast" plates to have comparatively small latitude and low contrast at their maximum development.

It is to be noted that the Hurter and Driffield measure of speed is bound up with the idea of correct tone rendering and with the use of the straight line portion of the characteristic curve. Other criteria of speed exist. For instance, the exposure necessary to produce a just noticeable action (threshold value); and the exposure necessary to give a chosen useful density in the high-lights when development is pushed to the limit set by the growth of fog.

As has already been pointed out, correct tone rendering is not necessary or even indicated as desirable in aerial views. It is, moreover, a matter of experience that the majority of aerial exposures with existing plates fall in the "under exposure" period, where contrasts with normal development are less than in the subject. This being the case, the problem

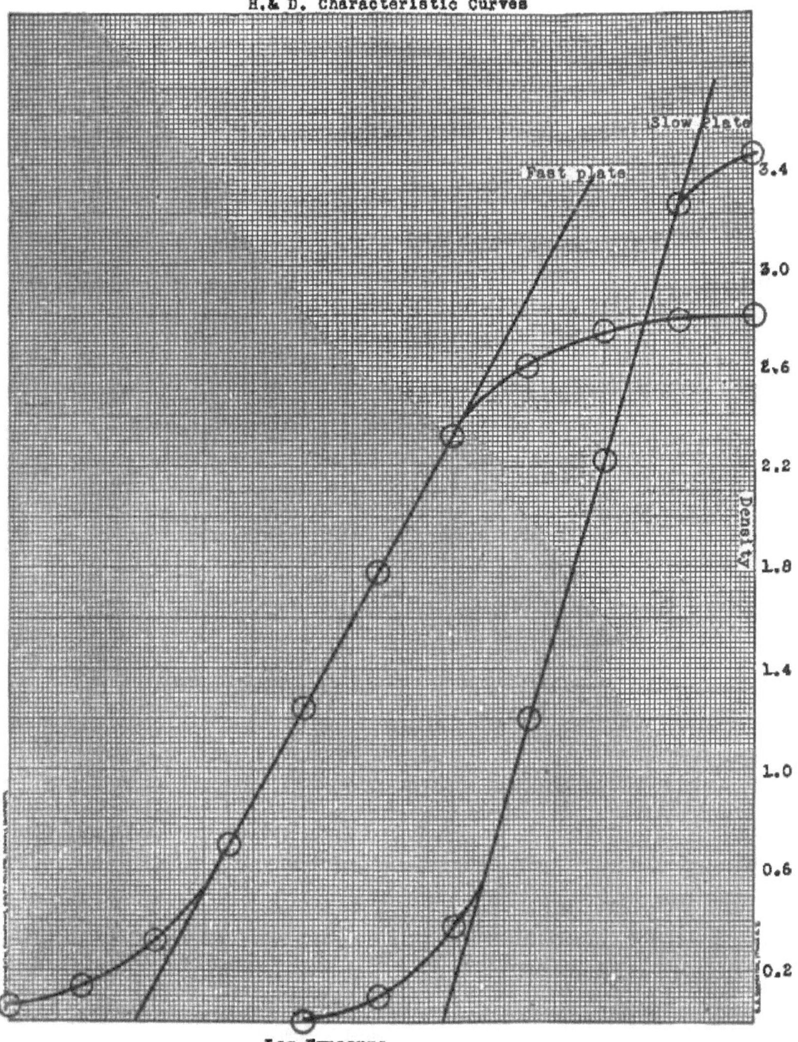

Fig. 104.—Characteristic curves of fast and slow plates, developed to maximum contrast.

is to select not necessarily a fast plate, by the H & D criterion, but a plate which will develop up workable densities in the under exposure region. A plate of medium speed will sometimes develop to greater densities in the short exposure region, if development is forced, than will a fast plate. The contrast in the normal exposure region will be excessive, but this is of no significance if no exposure falling in this region is present on the plate.

In addition to its capacity for developing density, the plate should have as low a threshold as possible, thus meeting to some extent the requirements of both the alternative criteria of speed given above. At the same time it is true that low threshold and good density for short exposures are not to be found in really slow plates. Consequently, while high speed, as ordinarily understood, is undoubtedly the first requirement, we may expect the complete specification for the best aerial plate to be a rather complicated thing, describing the characteristics of a workable "toe" of the curve, in terms of which several (*e.g.*, contrast and speed) are derived from another and quite different exposure region.

Effect of Temperature on Plate Speed.—It has been found by Abney and Dewar that very low temperatures materially decrease the speed of photographic emulsions. This decrease may amount to as much as 50 per cent. in the temperature range from 30 degrees Centigrade above zero to 30 degrees below zero, which is the range over which aerial photographic operations will have to be carried on in war-time. This effect has not been at all fully studied, and it is not known whether it is general or only found in certain kinds of plates. The remedy indicated is to provide means for heating the plates or films when low temperatures are encountered. This is fairly easy in film cameras, or in plate cameras like the deRam, where the entire load of plates is carried in the camera body. Plates carried in magazines present a more difficult problem. The heating coil incorporated in the German cameras is perhaps partly for this purpose.

Color Sensitiveness.—Complete specifications for an aerial plate cannot be made solely on the basis of its speed, contrast, latitude, threshold, and other sensitometric values which have to do only with the intensity of the light acting on it. These in general apply to photography from low altitudes, where the illumination and natural contrast of the subject are the only factors to consider. When higher altitudes are reached the interposition of haze decreases the already deficient contrast, calling either for the development of more contrast in the plate, or for the use of color filters to cut out the action of the blue and violet light predominant in haze. Along the lines discussed in the last section, it is not surprising to find that some plates are better than others for bringing out gradations masked by haze, even though no filters are used and though the plates are similar in color sensitiveness. But the limitations to securing contrast by manipulating the characteristic curve of the plate are soon reached, and it becomes necessary to resort to haze-piercing color filters, used with *color sensitive* plates.

Roughly, two general types of color sensitive emulsions may be distinguished: first, those in which sensitiveness to green and yellow is added to the natural blue sensitiveness, and second, those sensitive in a useful degree to all colors of the spectrum. The former are called *iso-* or *ortho-chromatic*, the latter *panchromatic* emulsions. Spectrograms exhibiting the distribution of sensitiveness throughout the spectrum for several representative plates are shown in Fig. 105. Orthochromatic plates are adequate for use with light yellow filters and have the slight practical working advantage that they can be handled by red light. Panchromatic plates are necessary for use with dark orange or red filters. They must be handled in total darkness or in an exceedingly faint blue-green light, taking advantage of the common drop in sensi-

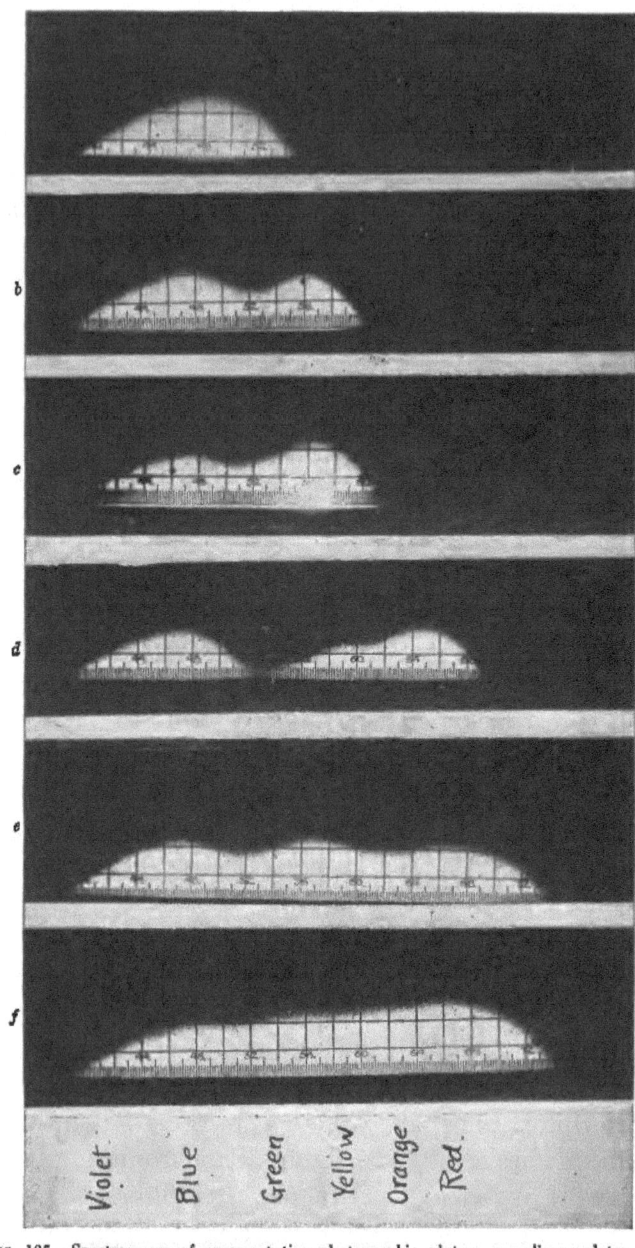

Fig. 105.—Spectrograms of representative photographic plates: *a*, ordinary plate; *b*, orthochromatic plate; *c*, specially green-sensitive plate; *d*, red sensitive plate, insensitve to green; *e*, panchromatic plate; *f*, specially red-sensitive panchromatic plate.

bility in that region of the spectrum. Plates can, indeed, be sensitized for the red alone, leaving a gap of almost complete insensibility in the green, as shown in the fourth spectrogram of Fig. 105. When used with a yellow filter these plates behave as do panchromatic plates with a red filter.

A rougher idea of color sensitiveness than is given by spectrograms is furnished by the *tri-color ratio*, which is the ratio of exposure times necessary with white light to give equal photographic action through a certain set of red, green and blue filters, expressed in terms of the blue exposure as unity. In an excellent panchromatic plate the three exposures would be equal. In an orthochromatic plate the red exposure will be too large to be figured. In interpreting either spectrograms or tri-color ratios care must be taken that the *absolute* exposures necessary are known. Thus a relatively high red sensitiveness may mean merely low absolute blue sensitiveness.

Two methods are used in imparting color sensitiveness. Either the sensitizing dye is incorporated in the plate emulsion before it is flowed; or the plate is bathed in a dye solution not long before using. The latter method gives higher color sensitiveness but poorer keeping quality, and is not a practical method for field operations. Greatly enhanced sensibility may be given by treatment with ammonia, but this again is a method for laboratory rather than field use.

Resolving Power.—A question which arises in connection with all photography of detail is the size of the grain of the photographic emulsion. Dependent on the size of the grain is the *resolving power*, or ability to separate images of closely adjacent objects. This varies with the speed, fast plates being of coarser grain than slow ones; with the exposure; and with the method and time of development. In general, it may be said that the resolving power of the plate does not enter practically into aerial work, because the resolving power of all plates so far found usuable corresponds to a smaller distance than the size of a point image as limited by the performance of the camera lens and the speed of the plane. Remembering that $\frac{1}{10}$ mm. is a fair value for the size of a point image as rendered by the lens, the rôle of plate-resolving power is shown by consideration of the following table. Resolving powers are given in terms of lines to the millimeter just separable.

Emulsion.	Resolving Power.
Seed Graflex	25
Eastman Aerial Film	37
Hammer Ortho	44
Cramer Isonon	48
Cramer Spectrum Process	57
Eastman Portrait Film	61

Tabulation of Requirements for Aerial Emulsions.— In terms of the sensitometric quantities just discussed the general requirements for aerial plates may be listed as follows:

1. *Speed.* The speed usually connected with the contrast and density required for the exposure times available is about 150 H & D. Faster plates in general have too low contrast, but the highest speed that will give the necessary contrast is desired.

2. *Contrast.* The contrast capable of development without fog should be from 1.5 to 2. This contrast should be produced by light of daylight quality, and, in orthochromatic and panchromatic plates, with the yellow or orange filters intended to be used with them. This contrast means a gamma infinity approaching 2.5.

3. *Speed of development.* A gamma of nearly 2 should be developed in 2½ minutes at 20 degrees C. in the developers recommended below.

4. *Fog.* Not over .25 for this degree of development, and not over .40 for six minutes development.

5. *Color sensitiveness.* This should in general be as high as possible. In terms of certain representative filters (described in a subsequent chapter) color sensitiveness should be such that with the white light speed above specified the relative exposures through the filters shall not be greater than as follows:

	No filter	Aero 1	Aero 2	#21	#23a	#25
Panchromatic plate	1	3	4.5	7	9	12
Ortho plate	1	2.5	3.5	6		

Relative Behavior of Plates and Films.—The advantages of film from the standpoint of weight and bulk have been discussed in connection with aerial cameras. Were there no other considerations film would unquestionably be the most appropriate medium for aerial photography. There is, however, the question of ease of handling, to be treated in a subsequent chapter, and the question whether the purely photographic characteristics of film are satisfactory. Can the same speed, contrast, and color sensitiveness be obtained on film as on glass? Is the picture so obtained as permanent or reliable as the plate image?

It must be confessed that up to the present emulsions on film have not proved the equal of those on glass. It has been found by emulsion manufacturers that the same emulsion flowed on film and on glass gives better quality on the glass. Emulsions specially prepared for film fall somewhat short of the best plate emulsions. It has also been found harder to color-sensitize film, and to insure good keeping quality in the color sensitized product.

In addition to the question of photographic quality there arises the matter of shrinkage and distortion. These are negligible with plates, but are a more or less unknown quantity in film. Irregular shrinkages of as much as two per cent, are found on experiment. This defect, of course, would be an obstacle only in exact mapping work.

Positype Paper.—The need sometimes arises in military operations to secure prints ready for examination within a few minutes after the receipt of the negatives. Even the 15 or 20 minutes within which a negative can be developed and a wet print taken may be considered too long. While such occasions are probably more apt to occur in popular magazine stories than in actual warfare, it is important to have available methods of producing prints with an absolute minimum of delay. This need is met to some degree by a direct print process, commercially exploited under the name of "Positype."

In this process the exposure is made directly on a sensitized paper or card, which is developed, the image dissolved out, the residue exposed, and again developed; thus furnishing a positive picture (reversed right and left). The time necessary to develop a print ready for examination need not be more than three minutes. Only a single print is available, but this is all that would be called for under the extreme conditions suggested. If later, copies are desired they may be made by the same process.

Plates and Films Found Satisfactory for Aerial Work.—The following plates and films have been found particularly good for aerial photography. The list is not intended to be complete. Furthermore, it may be expected to be soon superseded, as the efforts of various manufacturers are directed toward developing special aerial photographic plates.

Among orthochromatic plates: The Cramer Commercial Isonon, the Jougla Ortho.

Among panchromatic plates: The Ilford Special Panchromatic, the Cramer Spectrum Process.

Film: Ansco Speedex, Eastman Aero.

CHAPTER XIX

FILTERS

The Function of Filters in Aerial Photography.—The use of color screens or filters has been very common in ordinary landscape photography, for the purpose of securing approximately correct renderings of the brightnesses of colored objects. Plates of the non-color-sensitive type have their maximum of sensitiveness in the blue of the spectrum (Fig. 105) and in consequence blue skies photograph as white, while other colors are likewise reproduced on a totally wrong scale. Filters for correct brightness rendering are calculated for a given color sensitive plate so that the resultant reaction to the light of the spectrum copies the sensitiveness of the eye, which is greatest in the yellow-green. Such filters for use with the common orthochromatic plates are of a general yellow color.

Filters for aerial work are meant to serve quite a different purpose. Correct tone or color rendering is of quite secondary importance to another use of filters, namely, to cut or pierce aerial haze. It is quite a matter of accident that the same general color of filter is called for both to give correct color rendering and to pierce aerial haze, namely, *yellow*. Yet on closer analysis it is found that quite different types of yellow filter are demanded, spectroscopically considered.

Figure 106 (K_1 and K_2) shows the spectral transmission curves of the Wratten K_1 and K_2 filters, intended for correct color rendering with orthochromatic plates. The absorption increases *gradually* toward the blue. In the same figure is shown on an arbitrary scale the spectroscopic character of typical haze illumination, increasing in brightness inversely

239

as the fourth power of the wave-length, that is, with great rapidity in the blue and violet. It is evident from this that a much more abrupt absorption than that of the K_1 or K_2 filter is desirable, because in the green of the spectrum the haze light is comparatively weak, and more will be lost by any absorption in this region through decreasing useful

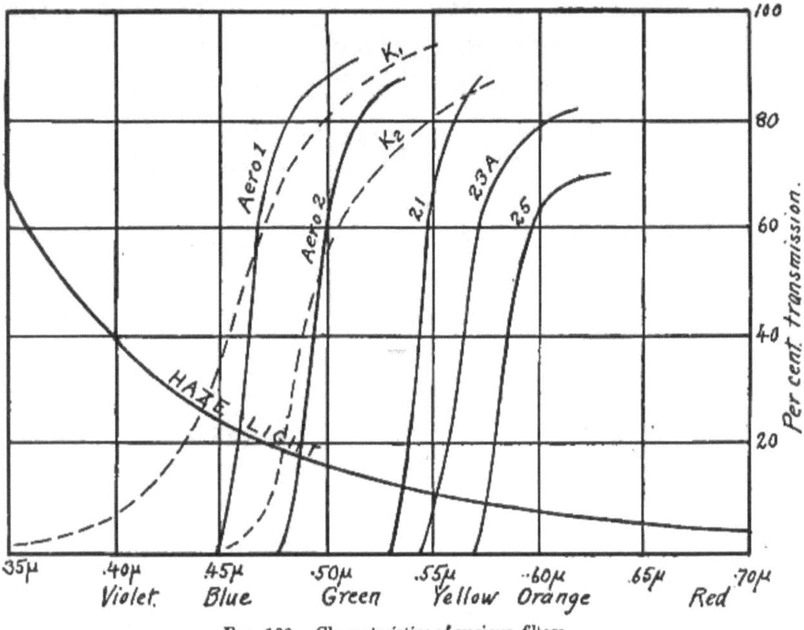

Fig. 106.—Characteristics of various filters.

photographic action than will be gained by cutting out the haze. This latter consideration is important. The use of any filter means an increase of exposure; the use of yellow filters multiplies it several times. Careful experiment has shown that no filter of depth less than K 1½, to use the Wratten filters as a basis for discussion, are of real value in haze piercing. The *filter ratio*, or ratio of exposures with and without filter, is 4.7 for the K 1½ with the Cramer Isonon plate—a figure which shows the importance of securing the necessary haze-piercing character with the minimum absorption of useful photographic light.

Practical Filters.—Since the character of the absorption of the "K" filters is not all that could be desired, new filters, both of dyed gelatin and of glass, have been produced. The glass, a Corning product having a very sharp-cut absorption, has not yet been produced on a commercial scale with the high transparency in green, yellow and red that selected samples have shown. The United States Air Service has adopted filters of a new dye, called the EK, from the name of the company in whose laboratory it was produced. These filters are standardized in two depths of staining, called the "Aero No. 1" and "Aero No. 2." Their spectral transmission curves appear in Fig. 106, along with those of certain darker filters useful only with panchromatic plates for exceptionally heavy haze. The characteristic of these Aero filters is their great transparency through all the spectrum except the blue, whereby the greatest haze-cutting action is attained together with a low filter factor. The filter factors of the Aero No. 1 and No. 2 with Cramer Isonon plates are 3 and 5, respectively.

Effects Secured by the Use of Filters.—The efficiency of yellow filters for haze-cutting is best shown by photographs taken at high altitudes with filters and without. Such illustrations are given in Figs. 107 and 108, where the first photograph is one taken at 10,000 feet without a filter, the second taken at the same altitude under the same conditions, but with an orange filter. Both are on panchromatic plates, and it will be seen that even with these plates the filter makes all the difference between a useless and a useful picture. But it must be clearly understood that the difference here lies between a plate sensitive chiefly in the blue and violet, and a plate affected only by the yellow, orange and red. The difference is not between what the eye sees and what a plate with a filter sees, as is sometimes supposed. As shown in Fig. 108, a filter enables the plate to photograph through the haze between clouds, but not through the clouds themselves. In general, no filter and plate combination which is feasible

Fig. 107.—A photograph taken at 10,000 feet, without a filter.

for aerial exposures is capable of showing more than the eye can see if yellow or orange goggles are worn. To do this it would be necessary for the photographic action to take place by deep red or infra-red light, which would demand exposures now out of the question.

Filters are almost always necessary in photographing from high altitudes or in making distant obliques. At times, particularly after a heavy rain, the air is clear enough so that filters may be dispensed with. Clearing weather was therefore chosen whenever possible for making obliques of the battle front.

Filters for the Photographic Detection of Camouflage.— In the photographic as in the visual detection of camouflage, the problem is to differentiate colors which ordinarily look

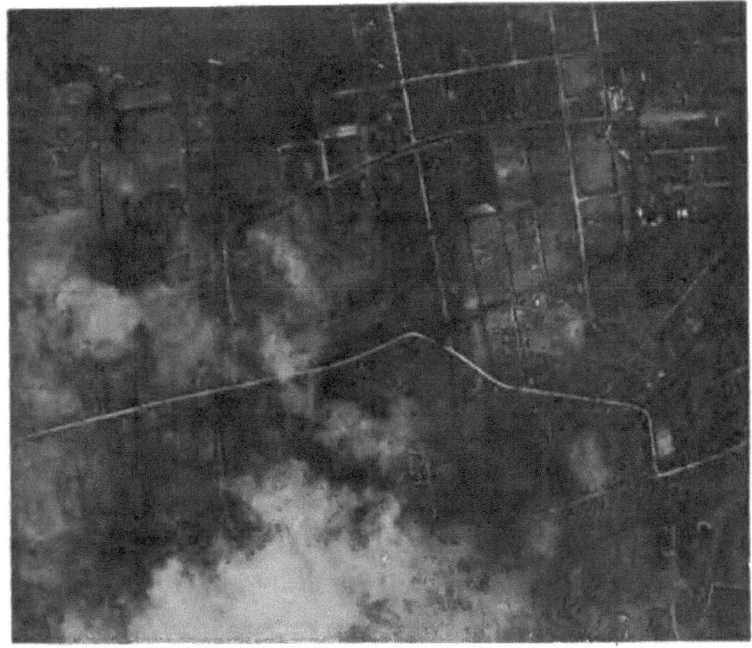

FIG. 108.—Photograph taken at same time and over same neighborhood as Fig. 107, but with an orange filter.

alike, but which are actually of different color composition. Particularly important are the differences between natural foliage greens and the paints used to simulate them. If these differ in their reflection spectra, a proper choice of filter will show up the two greens as markedly different. Two kinds of difference may be produced; either the two colors may be changed in relative brightness, or they may be altered in hue. Thus foliage green, due to its possessing a reflection band in the red of the spectrum, which is absent in most pigments, may be made to appear *red* while the camouflage remains green or turns black. Filters which cause changes of color are of course of no use for photographic detection of camouflage, since the photographic image is colorless. Brightness differences are alone available.

Those same filters which have been worked out primarily for producing brightness differences in visual detection of camouflage could be used photographically, provided the plates employed were color sensitive, and were as well screened to imitate the sensibility of the eye. But the most useful visual filters are those causing color differences to appear; more than this, the visual camouflage detection filters as a class have low light transmissions, so that their usefulness in photography is doubtful. Little work has actually been done with camouflage detection filters for photography. Yet in spite of this photography has been of real service in this form of detective work. Its utility for the purpose comes from the fact that the natural sensitiveness of the plate to blue, violet and invisible ultra-violet acts to extend the range of the spectrum in which differences between identical and merely visually matched colors may be picked up. Consequently the plain unscreened plate has proved a very efficient camouflage detector—so efficient in fact that all camouflage materials have had to be subjected to a photographic test before acceptance. Fig. 171 shows how an ordinary photograph reveals the unnatural character of the camouflage over a battery.

Methods of Mounting and Using Filters.—The most primitive way of mounting a gelatin filter is to cut a disc from a sheet of dyed gelatin and insert it between the components of the lens. For this purpose the gelatin must be perfectly flat, which is insured by its method of preparation and test. One disadvantage of this method is that the filter can be inserted and removed only upon the ground. It is less satisfactory the larger the diameter of the lens, and the wastage of filters due to insertion and removal is apt to be high. The camera should be refocussed after filters of this kind are inserted.

Glass filters, ground optically true, or gelatin filters, mounted between optically flat glass plates, are the most convenient and satisfactory. They may be mounted in circular cells to screw or attach by bayonet catches to the front of the lens. Or they may be mounted in rectangular frames to slide into transverse grooves in the camera body. Fig. 44 shows the mount of this latter form adopted in the larger United States Air Service cameras. This is particularly convenient if it is desired to insert or change the filter while in the air—a practice not generally considered feasible in war work with the photographically inexperienced observer, but likely to be common with the employment of skilled photographers for peace-time aerial photography.

German cameras are reported in which the glass filter is carried behind the lens, on a lever which also carries a clear glass plate of the same thickness, to be thrown in when no filter is needed, thus maintaining the focus. The performance of the lens will be impaired by this scheme, unless it is specially calculated to offset the effect of the glass introduced in the path of the rays behind the lens—optically true glass has no effect on definition if placed in front of the lens. Glass filters may also be placed in close contact with the plate or film, in which case they must be much larger, but do not need to be of as good optical quality.

Self-screening Plates.—Mention must be made of a quite different mode of realizing the filter idea, a method available where the sensitive plate is always to be used with a filter. This is to incorporate a yellow dye in the gelatin of the plate itself. The dye must be one which has no direct chemical effect on the plate, but which acts simply as a coloring agent for the gelatin. "Self-screening" plates, as they are called, have been produced by the use of the dye called "filter yellow" and have found some use in orthochromatic photography. They effect a useful saving of light through the elimination of the reflection losses at the surfaces of glass and gelatin filters. The filtering action of the dye in the plate is somewhat different from its ordinary one, since the deeper portions of the sensitive film are subject to greater action than the surface, and this tends to diminish contrast.

CHAPTER XX

EXPOSURE OF AERIAL NEGATIVES

The principal factors governing the length of exposure in the airplane camera have already been discussed under various headings. These are briefly, the nature of the aerial landscape, the practically attainable lens apertures, the form of the camera support, the speed of the plane, and the characteristics of plates, films and filters. It is convenient however, to re-assemble this information in one place, in such form as to apply to the practical problem of determining the exposure to be given in any specific case.

Limitations to Exposure.—In the ordinary photography of stationary objects, exposure is a variable entirely at the operator's command. Plates of any speed may be selected, so that attention may be focussed on latitude, color sensitiveness, and other tone rendering characteristics. The exposure may be made of a length sufficient to insure all the useful photographic action lying in the "correct exposure" portion of the sensitometric curve. The exposure ratio of any filter it is desired to use is a matter of indifference—its effect on color rendering need alone be considered.

Airplane photography is sharply distinguished from ground "still" photography by its severe limitations as to the amount of the exposure. The actual duration is definitely restricted by the high speed of the plane. In peace work this can be offset in part by using slower planes or by flying against the wind. The practical limitation to $\frac{1}{100}$ second, set by war-time requirements as to definition of fine detail, may be increased to $\frac{1}{50}$ of a second, or even more, where mapping of grosser features is the object. A common, but entirely avoidable limitation, is that due to vibration of the camera. By proper mounting this may be entirely overcome, leaving the ground speed of the plane the only source of exposure-limiting movement. The amount of light reaching the plate constitutes a primary factor in exposure, and this is a matter of lens aperture. Generally, lens aperture is smaller the larger the plate required to be covered, and the greater the focal length. Because of their larger aperture, the short-focus lenses which will be favored for peace-time large-area mapping will permit more and longer working days than have been the rule in long-focus war photography. The necessary use of filters, particularly at the high altitudes which would be chosen in mapping, in order to economize in the number of flights needed to cover a given area, introduces an inevitable decrease in the amount of light available at the plate, as compared with surface photography under the same illuminations.

Broadly speaking, it may be said that all the demands made in reference to aerial photographic exposure work are to *decrease* the amount of light reaching the plate. Any surplus offered, as by the midsummer noon-day sun, must be immediately snapped up, either by decreasing the exposure to get greater sharpness, or by introducing filters to get greater photographic contrast. The absolute exposure of the plate tends to be kept at the irreducible minimum. As already stated, it lies, with present photographic materials, on the "toe" of the "H & D" curve, just reaching up into the straight line portion.

Estimation of Exposure.—According to the foregoing argument the problem of estimating an aerial exposure resolves itself largely into one of deciding how short this may be made. Or, if the light is strong, whether it is sufficient so that a filter may be introduced without demanding more than the $\frac{1}{100}$ second or thereabouts which is dictated by the motion of the plane.

Deciding upon exposures in the field has been largely a matter of experience and judgment. A majority of the cameras in use during the war were not furnished with shutters calibrated in definite speeds. Consequently, the sergeant upon whom the decision usually devolved became a storehouse of knowledge as to the slit widths and tensions appropriate to each individual camera. This knowledge had to be acquired from the results of actual photographic reconnaissances, or from special test flights, both of them wasteful methods. But the chief objection to this state of affairs lies in the fact that the knowledge thus acquired is of no use to anyone else, nor is it applicable to other types of camera.

The first essential to placing exposure estimation upon a sound basis is therefore an accurate knowledge of shutter performances. Either the shutter speeds should be placed upon the camera by the manufacturer and periodically checked, or a regular practice should be followed of calibrating shutters, either at a base laboratory or even in the field.

Assuming that the speeds of all shutters are accurately known, the process of estimating the requisite exposure becomes less a matter of mere guesswork and more nearly a matter of precision. For this purpose data on the variation of light intensity during the day and during the year (Figs. 101 and 102) should be taken as a guide. These data refer of course to visual and not to photographic light, but since it is always necessary to use color filters, which make the active light of approximately visual quality, this is no valid objection. The effects of clouds and mist must of course be learned largely by experience, but with the above daylight data at hand, anyone in possession of definite information on the correct exposure with a given plate for a known day and hour need not go far wrong in estimating exposures at any other time in definite fractions of a second.

Exposure data charts. Fig. 109 shows a chart, pre-

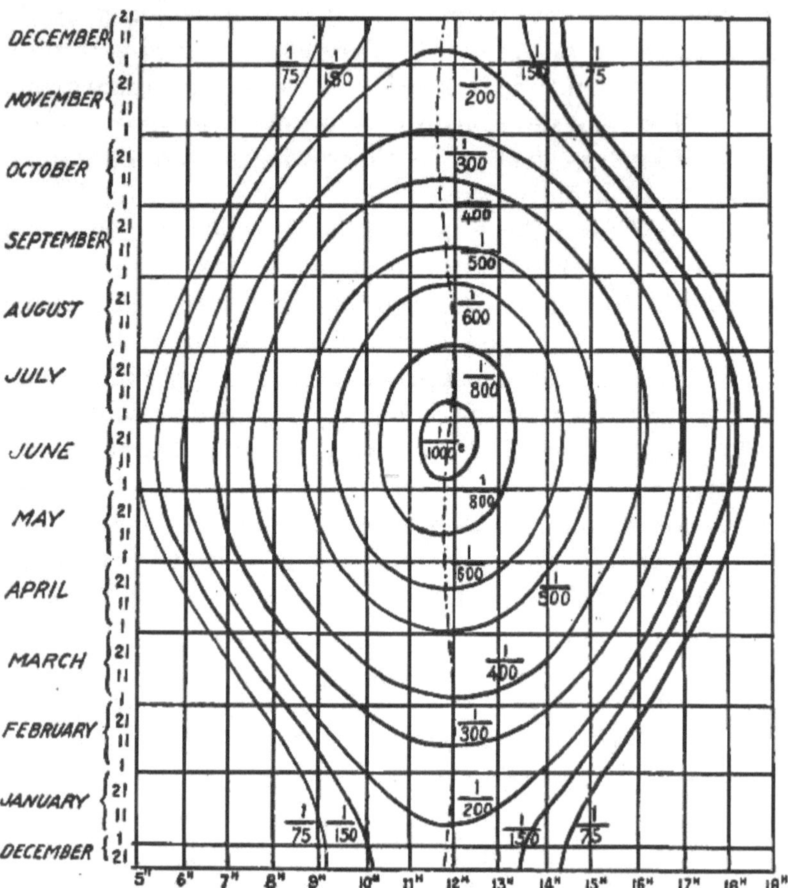

Fig. 109.—Chart showing aerial exposures for all times of the day and year. Data on basis of F/5.6 lens, Jougla orthochromatic plate, and clear sunlight, no filter. Exposures to be doubled and tripled for overcast and cloudy weather.

pared in the French service, indicating aerial exposures for all hours of the day throughout the year. These are for clear sunlight, for a lens of aperture F/5.6 and for "ortho" plates without a filter. They are based on what is probably an over-estimate of the actual speeds given by the French shutters. For "light" clouds the exposures are to be doubled, for "heavy" clouds quadrupled, and for forests and dark ground "lengthened." Charts of this form should be extremely useful, but they were actually not of great service because of the prevalent lack of knowledge of true shutter speeds.

Exposure meters. Aerial photography offers an excellent opportunity for the use of exposure meters, particularly those of the type in which a sensitive surface is exposed to the light for a measured time sufficient to darken a predetermined amount. The sensitive paper of the meter may either be exposed from the ground to the direct light of sun and sky, or from the plane to the light reflected from the ground. The first method will give figures subject to some correction for the character of the ground to be photographed—whether fields, forests, or snow. The second method is to be preferred where the shutter speed can be adjusted in the air, according to the indications of the meter, or where the filter can be selected and put in place during flight. Trials with a commercial Wynne exposure meter, used in the latter manner, give as a working figure an exposure of .001 second for each 4½ seconds taken to darken the sensitometer strip to match the darker comparison patch. This relation applies to a lens of aperture F/4.5, on Cramer Commercial Isonon plates without filter.

CHAPTER XXI

PRINTING MEDIA

Skilled photographers can examine a negative and can interpret its renderings with practically as much satisfaction as they get from a print, whereby a considerable amount of time can be saved in an emergency. The original glass negative should always be used when accurate measurements are to be made. These and a few other cases constitute the only use of a negative apart from its normal one, namely, for producing positive prints, usually in large numbers. The commonest form of print is on paper, although the most satisfactory print from the photographic standpoint is the transparency on glass or celluloid film.

Transparencies.—Transparencies are made by the regular photographic processes of exposure and development, on glass plates or films placed in contact with the negative, or in the appropriate position in an enlarging camera. The sensitometry and the terms used to describe the qualities of plate or film for this purpose are those already given in connection with the general discussion of plates and films. But the kind of emulsion to be selected is quite different from the aerial negative emulsion. There is here no practical limitation to the speed, contrast or latitude. Consequently, we can choose a positive emulsion on which the exposure through the aerial negative falls entirely on the straight line portion of the characteristic curve, thus reproducing all of its tones, and the contrast of the negative may be increased to any desired extent. The possibilities of positive emulsion are indeed rather greater than the usual aerial negative can utilize. A range of clearly graduated opacities of two or three hundred to one is possible, so that not only can detail be well rendered in the high-lights, but also equally well in dark shadows where, indeed, an increase of illumination is necessary for it to be made easy to examine. This range is to be contrasted with the 1-to-7 range in the aerial landscape, which may be doubled by a contrasty plate. In resolving power, the positive emulsion, which is slow, exceeds the negative emulsion. It easily bears examination through a magnifying glass, thus making any enlargement unnecessary in the printing process.

Glass transparencies are of course impractical for general distribution, on account of their fragility. Heavy film transparencies are not open to this objection, and, especially in the form of stereos, constitute the most beautiful form of aerial photographic print.

Paper Prints.—Prints on paper suffer by comparison with transparencies, in the range of tones which they exhibit. This lies between the white of the paper, which never has more than 80 per cent. reflecting power, and its darkest black, which differs with the kind of paper. In dull or mat papers the blacks will reflect as much as 5 per cent.; in glossy papers, ordinarily used for aerial negatives, the reflection from the black may be as low as one per cent., but in order to get the benefit of this the paper must be so held as not to reflect any bright object to the eyes. This deficiency in the range of paper gradations is not so serious with aerial negatives as with ordinary properly exposed negatives because of the small range of brightness in the aerial view.

The sensitometry of papers is similar to that of plates, with the difference that reflecting powers take the place of transparency. As in the case of transparency emulsions there is in papers no dominating requirement for extreme speed, to which other characteristics must be subordinated. Yet speed is of sufficent importance in handling large quantities of prints so that ærial negative printing for military purposes has been done almost entirely on the rapid enlarging papers, rather than on the true contact printing papers, which are slower.

The two principal types of rapid enlarging papers, the

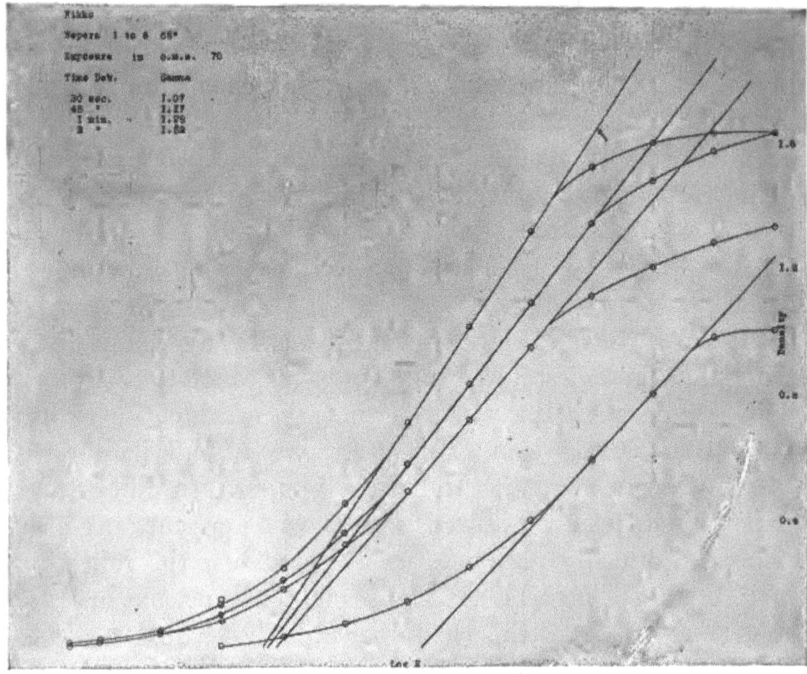

Fig. 110.—Characteristic curves of bromide paper.

bromide and the "gas light," exhibit certain characteristic differences which are important to bear in mind in seeking to obtain any particular quality of print. Bromide papers, of which "Nikko" is a good example, show sensitometric curves rather like those of plates. That is, they increase in contrast with continued development. At the same time, as is shown in Fig. 110, they increase somewhat in speed with development; that is, under exposure can be compensated for to a small degree by protracted development. These characteristics of bromide paper can be utilized to secure prints of a quality quite different from that of the negative. Thus, if the negative has a long range of tones, a

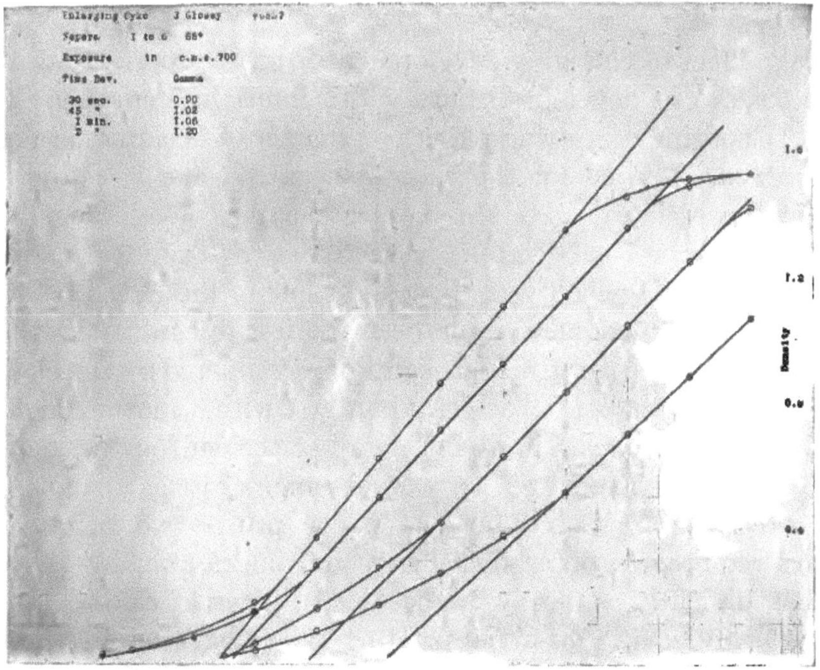

FIG. 111.—Characteristic curves of gas-light paper.

flat print can be secured by full exposure and short development. If, as is apt to be the case with aerial negatives, a print of greater contrast than the negative is desired, a short exposure with long development is called for.

The sensitometric curves of a typical gas light paper "Contrast Enlarging Cyco," are shown in Fig. 111. Here the contrast is a fixed characteristic of the paper, and the only effect of changing development is on the speed; that is, exposure and development are, within limits, interchangeable.

Choosing a printing paper is a matter of deciding on the contrast required for the class of negative, and selecting a paper which will give this contrast with a good range of tones from a clear white to a deep black. The ideal paper would be one which was all straight line in the H & D plot. In such a paper there would occur no loss of contrast in the lighter tones when the high-lights were rendered by the clear white of the paper. Too great contrast with a short straight line portion, results in loss of detail at the ends of the scale. A negative possessing a very great range of tones cannot be correctly represented on one paper print—two printings are required, one for high-lights and one for shadows, but this difficulty is rarely to be faced in aerial views. The greatest demand for aerial printing papers has been for those of considerable contrast, because of the flat character of the negatives.

CHAPTER XXII

PHOTOGRAPHIC CHEMICALS

General Considerations.—Developing, fixing and other chemicals for aerial work differ in no essential respect from those used in ordinary photography. Full discussions of these are to be found in numerous texts and articles. The aerial photographic problem is to select those most suited for the under-exposed flat negatives characteristic of photographs from the air. At the same time selection from among the chemicals of appropriate quality must be governed by considerations of the conditions surrounding work in aerial photographic laboratories. These laboratories, especially in war-time, are apt to be most primitive in their facilities.

Characteristics of Developers for Plates and Films.—From the standpoint of practicability, aerial negative developers should have good keeping power, be slow to exhaust, and work well over a considerable range of temperatures. From the standpoint of the photographic quality desired in the negative, the developer should bring up the maximum amount of under-exposed detail. This means that it should impart the highest possible speed to the plate, with good contrast, and low fog or general reduction of unexposed silver bromide.

There are many characteristics to study in a developer: its effect on inertia or speed, gamma infinity, fog, time of appearance, "Watkins factor," speed of development, temperature coefficient, dilution coefficient, keeping power, exhaustion, length of rinsing, stain, color coefficient and resolving power. These are defined and described as follows:

Effect on inertia. The meaning of inertia has already

been given under the discussion of plate speed. While this is a constant, independent of time of development, for any one developer, it is altered appreciably by change of the latter.

Time-gamma relation. Contrast, symbolized by γ, has likewise been discussed under plate sensitometry. Viewed from the standpoint of the developer, the point of interest is the rate at which γ varies with development, and the maximum contrast which can be reached or γ infinity. Speed of development is commonly defined by the *velocity constant,* symbolized by κ, which is arrived at mathematically from a consideration of the time of development to produce two different contrast values. High γ infinity is desired for aerial negatives, and for rapid work κ must also be high.

Fog. The opacity due to chemical fog is to be kept at a minimum in aerial negatives, as it is chiefly prejudicial to under exposures.

Time of appearance and Watkins factor. The time of appearance is measured in seconds. The Watkins factor is a practical measure of the speed of development, and is determined by the ratio of the time of development required for a definite contrast, to the time of appearance. It is useful also as a guide to development time.

Temperature coefficient. This is the factor by which the time of development at normal temperature (20 Cent.) must be increased or decreased in order to obtain the same quality negative, for a change of seven degrees either side of normal.

Temperature limits are the temperatures between which development can be carried out with any degree of control or without serious damage to the negative. These factors are of great importance where climatic or seasonal changes have to be endured.

Dilution coefficient. This is the factor by which the

development time is increased in order to maintain a given quality negative in different dilutions of the developer. It is useful in tank development.

Keeping power. The keeping power of a developer, mixed ready for use, is determined by its ability to resist aerial oxidation. A developer of poor keeping power, which must be made up immediately before use, causes delay and waste of time whenever emergency work has to be done, whereas a developer of good keeping power may be left in its tank ready for instant use.

Exhaustion of a developer is the rate at which it becomes useless for developing, due both to aerial oxidation and to the using up of its reducing power by the work done in developing plates. It is conveniently measured by the area of plate surface developable before the solution must be renewed.

Length of rinsing. The time required for rinsing between development and fixing bath plays a not unimportant part in total development time. Dichroic fog is caused with some developers if, due to insufficient rinsing, any of the caustic alkali is carried over to the fixing bath. Stains develop also if the fixing bath is old, or if light falls on the unfixed plate while any developer remains in the film.

Color coefficient. The function of the sulphite, which forms a constituent of all developing solutions, is two-fold. It acts partly as a preservative, and partly to prevent the occurrence of a yellow color in the deposit. The yellow color, if present, increases the photographic contrast. This phenomenon has been purposely utilized, particularly in the British service, to give "stain" to negatives which otherwise would show insufficient printing density. The color index or coefficient of a negative (with a given printing medium) is the ratio of photographic to visual density. If we take a pyro developer containing five parts of pyro per thousand

and ten parts of sodium carbonate, and then vary the amount of sulphite from none to fifty parts per thousand, the color index varies as follows:

Sulphite Parts per Thousand	Color Index
50	1.16
25	1.24
15	1.30
10	1.45
5	1.80
0	2.75

The color index is somewhat different with various kinds of printing media.

This staining effect is a variable one, depending upon length of development, dilution of the developer, length of rinsing, temperature, the fixing bath used (plain hypo being necessary for a maximum effect), the length of washing after fixation and the properties of the water used. Standardization of these conditions in the field is difficult; hence any developer which will give the same effective contrast without resorting to stain is to be preferred.

Resolving power. Some developing processes and conditions will introduce bad grain into the negative. Hence the resolving power which a developer brings up must be investigated among its other characteristics.

Practical Developers for Aerial Negatives.—In the English service a pyro metol developer was generally used, producing stained negatives. The French, American and Italian practice was to use metol-hydrochinon, without staining. A special chlor-hydrochinon developer, worked out by the Eastman Research Laboratory for the United States Air Service, has probably the greatest merit of any yet tried. A comparison, given below, between it and a pyro metol formula used on a representative plate, illustrates the use of the various bases of study given above.

Pyro Formula

Solution A	Solution B
Pyro, 3.75 grams	Sodium carbonate, 53 g
Potassium metabisulphite, 3.75 g	
Metol, 3.05 g	
Potassium bromide, 1.5 g	
Water, 500 c.c.	Water, 500 c.c.

Use 1 part of A to 1 of B

Chlorhydrochinon Formula

Solution A	Solution B
Chlorhydrochinon, 25 g	Sodium carbonate, 30 g
Metol, 6 g	Sodium hydrate, 10 g
Sodium bisulphite, 2.5 g	Potassium bromide, 3 g
Sodium sulphite, 25 g	
Water to 670 c.c.	Water to 330 c.c.

Use 2 parts of A to 1 of B

	Pyro	Chlorhydrochinon
H & D speed	150	180
Gamma infinity	1.45	2.12
Fog (at maximum gamma)	.32	.60
Time of appearance	5 seconds	5 seconds
Watkins factor	25	10
Velocity factor "κ"	.320	.400
Temperature coefficient	1.40	2.0
Temperature limits	4° to 32° C	4° to 32° C
Keeping power	45 minutes	8 days
Exhaustion (100 c.c.)	30 sq. in.	300 sq. inches
Dilution coefficient	2	2
Color coefficient	1.50	1.00
Resolving power	47	53

Owing to the difficulty of securing pure chlorhydrochinon a metol hydrochinon of very similar properties has been worked out. Its composition is

Metol	16 grams
Hydrochinon	16 "
Sodium sulphite	60 "
Sodium hydroxide	10 "
Potassium bromide	10 "
Water to	1 litre

To keep the ingredients in solution in cold weather, 50 c.c. of alcohol should be included in every litre of solution. All things considered this is probably the most practical and satisfactory developer for aerial negatives.

Developers for Papers.—The following formula has been found very satisfactory for papers:

Metol	.9 gram
Hydrochinon	3.6 "
Sodium carbonate	20.0 "
Sodium sulphite	14.0 "
Potassium bromide	.5 to 1.0 "
Water to	1 litre

Fixing Baths.—For plates the following fixing and hardening bath is recommended:

Sodium thiosulphate (hypo)	350 grams
Potassium chrome alum	6 "
Sodium bisulphite	10 "
Water to	1000 c.c.

During hot weather, the above quantities of chrome alum and bisulphite are doubled.

For papers the following:

Hypo, 35 per cent.	100 volumes
Acid hardener	5 "

The acid hardener is constituted as follows:

Alum	50 grams
Acid acetic 28°	400 c.c.
Sodium sulphite	100 grams
Water to	1 litre

Intensification and Reduction.—These processes have been little employed in air work. Reduction is rarely necessary, for obvious reasons. Intensification would often be of value, but the common practice, which saves some time, is to use printing paper of strong contrast for those negatives which are deficient in density and contrast. When intensification is desirable or permissible, either the ordinary mercury or uranium intensifier may be used.

Water.—In the field it is found necessary in many cases to purify the water that is to be used in mixing up chemicals.

Water may contain suspended matter or dirt, dissolved salts, and slime. It is important to remove the suspended matter, as it may cause spots on the plates and papers, while any slime would coagulate, forming a sludge in the developer which would also tend to settle on the plates and cause marks during development. The dissolved salts may or may not cause trouble. Two methods of purification are possible:

(*a*) Filter the water through a cloth into a barrel, add about one gram of alum for every four litres of water, and allow to settle over night. Draw off the clear liquid from a plug in the side as required.

(*b*) Boil the water and allow it to cool over night. If the water contains dissolved lime, boiling will often cause this to come out of solution.

V

METHODS OF HANDLING PLATES, FILMS AND PAPERS

CHAPTER XXIII

THE DEVELOPING AND DRYING OF PLATES AND FILMS

Field Requirements.—Developing, fixing, drying and printing in the field demand simple and convenient apparatus that may be carried about and installed with the least amount of labor. On top of these requirements military needs impose others that are more difficult. *Speed* is, on occasion, imperative. A print may be required within a few minutes after landing, and many thousands within a few hours. Quantity production must be achieved under the most primitive conditions. Nothing, in fact, shows the calibre of the photographic officer better than his choice of workplaces as the army moves forward. Ingenuity and practical judgment are at a premium. Cellars, stables, dog kennels, or huts hastily built from packing cases, must be equipped and in working order over night. All the facilities offered by a great city are urgently needed—water, electric light, power for driving fans—but must be dispensed with if the photographic section is to be convenient to the airdrome, whose portable hangars are most apt to be pitched in the open country. Water must be carried, electricity generated, and to the photographic problem is added the military one of concealment and protection. Dugouts and bomb proofs must be built for supplies, and "funk holes" for the men. Entire underground emergency extensions have sometimes been built in stations occupied for extended periods, for airdromes are a favorite bombing target.

For getting the exposed plates to the photo section, messengers, on motorcycles if possible, are employed. In some cases, where hangars and photographic hut are forced to be widely separated, recourse has been had to parachutes (Fig. 112), a device also employed to distribute prints to infantry during an advance.

For warfare of movement, especially in sparsely settled

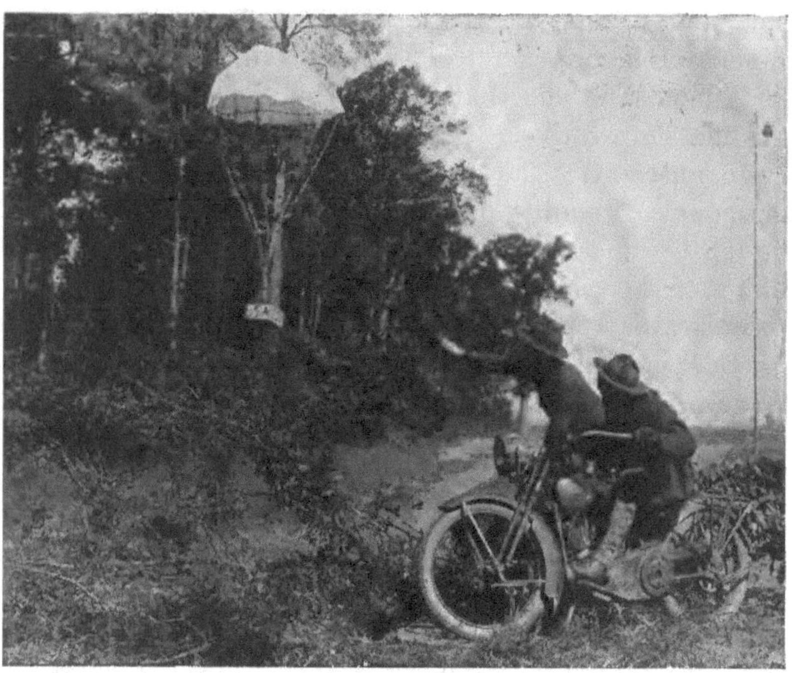

Fig. 112.—Receiving pictures from plane by parachute.

or devastated country, where cellars are unavailable, the dark room must be taken along. *Motor trucks and trailers* (Figs. 113, 114, 115), the former for hauling supplies and electric light generating plant, the latter fitted as a complete developing and printing laboratory, form the headquarters of each photographic section in the field. Usually altogether too small for the amount of work required, they were extended by tents and lean-to's, or ingeniously used as a nucleus for the organization of the favored stable or cellar.

Methods of Plate Development.—Where speed is not required the simplest and commonest mode of developing plates is in the tray, one plate at a time. Common practice is to examine the plate at intervals during development, and discontinue the operation on the basis of its appearance.

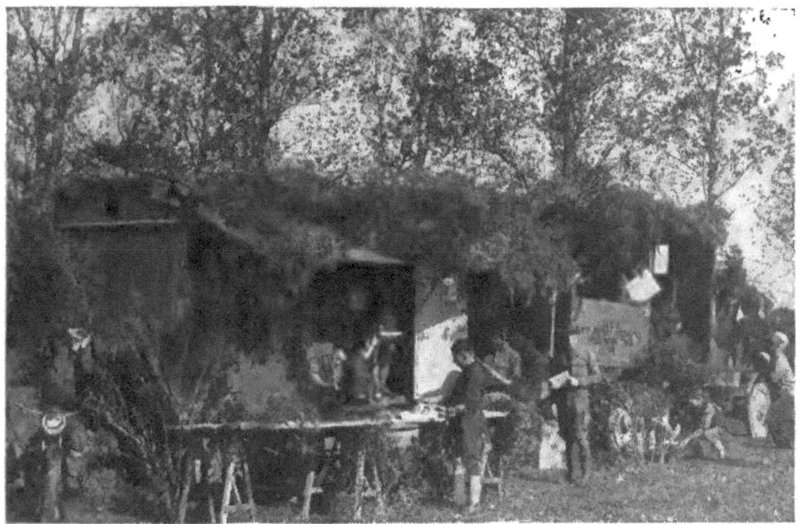

Fig. 113.—Mobile photographic laboratory.

This is only possible if the plates used are insensitive to some light by which the eye can see. Deep red light is suitable for ordinary and most orthochromatic plates. A faint blue-green may be used with some panchromatic plates. The best practice, however, is to develop by time in total darkness, whereby all chance of dark room fog is avoided. Development time for plates of the average exposure of the one to be developed is either known from previous experience, or is found by trial on the first one. Development by time results in negatives of densities varying with the exposures, but, as was brought out in the discussion of sensitometry, this difference can be compensated for by the choice of the paper used for printing, and by its treatment.

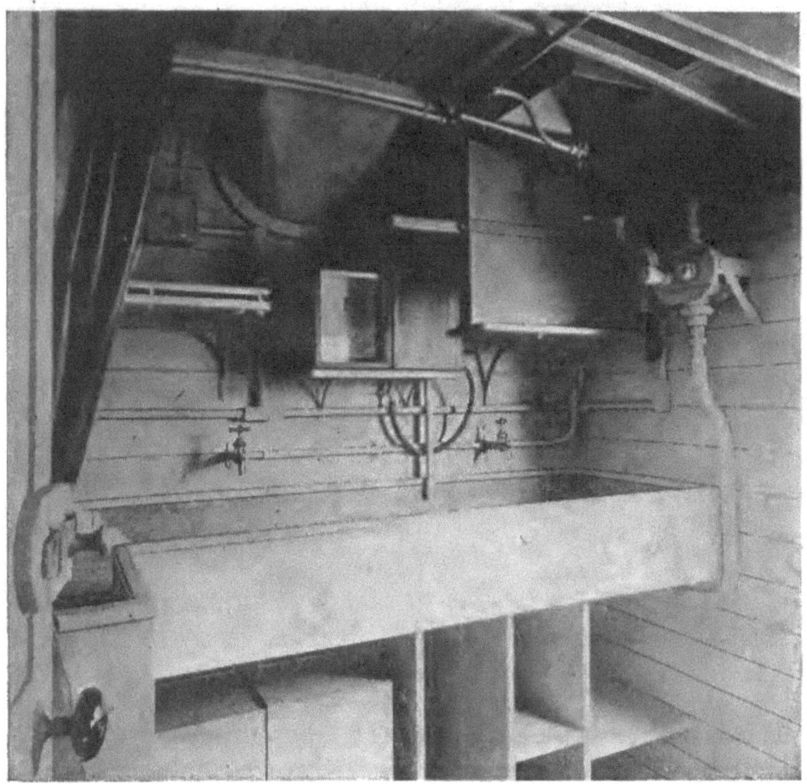

FIG. 114.—Interior of photographic trailer, developing room.

Where larger quantities of plates are to be handled *tank development* is adopted. In ordinary tank development the plates are placed in grooved tanks, into which is poured first the developer, next the rinsing water, and then the hypo. It has been customary in tank development as practiced for peace-time work to use dilute developer, requiring from ten to thirty minutes, but speed requirements in war-time aerial photography dictate the use of full-strength quick-acting developer. An improvement on the simple grooved tank is provided by metal cages or racks, each holding a dozen or

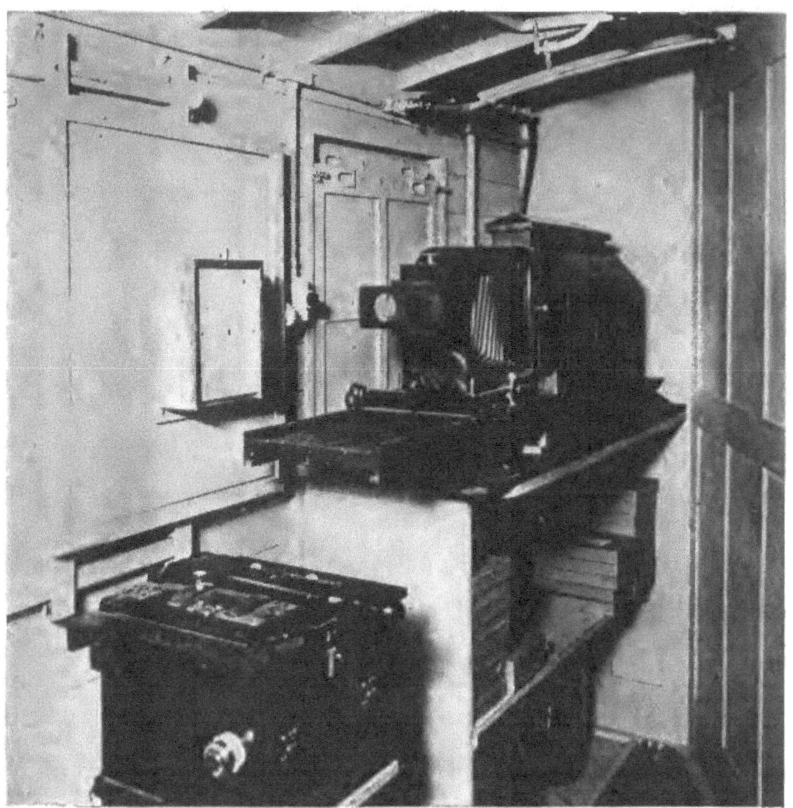

Fig. 115.—Interior of photographic trailer. Enlarging camera and printer.

more plates, which may be introduced or removed from the tank as a unit (Fig. 116).

The *core rack* system combines certain of the features of both tray and tank development. Each plate is inserted in a separate metal frame with projecting lugs to rest on the top of the tank and so suspend the plate in the solution. The process of development is the same as in the tank system, but any individual plate may be examined and removed.

Film Developing and Fixing.—The problem of quicky handling roll film of large size is one upon whose solution depends in large degree the feasibility of film cameras for

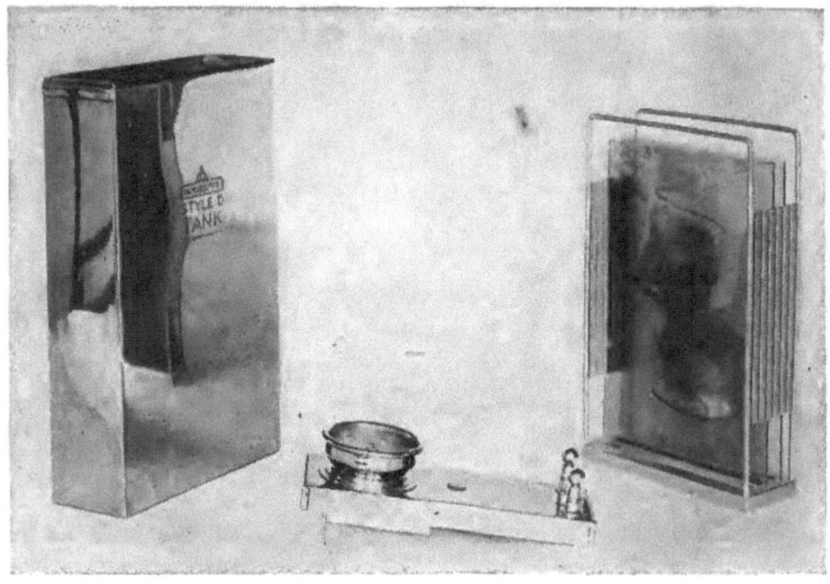

Fig. 116.—Tank and rack for tank development.

aerial work. It presents many difficulties: a long film is unwieldy, is inherently subject to curling, and takes up much space if it is handled entire. For small scale operations roll film can be cut into short strips and developed either by drawing through a tray or, if cost of developer is no object, in a deep tank. In order to make the cutting apart of exposures easy in the dark, film cameras should make some form of punch mark in the film between the exposed parts, or the space between exposures should be uniform, so that a print trimmer set to a definite mark may be used. Racks for holding two or three feet of film, folded back on itself and clasped by spring clothes-pins, are fairly practical. One object of the use of film, however, is to greatly increase the number of possible exposures; and where hundreds instead of dozens of exposures are to be developed, this method takes up entirely too much time.

Following the practice in moving picture development, *film developing machines* of various designs have been devised. Among these may be described the G. E. M. machine; the Ansco machine; the Eastman apron machine; the Brock frame and tank apparatus; the Eastman reel machine; and a modification of the latter by the United States Air Service.

The G. E. M. film developing apparatus, similar in idea to the Eastman "apron" method of film developing, as exemplified in the familiar amateur film developing machines, has the film wound in a spiral on a long linked metal frame or chain. After being wound it is placed in a tub of developer, from that to a tub of water, thence to a tub of hypo, and finally to a tub of water, where it is washed in several changes. The objections to the method are that it takes up much floor space for the various tubs, and that it requires such large quantities of solution. To develop a thirty-five foot length of 18×24 centimeter exposures requires approximately 28 gallons of developer; for the rinsing, 28 gallons of water, and the same for hypo, and at least three times that for washing. In all 168 gallons of water must be brought to the developing hut or lorry.

The Ansco machine makes use of an idea frequently applied in the moving picture industry. The film is carried spirally, upon two cross-arms which bisect each other at right angles, and which contain vertical pins around which the film is looped, beginning at the center and working out.

18

After it is wound it is placed in a tub of developer, as in the G. E. M. machine. It has an advantage over this apparatus in that the shape of the tubs or tanks is square instead of round. But it is equally extravagant of space and water.

This same criticism may be made of the Eastman apron apparatus for film developing. This is similar to the G. E. M. machine, but differs from it in using a perforated celluloid apron to support the film during the various operations, instead of a metal chain.

The Bröck developing outfit consists of a rectangular wooden frame and a three-compartment tank. The frame, which is approximately 3 by 4 feet in size, is used as a support for the 4 inch wide film, which is wound spirally around it, between guiding pins. A special support is provided, on which the frame may be rotated as the film is fed off the camera spool. The frame, with the film on it, is lowered successively into the three narrow but deep compartments of the developing tank. The first compartment holds developer, the next water, the next hypo. The amount of developing solution required is rather large (96 gallons of water in all for a strip of 100 4×5 inch exposures), but because of the small surface exposed to the air, it keeps for a considerable period. The chief demand for floor space with this apparatus is for feeding the film on to the frame.

In the Eastman twin reel machine the film is wound on a wooden drum or reel of large diameter, to form a helix. The drum is suspended so that the bottom edge touches the developing solution, and, upon revolving the drum, every portion of the helix of film is brought into contact with the developer. By shaping the developing tank so that it closely conforms to the shape of the reels, a high economy in quantity of developing agent can be achieved. When developing action is finished, the developer is emptied out, rinse water put in;

hypo follows, and then comes the final washing with water. With this apparatus the whole cycle is completed, for the 35 feet length of film above considered, with seven gallons of water.

The Air Service apparatus differs from the above only in the drying method, which will be described below.

Heavy cut film, such as is marketed under the name of Portrait Film, has not thus far been used in aerial work, except for printing transparencies. It is conceivable, however, that film in the cut form may be used in some future design of camera. This may be developed expeditiously in a tray, six or eight films being handled at a time, in a pile, pulling out the lower one frequently and placing it on top. The core rack system is also available for film in this form, special racks with clips to hold the film being necessary.

Plate Drying.—The drying of negatives on glass is a comparatively simple matter, owing to the rigid nature of the emulsion support. A large number of plates may be placed in a compact mass in the ordinary plate racks of commerce with the wet sides accessible to a draft of air. Two dozen plates separated from each other by a quarter of an inch and left to dry spontaneously in a room of ordinary humidity and living temperature will dry in two hours and a half. If the surface be wiped with soft cheese-cloth or chamois, so as to absorb all the surface moisture before the plates are placed on the rack, this time may be appreciably reduced. By placing the plates in a forced draft of air, from an electric fan, this time may be reduced to an hour.

Extra rapid drying of plates may be accomplished by placing them in a bath of alcohol before putting them in the racks. The alcohol displaces all the water in the film, and is itself very quickly dissipated into the atmosphere when the plate is taken from the tray. The plate must be left in the alcohol tray long enough for the substitution of the alcohol for the water in the film to take place. Five minutes is long enough. The alcohol before use must be as nearly free from water as possible. The best way to make sure of this is to place in the bottle of alcohol some lumps of calcium oxide, which will take up the water and form calcium hydroxide, which settles at the bottom of the bottle.

Another method of quick plate drying takes advantage of the extraordinary greediness of potassium carbonate for water. The wet plates are placed in a saturated solution of potassium carbonate and left for a minute. If a plate be now taken from the solution and its surface wiped with a soft cloth, it will be found that the film has a greasy, slippery feeling, but that it contains no water and can be printed from at once. Plates so treated should be washed, however, at some time in the succeeding four months, or the traces of potassium carbonate left in the film cause deterioration.

Film Drying.—Unlike the drying of plates, drying of film negatives is a very puzzling problem, and may be considered as the crux of the successful use of film in aerial cameras.

Apron and similar machines have very poor drying efficiency if the film is left in place, for not only the film but the apron or chain must be freed of water. This may be hastened, as in the G. E. M. machine, by blowing air through with fans, but even with their help drying a 35 foot film is a matter of two hours or more. Passing the film through wringers or a squeegee to remove excess water is a considerable aid; the film may either be re-wound on a dry reel, to be put in a forced draft of air, or may be hung up in short lengths or festooned, either method taking up a great deal of space. The use of alcohol is not advisable as it may abstract camphor from the celluloid and cause the film to become distorted.

The Eastman twin reel machine had an upper reel joined to the lower or developing reel, with a chain and sprockets, so that the upper reel revolved at the same time and rate of revolution as the lower, when the lower was being revolved at the gentle speed appropriate to the developing process. Fans blew a draft of air over the upper reel. This method necessitated over an hour for drying.

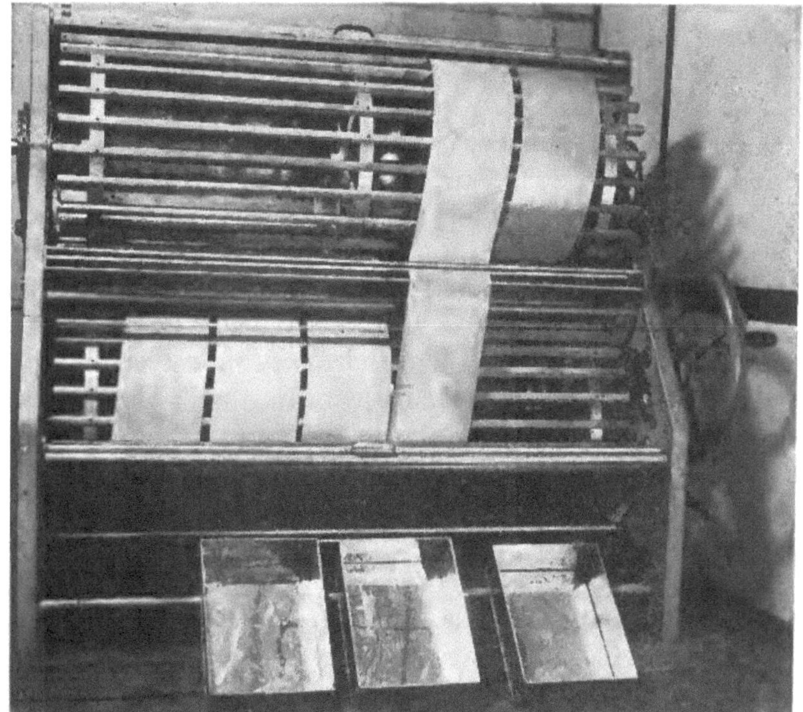

FIG. 117.—U. S. Air Service film developing machine for film 24 centimeters wide.

The Air Service model of film developing and drying machine (Fig. 117) introduces an essential modification in the drying scheme of the Eastman apparatus. The upper reel is quite independent of the lower reel and is revolved at a high rate of speed, so that a whirling action is introduced into the drying. Large rotating fans at the same time drive a considerable volume of air across the film surface, and the combination of the two agencies makes it possible to dry 35 feet of 18×24 centimeter film in 20 to 30 minutes. This for large numbers of pictures makes the use of film even quicker than that of plates. The only practical drawback to the apparatus is its bulk, which calls for a separate room or trailer. This, however, seems to be inevitable in the use of large roll film.

Cut film can be dried with speed only if placed in a draft of warm air. Drying boxes, with a chute or chimney and with fans to drive the air through from an alcohol stove, will dry several dozen films in an hour. The films must not be closer together than about one inch, which makes the drying boxes rather cumbersome.

Marking Negatives.—After development and drying, and before filing or printing, each plate should be marked with data for purposes of future identification. This is most easily done with pen and ink on the film side (in reversed lettering) either along an edge in the unexposed portion covered by the sheath or in a corner, so as to lose as little of the photograph as possible. Just what data shall be inscribed is dictated by the purpose for which the negative was made. The date, altitude, time of day, true north (from known permanent features or from shadow direction and time of day), number of the camera used, the focal length of the lens. Other records, such as the plane and squadron numbers, or even the pilot's and observer's initials, may be called for (Fig. 75). For mapping work the scale of each of a set of negatives, once found, may be marked, either in figures or by means of a line of length corresponding to a fixed distance on the ground. Rectifying data can similarly be inscribed, so that the negative can be printed in the enlarging and rectifying camera with the minimum of delay.

CHAPTER XXIV

PRINTING AND ENLARGING

Contact Printing.—Single prints are made most simply in a printing frame held at a short distance from a light source. When any quantity must be made, as in turning out prints at high speed for distribution to an army before an attack, *printing machines* are employed. These consist essentially of a light box, a printing frame of plate glass, and a pressure pad. In the commercial models, such as the Crown and the Ansco, which are equipped with electric light, merely bringing the pressure pad down and clamping it automatically turns on the light, while release of pressure terminates the exposure.

The question of regulating the distribution of light is of considerable importance with negatives taken by focal-plane shutters of non-uniform rate of travel. In the McIntire printer (Fig. 119), the separate electric bulbs are on long necks in ball and socket joints, so that they can be brought individually closer to the printing surface or farther away from it, thus permitting a wide range of "dodging." This printer also has an automatic time control for the light, a valuable device where many prints from the same negative are desired.

These machines are well suited for printing aerial negatives, either plate or cut film, if used where a source of electric current is available. The chief defect, which may be caused by faulty construction, is imperfect contact between paper and negative, a cause of serious unsharpness on prints destined for minute study in interpretation.

The printing of aerial negatives may be done either on

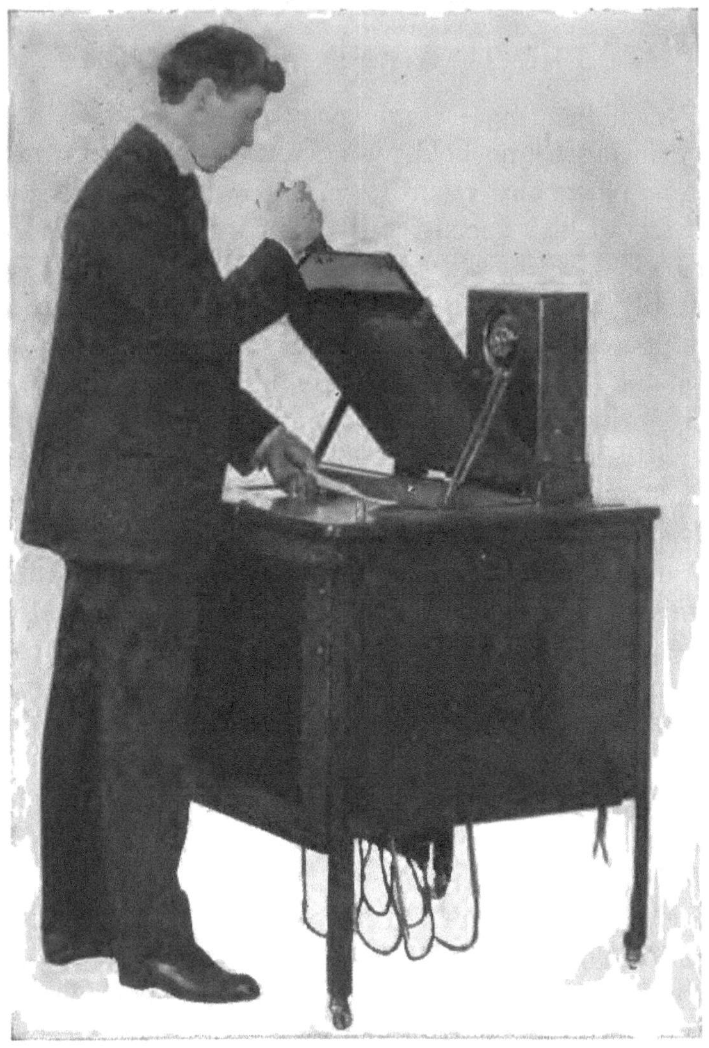

Fig. 118.—Printing machine.

roll or cut paper, and if films are used, a further alternative is offered of handling it either in the roll or in cut form. Where many prints are to be made from one negative roll paper has some advantages, particularly if a developing and drying machine is available. But for moderate numbers the advantage is small, since cut prints can be developed quite

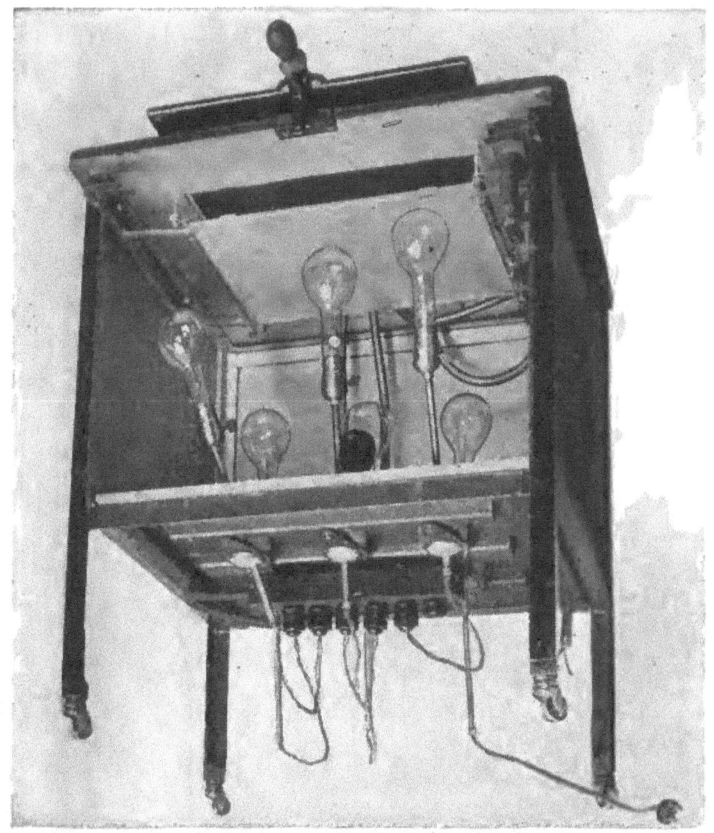

FIG. 119.—Interior of McIntire printer, showing lamps adjustable in position for "dodging."

conveniently in goodly numbers in the ordinary trays. But the advantages of keeping film in the roll form are very great, both in respect to storage and in respect to handling during printing, as the rollers provide the necessary tension and prevent the film "getting away."

For the American Air Service, cut paper has been used exclusively. For film printing, the Ansco machine has been

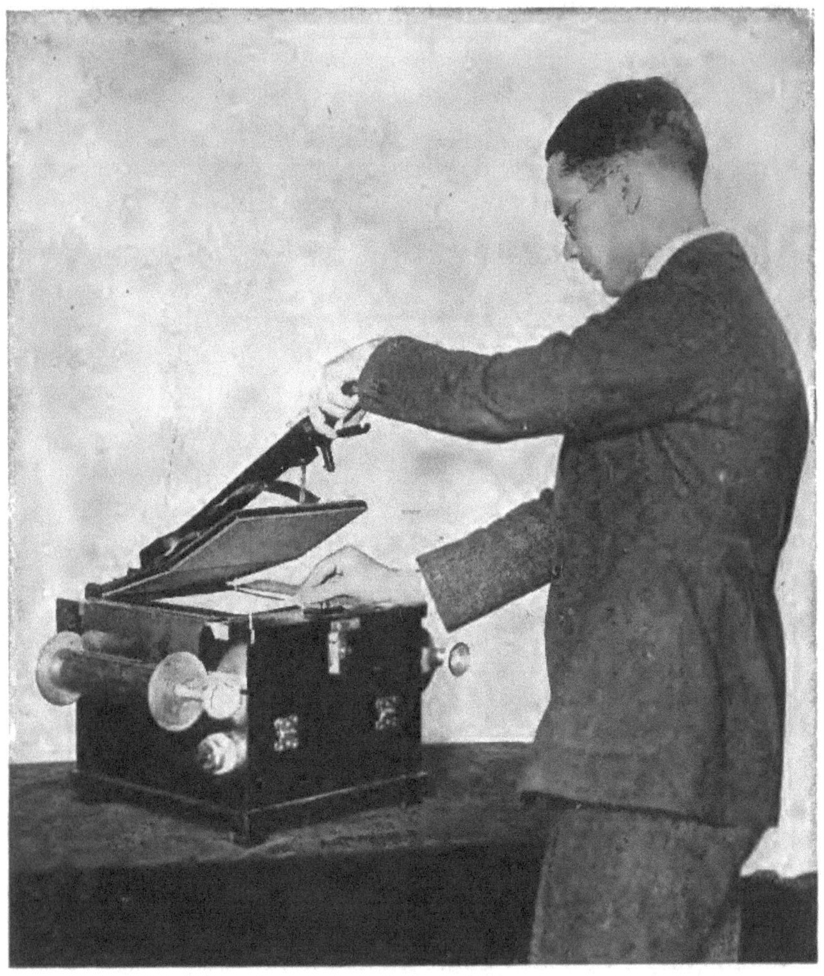

FIG. 120.—Film printing machine.

equipped with roll pivots to take film 24 centimeters wide which may be advanced in either direction by turning large milled heads (Fig. 120). If we put rollers on the two remaining sides of the box to handle paper we transform the printer into the same form as a French machine, in which paper and film are moved at right angles to each other. A disadvantage of this modification, however, is the difficulty of examining the negative to be printed.

Stereo Printing.—To make separate prints from the two elements of a stereoscopic pair and mount them side by side after proper orientation is too slow a process if quantities of prints are needed. One method of multiple production is to make a master stereogram, and then produce photographic copies of it, but there is inevitable loss of quality in this copying process. An intermediate method is to print from both negatives on the same sheet of paper. In order to do this the negatives must be placed in rather large frames, with mats properly located to guide the placing of the paper. The Richard double printing frame is a practical device which simplifies the necessary manipulations. It consists essentially of a platform pierced with three illuminated openings. The two negatives are compared, superposed, and orientated over the central opening and then shifted laterally, one to each of the two side openings, which serve both as printing frames and masks. The printing back slides on a rod, permitting the paper to be lifted up and moved between exposures. Once the negatives are properly placed, stereo prints can be turned out quickly and easily.

Enlarging.—In the French service contact printing was the rule during the war. The English practice, on the other hand, was to take small negatives—4×5 inches, with 8 to 12 inch lenses—and enlarge them, usually to 6½×8½ inches. For this purpose a regular part of the English photo section equipment was the *enlarging camera* (Fig. 115). This may be briefly described as a short focus camera in which the subject to be photographed is a negative, illuminated by transmitted light, whose image is thrown by the camera lens on the paper or other sensitive surface. By making the distance between negative and lens less than that between lens and paper, the resulting print is an enlargement, and *vice versa*. The scale of enlargement or of reduction is varied over limits set only by the length of the camera and the amount of light available.

The lens employed must of course possess sufficiently high quality to preserve all the sharpness of the negative, and focussing must be done with accuracy. Next to the lens the most important element is the light source. This may be of the point form, such as a concentrated filament electric lamp, an oxy-acetylene lime light, or an acetylene flame. The latter was extensively used in the English service, while acetylene generators for emergency purposes formed part of each American photo truck equipment. With point light sources we must use *condensers* to focus the light into the projecting lens. Much less efficient, but the only recourse where large condensers are not available, is a diffusing glass behind the negative, illuminated either by a bank of electric lamps with mirrors or by a U tube mercury vapor lamp, where proper current can be got.

The device for holding the printing paper must permit quick changing, but insure good contact. We may use either a spring plate to hold the paper against plate glass from behind, or else a weight acting on a lever arm of sufficient length.

The need for some automatic means of focussing an enlarging camera has been very generally felt. An illustration of such an enlarging camera is that put out by Williams, Brown & Earle, of Philadelphia, known as the "Semper-focal" (Fig. 121). In this camera the movements of the lens, paper easel and negative are so inter-related and actuated with respect to each other that the correct focus of the instrument is maintained for any degree of enlargement or reduction. This feature is a great help in making up mosaic maps, where prints of continuously varying scale ordinarily occasion serious delay for individual focussing.

Determining the correct enlargement for each negative of a mosaic is perhaps the most important problem in the use of the enlarging camera for aerial work. The correct

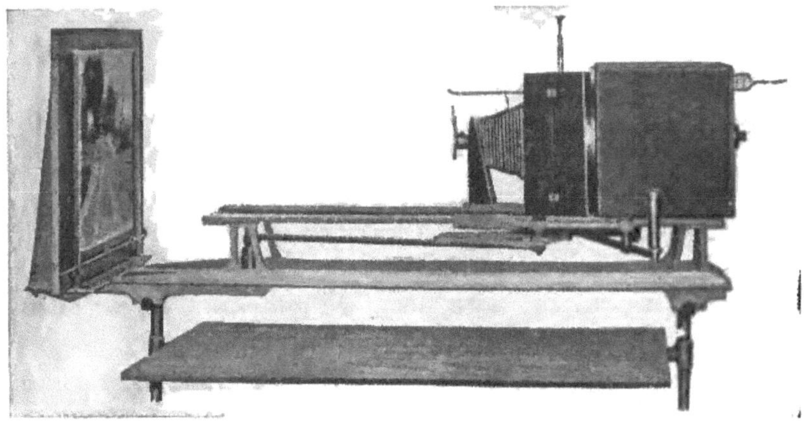

FIG. 121.—"Semperfocal" enlarging camera, with mechanism for holding image in focus at any enlargement.

setting of the camera may be found by either of two methods: the negative may be previously scaled and marked with a line on its edge, which must be projected to a definite size; or the true location of several points in the picture as obtained from an accurate map may be marked on the enlarging camera easel according to the desired scale, and the negative image projected to coincide with these. In either case, if an exact scale is desired, allowance must be made for paper shrinkage, a matter which must be determined by previous experiment.

Rectifying.—Negatives taken when the plane is not flying level will be distorted (Figs. 134 and 135). Contact prints from these will not fit into a mosaic, and no mere enlargement or reduction will make them available. It is necessary with these negatives to resort to a *rectifying camera.* This is an enlarging camera built so that the negative and print easel may be inclined about vertical and horizontal axes, thereby purposely introducing a distortion sufficient to offset the distortion of the negative. Thus, if the bottom of the printing surface is moved away from the lens, that part of the picture will be enlarged; if moved toward the lens, reduced.

For small rectifications the common practice is to tilt the printing surface alone, a method that is practical as long as this tilting does not affect the focus so much as to require prohibitive stopping down of the lens. For great distortions, such as that inherent in the principle of the Bagley camera, it is necessary to tilt both negative and print in order to preserve an approximate focus, a given portion of the negative moving toward the lens as the corresponding portion of the print is moved away. Both schemes for rectification are shown diagrammatically in Fig. 122.

Developing and Drying Prints.—The developing of prints follows closely that of cut or roll film, and so need not be treated separately.

The drying of emulsions on paper is more easily accomplished than the drying of emulsions on glass, for two reasons: the emulsions on paper are much more thinly coated, and there is diffusion of moisture into the atmosphere from front and back of the printing medium. In the field a common method has been to soak the prints in water-free alcohol and then burn off the alcohol, thus securing a dry print within two or three minutes after the conclusion of washing. A later method very generally employed is to cover wooden frames three or four feet above the ground with chicken wire or muslin, and on these lay the prints after soaking them in alcohol. Below the frames currents of warm

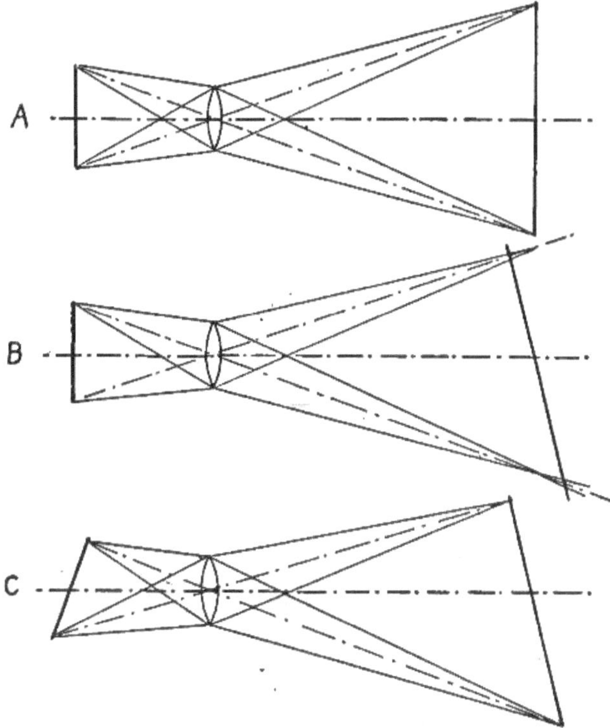

Fig. 122.—Diagram showing enlarging with and without distortion: *A*, enlarging without distortion; *B*, distortion for rectification of print, by inclining printing surface; *C*, distortion, for rectification of print, by inclining both negative and printing surface.

air rise from pans of burning alcohol, previously used to soak the prints and now useless as alcohol because of their high water content.

Before putting them in alcohol it is advisable to squeegee all the surface water from the prints. This may be expeditiously done by removing them in mass from the final wash water upon a large ferrotype plate, and either running the plate and prints together through a wash wringer with light pressure, or covering the whole with a sheet of blotting paper and pressing out the water underneath by means of a rubber squeegee vigorously applied.

For base work one of the modern automatic print-drying machines used in commercial photography would be desirable. Glossy surfaces are given prints by the usual ferrotype plate method. But this is too time-consuming for war practice, and besides has but doubtful advantage where papers of the glossy type are chosen.

VI
PRACTICAL PROBLEMS AND DATA

CHAPTER XXV

SPOTTING

"*Spotting,*" as distinct from mapping or from the photography of continuous strips, is the photography of a definite individual objective. In military work spotting or "pin pointing" includes the photography of particular trenches or pivotal points in a trench system before an attack (Fig. 123), of roads or bridges along which an advance must pass (Fig. 124), of batteries or big guns which are the subject of artillery fire (Fig. 125), both before and after their bombardment (Fig. 126), of gun puffs or exploding bombs (Fig. 131).

The technique of spotting consists largely in getting properly over the target and then securing the exposure at just the right moment. This is chiefly a question of proper piloting; but the aid which can be offered to the pilot by camera auxiliaries designed particuarly for spotting needs is very large.

Discussion of the task of the pilot who must steer a photographic plane accurately over a previously selected point of interest cannot be undertaken without raising the question of who should take the picture, pilot or observer? In the English service the most general practice was for the pilot to be charged with the responsibility both of covering the objective and of exposing. If a propeller drive was used on the camera, this left to the observer only the task of changing magazines. If the camera was hand operated the plates were changed either by the observer, or else, as was frequently the case, distance operating devices were attached, so that the pilot even then did everything except change the magazines, and the observer was kept free to watch the sky

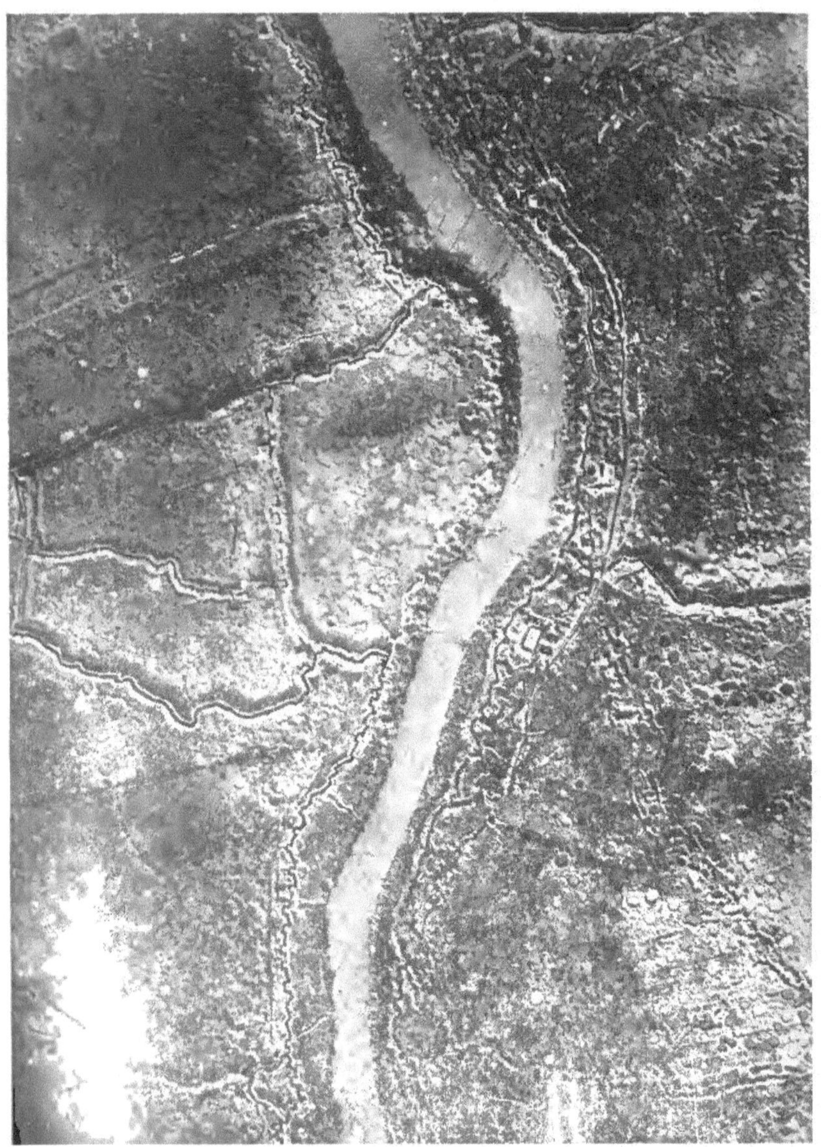

FIG. 123.—Low view of trenches on the Yser, showing concrete structures undamaged by bombardment.

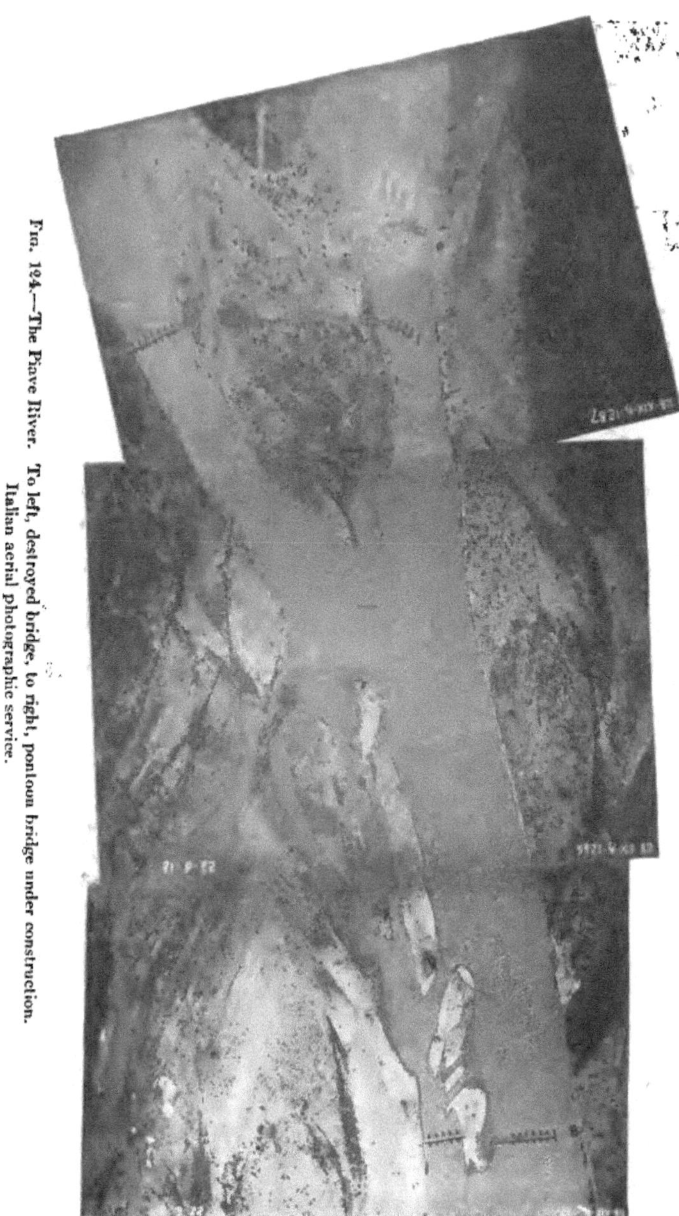

Fig. 194.—The Piave River. To left, destroyed bridge, to right, pontoon bridge under construction. Italian aerial photographic service.

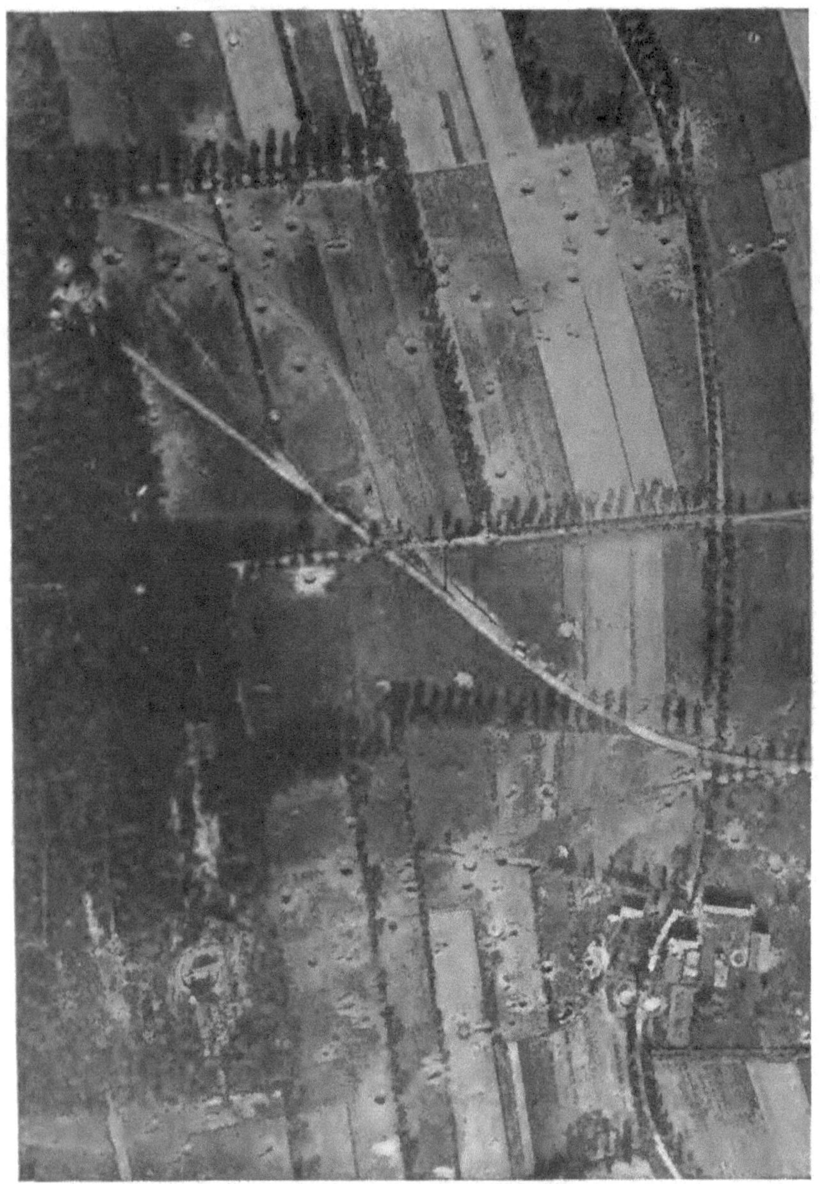

Fig. 125.—Showing big gun hidden in forest. (Upper left-hand corner)

for enemy aircraft. A very desirable adjunct to the camera when plates are shifted automatically or by the observer is a distance indicator, to show the pilot when the shutter is set. Electrical indicators for this purpose have been devised.

If the camera is completely hand operated, as were most of those in the French and German services, there is little

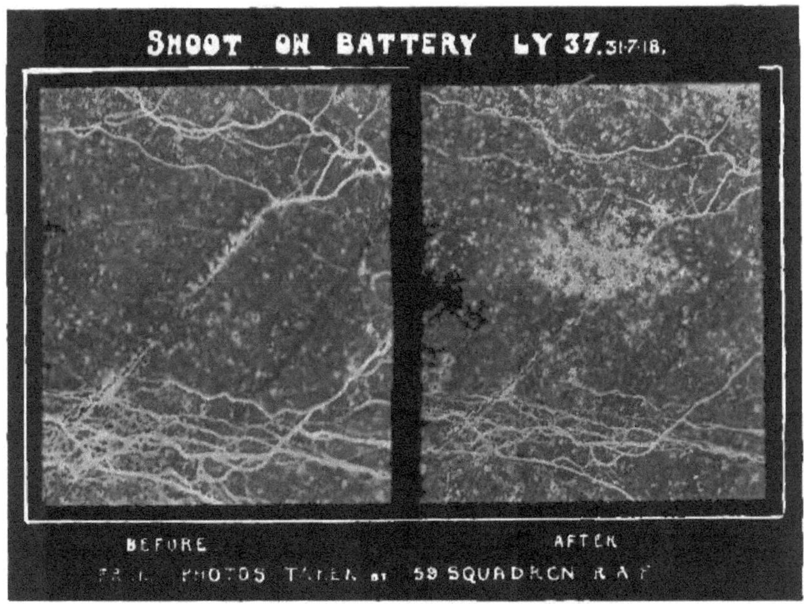

FIG. 126.—Example of spotting. Battery before and after bombardment.

choice but for the observer to perform the entire operation. The exposing operation could have been delegated to the pilot, but such was not the custom with the French or with the American squadrons using French apparatus. In this method of operation the observer depends on the pilot to get the plane over the target, while the pilot depends on the observer to get the picture when the target is covered. Ample opportunity is thus offered for misunderstanding and disagreement. This can be avoided only by excellent sights properly aligned, for both pilot and observer, and by some means of communication between the two men concerned.

The simplest *means of communication* is of course direct conversation. But this is only possible in those planes, such

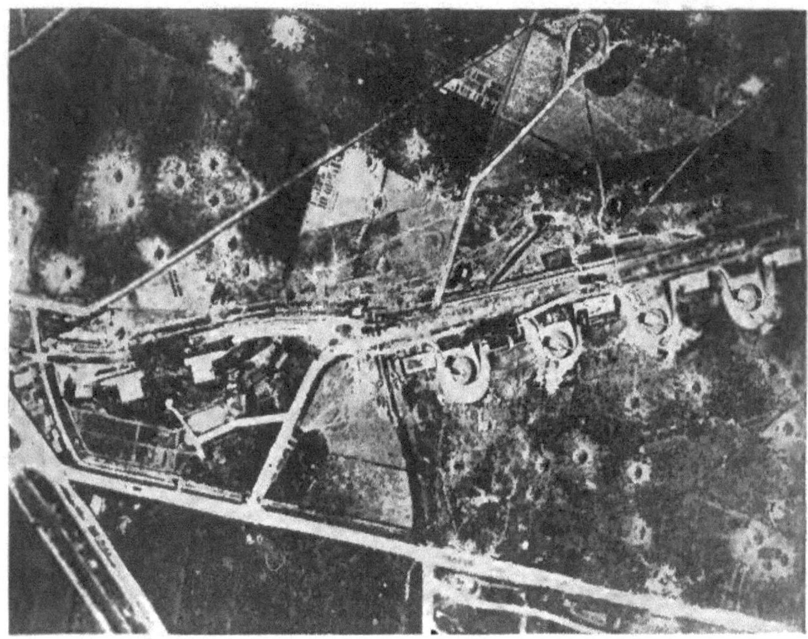

Fig. 127.—Photograph, made with long focus lens to determine the results of aerial bombing. The "Tirpitz" battery of long range naval guns directed on Dunkirk.

as the DH 9, in which pilot's and observer's cockpits are immediately together, so that, by shouting, any desired information can be conveyed with fair ease. When the distance is increased to four or five feet, as in the DH 4, the loudest shouts are totally lost in the roar of the engine and the blast of the wind. Speaking tubes and telephones are now fairly good, but are none too comfortable or convenient to have strapped on one's head and face. A primitive device used to some extent in the war was merely a pair of reins attached to the pilot's arms, by which he could be

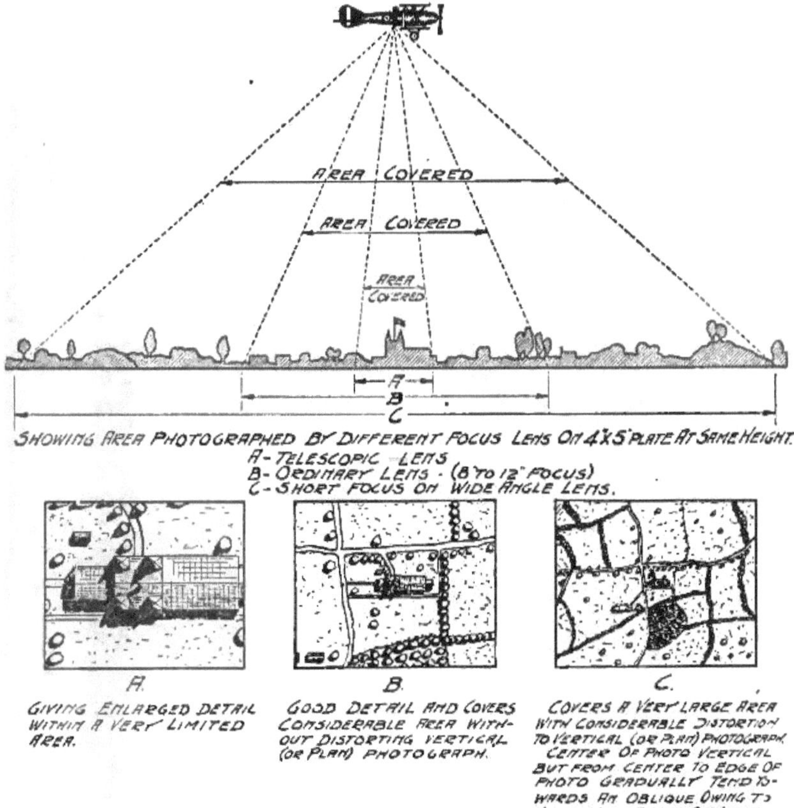

FIG. 128.—Diagram showing relationship between focal length and area covered by plate.

directed which way to steer. There is much to be said for a simple semaphore system, where an indicator in the observer's cockpit actuates a similar dial in front of the pilot, indicating "right" or "left," "picture obtained," "try again," etc. If the observer has a sight by which he can see far enough ahead to correct the pilot's error of pointing, the need for an accurate sight for the pilot is diminished.

In considering the question of sights, attention may again be called to the poor "visibility" from the pilot's seat in the present prevailing type of two-seater tractor plane.

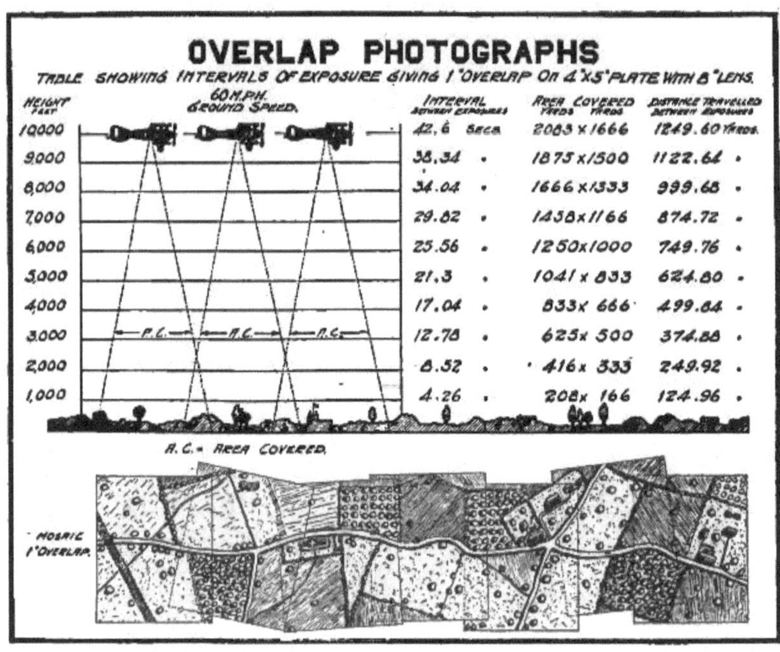

HEIGHT FEET	INTERVAL BETWEEN EXPOSURES	AREA COVERED YARDS YARDS	DISTANCE TRAVELLED BETWEEN EXPOSURES
10000	42.6 SECS.	2083 x 1666	1249.60 YARDS.
9000	38.34 "	1875 x 1500	1122.64 "
8,000	34.04 "	1666 x 1333	999.68 "
7,000	29.82 "	1458 x 1166	874.72 "
6,000	25.56 "	1250 x 1000	749.76 "
5,000	21.3 "	1041 x 833	624.80 "
4,000	17.04 "	833 x 666	499.84 "
3,000	12.78 "	625 x 500	374.88 "
2,000	8.52 "	416 x 333	249.92 "
1,000	4.26 "	208 x 166	124.96 "

FIG. 129.—Diagram giving data on area covered at various altitudes by representative lens.

Blind directly in front, beneath, and to either side (Figs. 7, 8 and 9), it is no unusual thing for a pilot to entirely miss an objective, such as a railway line, which he can only estimate to be beneath him by judging its distance from those objects to either side which he can actually see. The English practice of leaving a clear space of six inches to a foot between the fuselage and the beginning of the wing fabric, allows the pilot to look down over the side, a decided advantage. But for photographic purposes nothing can compare with a good negative lens carrying fore and aft lines or wires, so that the pilot can see his object ve in ample time to head directly for it. The lens should either be large enough so that its rear

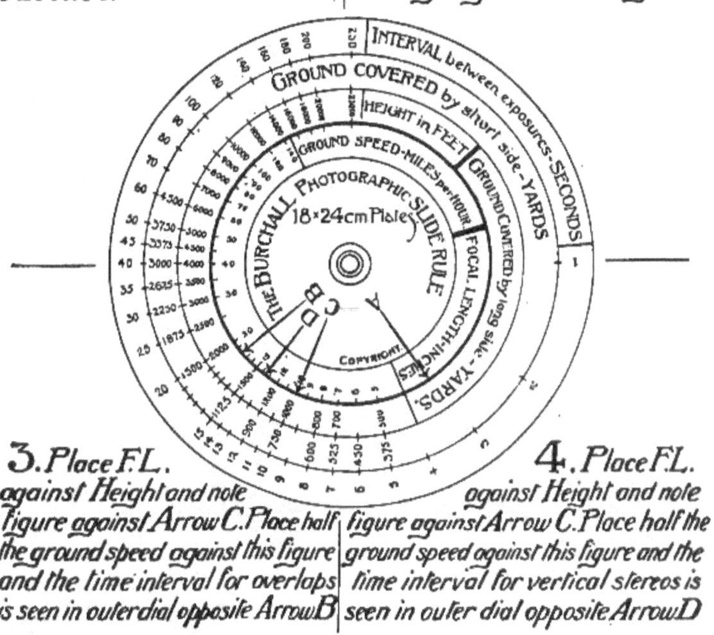

Fig. 130.—Burchall Slide Rule, for calculating intervals between exposures, and for other aerial photographic data.

edge gives the view directly downward, or supplemented by an additional lens pointing directly down, so that the covering of the target is assured. To locate such a lens in the front cockpit, free of all controls, is a very hard task; even so its view is likely to be badly interrupted by the landing gear. Nevertheless, so important is it, both in photography and in

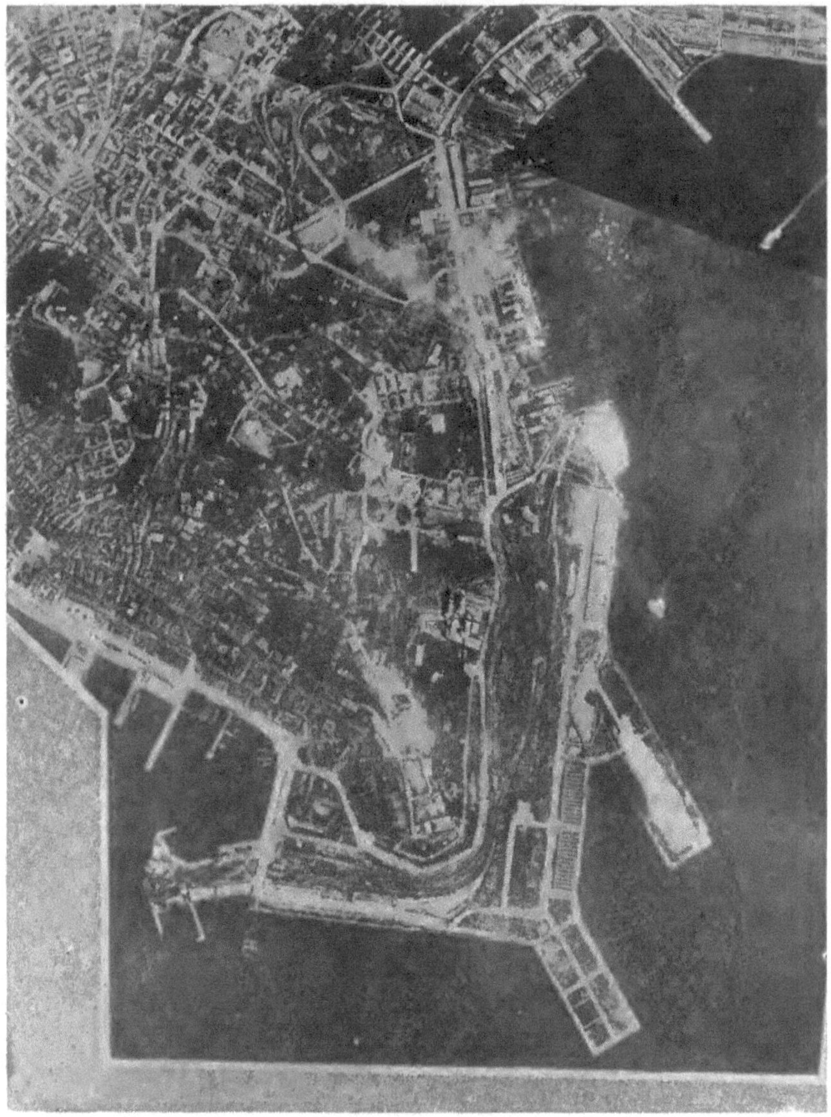

Fig. 181.—Aerial bombardment of Trieste. Note falling bombs in center of picture; and exploding anti-aircraft shells over the water.

Italian official photograph.

bombing, to have a sight by which the plane can be accurately directed that designers of planes should recognize this need and make every effort to provide a suitable location.

Sights for the observer have been discussed already. Here again the negative lens is to be preferred, but while the pilot's lens needs only directing lines in the axis of the plane (unless

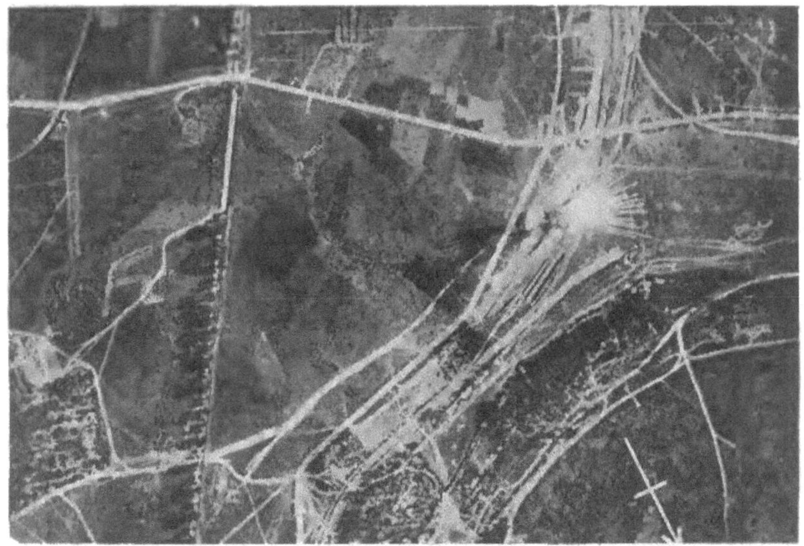

FIG. 132.—Example of spotting requiring exposure at exact instant. Explosion following burst of bomb in ammunition dump.

British official photograph.

he takes the picture), the observer's lens needs both an accurate center mark and an additional upper or lower sighting point. Accurate alignment of these marks with the camera axis must be arranged for in precise spotting.

Accurate spotting work requiring the delineation of fine detail calls for cameras of considerable focal length. The camera of longest focal length used in the war was the French 120 centimeter (Fig. 41). This was employed with great success in such work as regulating the fire of heavy railway guns brought into range only at night, to fire a few shots at chosen angles. Photographs taken the next day would then show the exact spot where each shell fell, and the damage it did, to serve as a guide for the next night's operations (Fig. 127). The field of these cameras is quite small—8 to 12 degrees—and so not only must sighting be exact but the

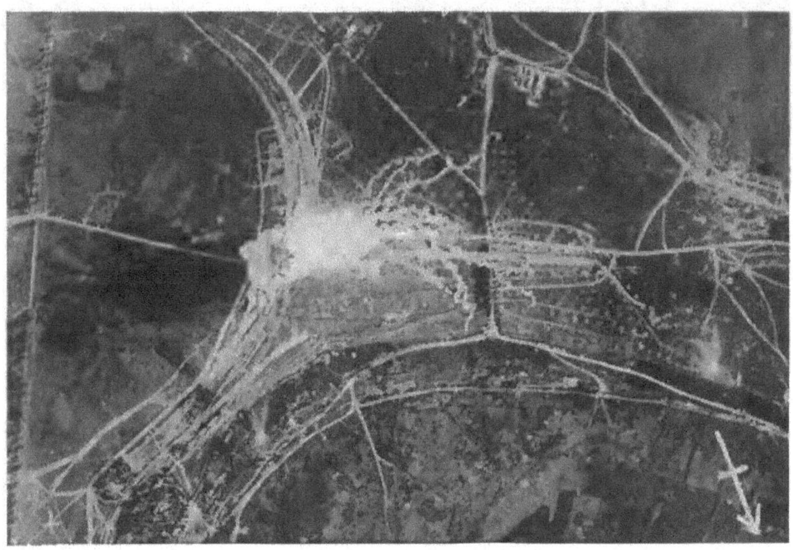

Fig. 133.—The same subject a few minutes later. Height of smoke shown by shadow. British official photograph.

area covered on the ground must be accurately known. This is to be calculated from the altitude, focal length, and plate size, by the relation—

$$\frac{\text{distance on ground}}{\text{plate length}} = \frac{\text{altitude}}{\text{focal length}}$$

Data derived from such calculations may be incorporated in tables, or graphically in diagrams such as Figs. 128 and 129.

These calculations and others required in mapping and stereo-work are simply and quickly made by slide-rule devices. One of these, the Burchell Photographic Slide Rule, developed in the English service, is shown in Fig. 130. This consists of two dials, the center one of which is mounted—usually by a pin pushed into a cork behind—so as to turn freely, to permit its being set for altitude, focal length, ground speed, plate size, etc., whereupon the area covered, or the appropriate interval between exposures may be read off.

Cameras for spotting work should be capable of exposure at the exact moment desired. For if the camera is ever to catch the gun as it discharges, the bomb as it falls (Fig.131), or the shell as it explodes (Fig. 132), the photograph must be taken within the instant. Automatic cameras, exposing at regular intervals, while adequate for mapping, are not fitted for many kinds of spotting.

CHAPTER XXVI

MAP MAKING

Technique of Negative Making.—Stated in its simplest terms, the whole problem of making a photographic map from the air consists in taking a large number of slightly overlapping negatives, all from the same altitude, with the plane flying uniformly level. When trimmed and mounted in juxtaposition, or pasted together so as to overlap in their common portions, the prints from these negatives constitute a complete pictorial map. There is thus furnished by a few hours' labor topographic information which would be the work of months to obtain by other means.

The making of map photographs involves all the special technique of spotting, with much in addition. The pilot's task is not merely to go over one object; he must navigate a narrow path, at a constant altitude, on an even keel. If he is to make not merely a ribbon, but a map of considerable width, he must take successive trips parallel to the first, each displaced just far enough from the previous course to insure that no portion is missed—a difficult task indeed.

It is the observer's duty to so time the intervals between exposures that they overlap enough, but not so much as to be wasteful of plates or film. He must also change magazines or films so quickly as to miss no territory, or if some be missed, his is the task of directing the pilot back to the point of the last exposure, where they begin a new series.

Level flying is entirely a pilot's problem. Its importance will be realized when we consider the accompanying diagrams (Figs. 134 and 135), where the effect on the resultant picture is shown of climbing, gliding, or banking to either side

Prints from negatives distorted in this way neither will be true representations of the territory photographed, nor will they match when juxtaposed. In fact, they can be utilized only if special rectifying apparatus is available for printing.

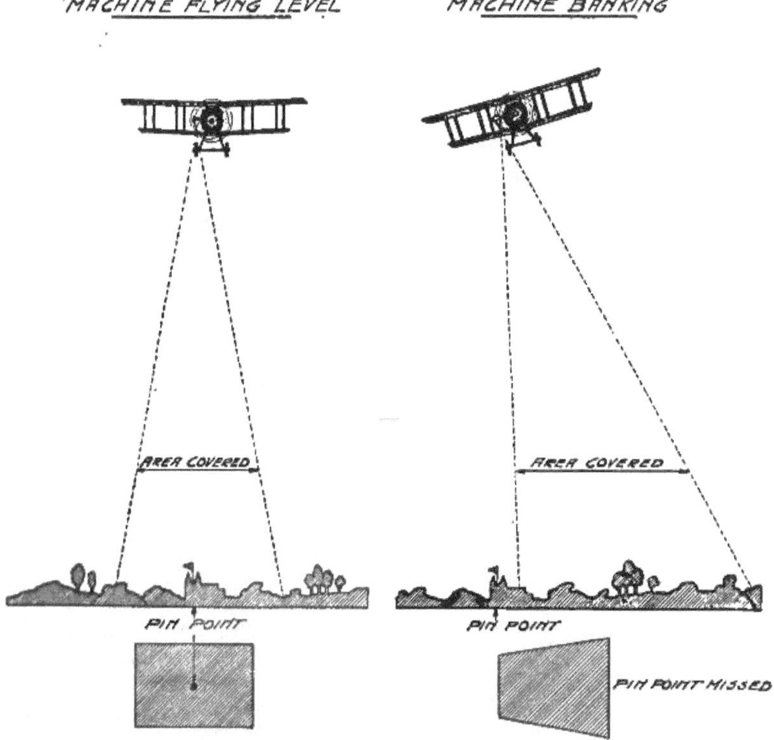

Fig. 134.—Diagram showing effect of banking on aerial photograph.

Flying at a constant altitude is similarly necessary if the prints are to be utilized without enlargement or reduction in order to make them fit.

Assuming a skilled pilot who will do his part, the next step is to calculate the exposure intervals in order to insure an adequate overlap. If a negative lens is installed which has been marked with a rectangle the size of the camera field, the simplest method is to estimate the proper instant for exposure by watching the progress of objects across the lens face. This of course requires constant attention, and

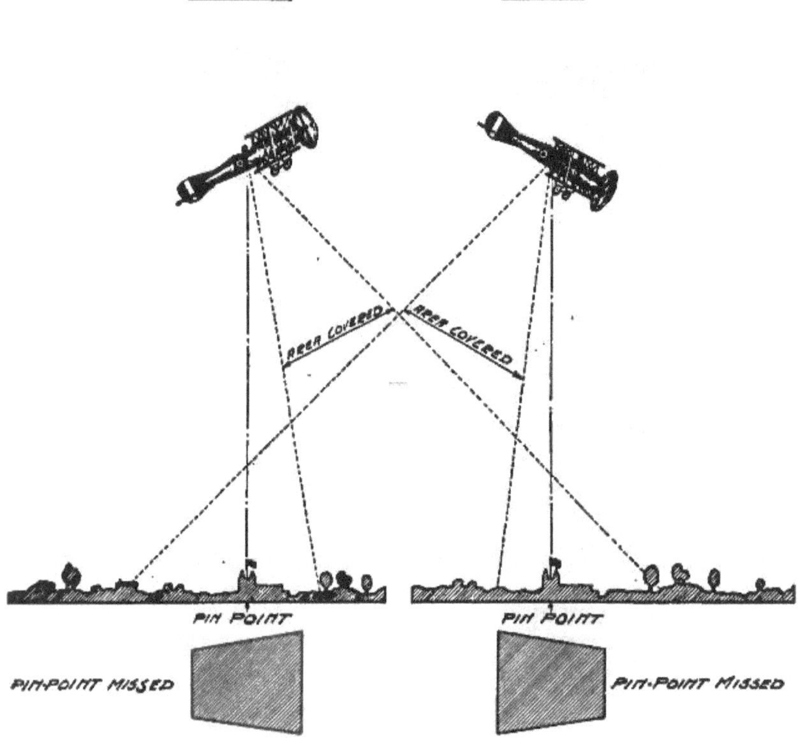

FIG. 135.—Diagram showing effect of climbing and diving on aerial photograph.

it is easier to do this only occasionally, in order to determine the ground speed in terms of camera fields traversed per minute. Thereafter exposures are to be made by time, as determined by a watch or clock. Any desired degree of overlap can be chosen, and either estimated, or more or less accurately fixed by lines marked on the negative lens at a shorter distance apart than the edges of the field. The most usual overlap is 20 per cent., except for stereos, which call for 50 to 75 per cent.

In the absence of a negative lens or some other sight to show the whole camera field, it is necessary to resort to calculation from the speed and altitude of the plane, the focus of the lens and the dimensions of the plate. If A is the altitude, a the focal length of the lens, d the diameter of the plate in the direction of travel (usually the short length is chosen for economy of flights to cover a given width), f the fractional part by which one negative is desired to overlap the next, and V the ground speed of the plane, then we have, by simple proportion, that the interval between exposures, t, must be—

$$t = \frac{Ad(1-f)}{aV}$$

If $A = 2000$ meters, $d = 18$ centimeters, $f = 1/5$, $a = 50$ centimeters, and $V = 200$ kilometers per hour, this relation gives—

$$t = \frac{2000 \times .18 \times .8 \times 3600}{.5 \times 200{,}000} = 10.3 \text{ seconds}$$

The principle of overlapping map exposures is shown in the accompanying diagram (Fig. 129), together with data calculated as above for a 4×5 inch plate.

It is particularly to be noted that it is the *ground speed* of the plane that is used. This may be calculated by knowing the air speed and the wind velocity and direction. Fig. 136 shows the method of doing this graphically. First an arrow is drawn representing the direction it is desired to fly. Next a second arrow is drawn of length to represent the wind velocity. This must be inclined toward the first arrow in the direction of the wind, and its head is to touch the head of the first arrow. Then with the farther end of this second arrow as a center, describe a circle of such a length as to represent the air speed of the plane, in the same units as the wind velocity. Connect the point where this circle cuts the arrow of flight direction to the center of the circle by a straight line. This line constitutes the air speed arrow, giving the direction it is necessary to fly, at the given air speed, to make the course desired. The length of the flight direction arrow between its head and its point of intersection with the air speed arrow gives the ground speed.

When the wind is ahead or astern this calculation reduces to the simple subtraction or addition of the wind velocity

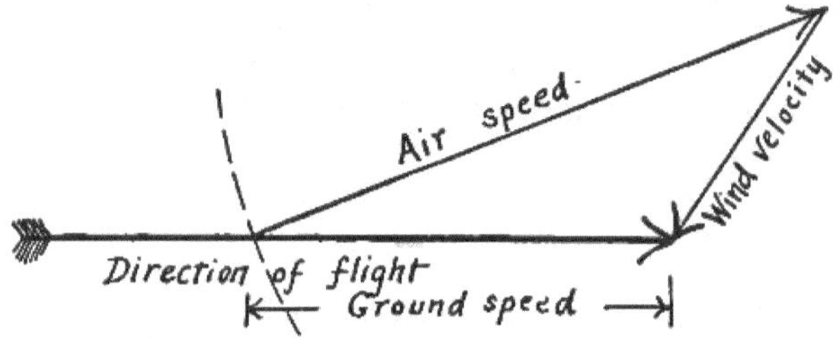

Fig. 136.—Diagram showing method of calculating ground speed from air speed and wind velocity.

to the air speed of the plane. Whenever possible, mapping should be done up and down the wind (Fig. 137). If the plane is "crabbing," the above calculations for overlap are only valid if the camera can be turned normal to the direction of travel over the ground. If the camera cannot be so turned the corners of the successive pictures overlap instead of their sides, with quite unsatisfactory results (Fig. 138).

Calculation of the distance apart of the parallel flights necessary to make a map of any width is done by the use of a formula similar to the longitudinal overlap formula above, distance figuring instead of time. Using the same symbols, and denoting the distance by D, we have—

$$D = \frac{Ad(1-f)}{a}$$

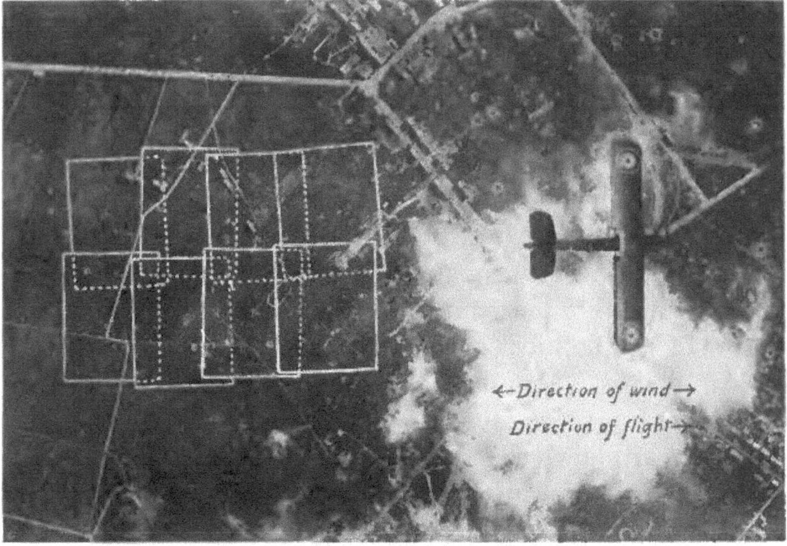

FIG. 137.—Overlaps made when flying with or against the wind.

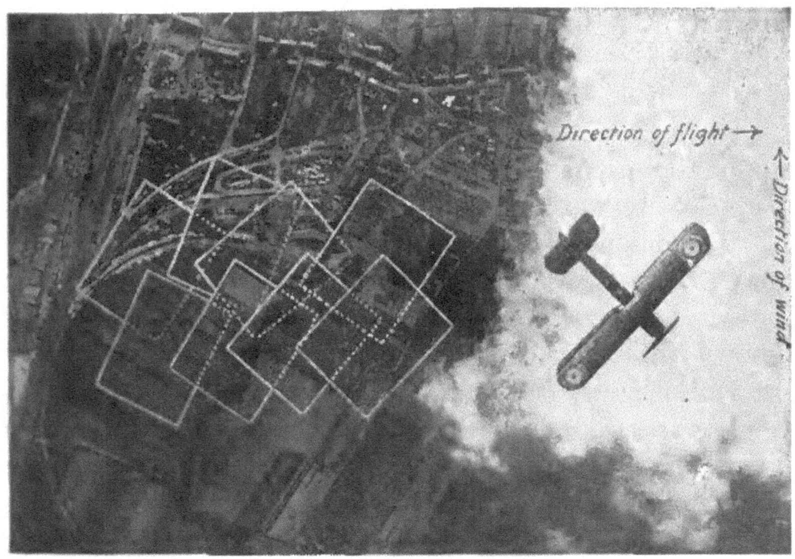

FIG. 138.—Unsatisfactory overlaps made when plane is "crabbing."

With the same figures as before, but substituting 24 centimeters for the plate dimension, this relation gives—

$$D = \frac{2000 \times .24 \times .8}{.5} = 768 \text{ meters}$$

It is of course largely a pilot's problem to steer the plane over parallel courses at a given distance apart, although the observer, noting conspicuous objects through a properly marked negative lens, may direct the pilot by any of the means of communication already mentioned.

An alternative method of securing parallel strips, which is to be highly recommended where enough photographically equipped airplanes are available, is for several planes to fly side by side, maintaining their proper separation (Fig. 139).

Cameras and Auxiliaries for Map Making.—Mapping can be done quite satisfactorily by hand operated or semi-automatic cameras, provided the observer has not too many other duties. On the other hand, the operation of exposing at more or less definite intervals of time, irrespective of the object immediately presented to the camera, is a largely mechanical one. It naturally suggests the employment of an automatic mechanism, whose speed of operation only is it necessary to watch.

If a non-automatic camera is used the timing of exposures may be done by watching a negative lens, as described above, or by reference to a clock, assuming that the ground speed is known through calculation. A very practical advance over the ordinary use of a clock is to attach a stop-watch to the shutter release, so that it is turned back to zero and re-started at each exposure (Fig. 70). In passing, it may be noted that if the stop-watch hand makes an electric contact which throws the shutter release, then the device constitutes an attachment for turning any semi-automatic camera into an automatic.

The most suitable cameras for mapping are unquestionably those of the entirely automatic type. The use of such cameras always demands a knowledge of the ground speed.

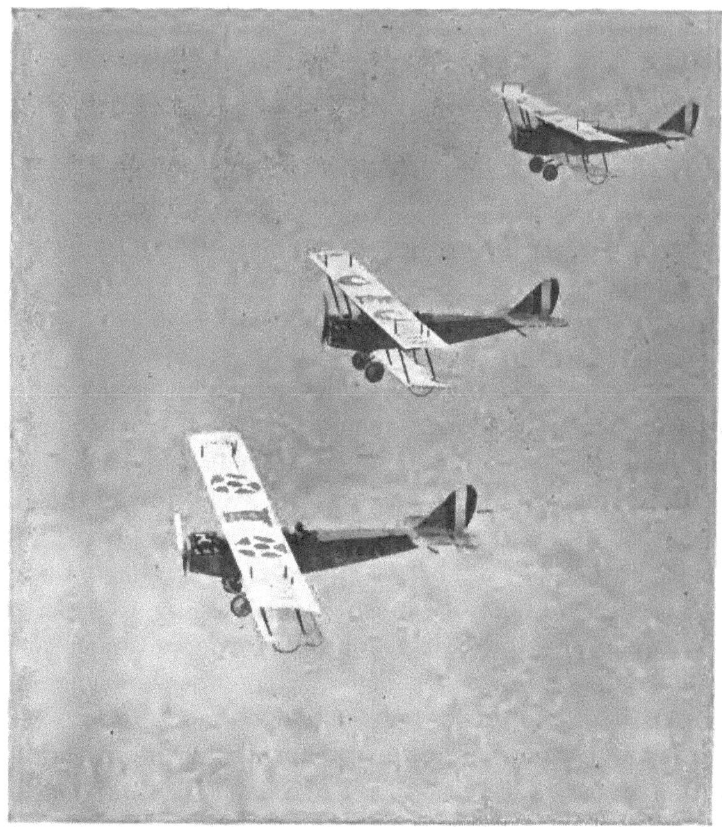

Fig. 130.—Planes starting out to make a map by flying in parallel.

This demand has led to many suggestions for *ground speed indicators.* The common idea of these is to provide a moving part on the plane—either a disc of large diameter, or a chain, or a revolving screw—whose speed may be varied until any point upon it appears to keep in coincidence with a point on the moving landscape below. The ground speed is then to be read off a properly calibrated dial. Or, as a further step, the frequency of the exposures may be directly controlled by the ground speed indicator mechanism. The entire control of the camera would then consist merely in occasional adjustment of the ground speed indicator.

While entirely possible in theory, these devices are not easy to work with in practice, because the plane is always subject to some pitching and rolling, which make it difficult to hold any object constantly on the moving point. This is especially true at high altitudes, where the apparent motion of the earth is quite slow compared to the swervings of the plane. This objection is in part removed if the ground speed indicator is carried by a gyro stabilizer.

Ordinary mapping does not demand such exquisite rendering of detail as does trench mapping. Nor is it necessary to fly in peace-time at such high altitudes as in war. In consequence, mapping cameras are preferably of the short focus, wide angle type, say, 25 centimeter focus for an 18×24 centimeter plate. Film is to be preferred over plates because of the greater number of exposures it is possible to make on a flight. The shutter of the mapping camera must be extremely uniform in its rate of travel so that the elements of the map may match in tone (Fig. 140). A mount which permits the camera to be turned normal to the direction of flight, such as the British turret mount (Fig. 87), is particularly desirable if flying across the wind is necessary, as will often be the case in mapping strips between towns or between flying fields. Devices to indicate compass direction and altitude are called for in new and poorly mapped territory, and may be expected to receive intensive study in the future. The question of their utility is, however, bound up with the whole question of the sphere of aerial photographic mapping.

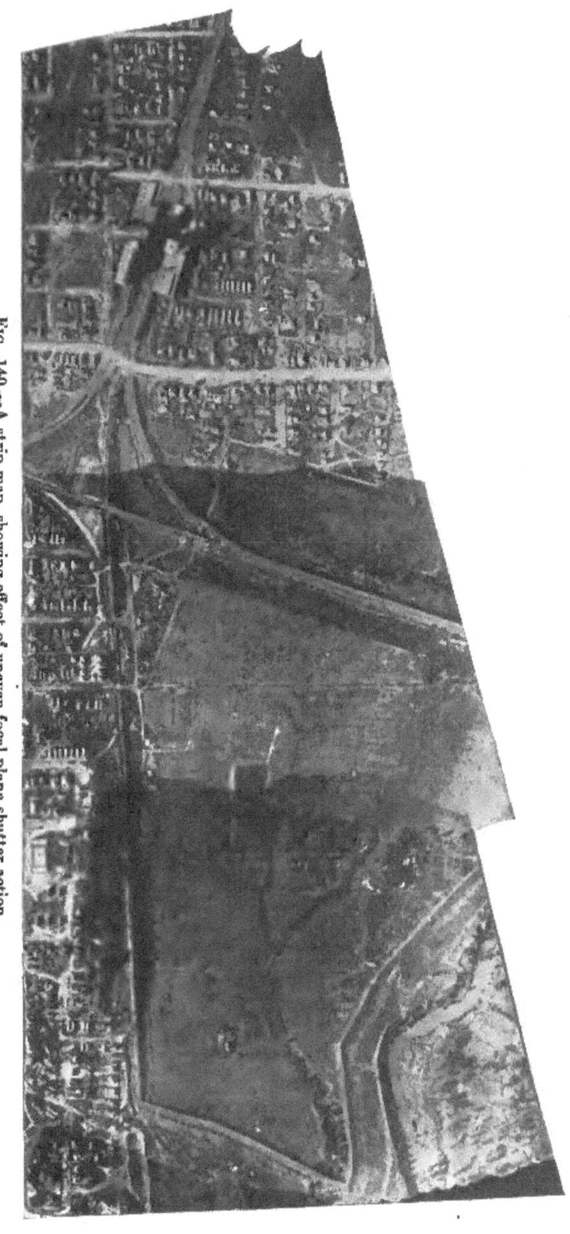

Fig. 140.—A strip map, showing effect of uneven focal plane shutter action.

Up to the present this has been almost entirely a matter of filling in details on maps obtained by the regular surveying methods, or of making pictorial maps for aviators. To what extent primary mapping can be done by the airplane is yet to be determined.

At this point mention must be made of special cameras for securing extremely wide angle views, thereby minimizing the number of flights. The *Bagley camera,* devised by Major Bagley of the U. S. Engineers, is an example. It has three lenses, a middle one pointing directly downward, and one to either side at an angle of 35 degrees. The pictures obtained with the side cameras are of course greatly distorted, and must be rectified in a special rectifying camera. The resultant definition is not good, but as the maps are made on a much smaller scale than the original pictures, this is not a serious objection. It is a matter for the future to decide whether the additional labor on the ground necessary for the rectifying process is to be more expensive than the extra flights which must be made with the ordinary types of cameras covering a smaller angle.

Printing and Mounting Mosaics.—With an ordinary set of overlapping negatives the first step toward producing a map is to *scale* the negatives. For this purpose one should be selected which by comparison with a map shows no distortion, and which is on the desired scale, or is known to have been made at the average altitude of flight. A sketch map of the territory should then be drawn, on this scale, based on available maps. This sketch is preferably made on a large ground glass illuminated from behind (Fig. 141). On this all the negatives should be laid, and their proper relative positions sought. When this is done it is evident at once whether all the territory has been covered, and whether there are any superfluous negatives. Each negative should then be examined as to its scale and distortion. If it can be made to fit the scale by simple enlargement or reduction, a line can be drawn on one edge of a length indicating its scale. This line will later be used as a guide in the enlarging camera. If the picture is badly distorted it must either be replaced by another negative, or if rectifying apparatus is available, it must be set aside for the making of a rectified print.

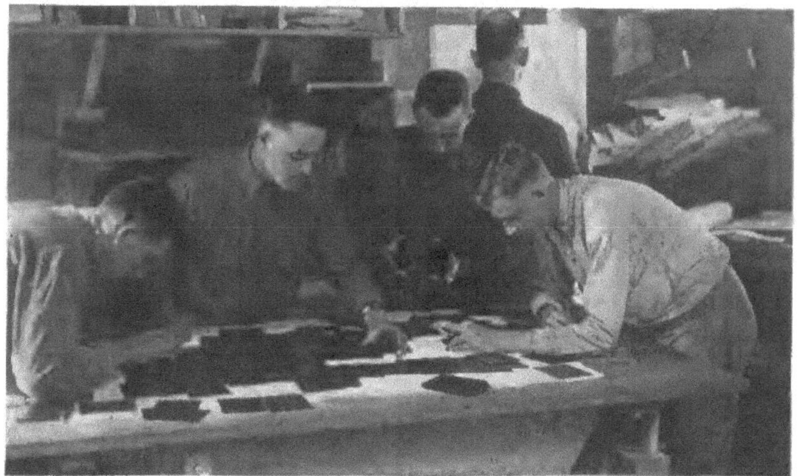

FIG. 141.—Scaling negatives for mosaic map-making.

The next step is to make prints from the negatives, which may be done either by contact, or, necessarily if differences of scale must be compensated, in the enlarging camera. If prints to an exact scale are required the shrinkage of the paper must be determined and allowed for. The prints must all show the same tone, and must be uniform from edge to edge. If the focal-plane shutter is not uniform in its travel, as is frequently the case, this means that the print must be "dodged," or exposed more at one edge than the other, by locally shielding the plate and paper during exposure. A case of the step-like effect caused by uneven shutter action is shown in Fig. 140. The effect due to uneven shutter action is of course absent with a between-the-lens shutter, which constitutes a strong argument in favor of that type for use in mapping cameras.

When the prints are made they must be mounted together on a large card or cloth background. For a very small

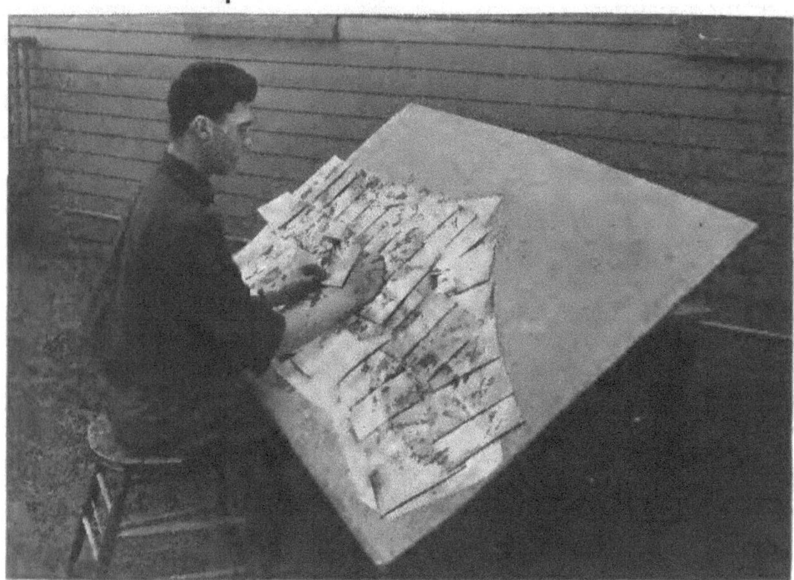

Fig. 142.—Arranging prints for a mosaic map.

mosaic they may be juxtaposed by simple examination, matching corresponding details in successive prints. For a mosaic of any size an accurate outline map must be drawn on the surface to which the prints are to be attached. The prints are then laid out on this outline, moved to their correct positions, and held down by pins (Fig. 142). When they are all arranged the final mounting may be begun. The excess paper, beyond what is necessary for safe overlaps, may be trimmed off, exercising judgment as to which print of each adjacent pair is of the better quality, and utilizing it for the top one at the overlapping junction. If one print shows serious distortion it may be placed under its fellows on all four edges, thus minimizing its weight. The edges are best made irregular by *tearing.* Straight edges are apt to force themselves on one's attention in the final mosaic and give an erroneous impression of the existence of straight roads or other features. Both forms of edging are shown in Figs. 124 and 143.

An alternative method of securing the final print mosaic, where film negatives are used, is to trim successive film negatives so that the trimmed sections will exactly juxtapose, instead of overlap. The sections are then mounted, by stickers at their edges, on a large sheet of glass, and printed together. Captured German prints show that this was the method commonly used with the German film camera (Fig. 62).

It will be noted that the procedure which has been described and illustrated by Figs. 142 and 143 assumes the previous existence of a map accurately placing at least the chief features of the country covered. This draws attention at once to the limitations and true sphere of aerial photographic mapping at the present time. With the cameras thus far it is not possible, nor is it attempted, to do primary mapping of unknown regions. Distortions due to lens, shutter, film warping and paper shrinkage considerably exceed the figures permitted in precision mapping. From the standpoint of geodetic accuracy the cumulative errors of deviations in direction, altitude and level, peculiar to flying, would soon become prohibitive.

The great field for aerial photographic mapping in the near future lies in filling in detail on maps heretofore completed as to general outlines, or, as in the war, on maps far

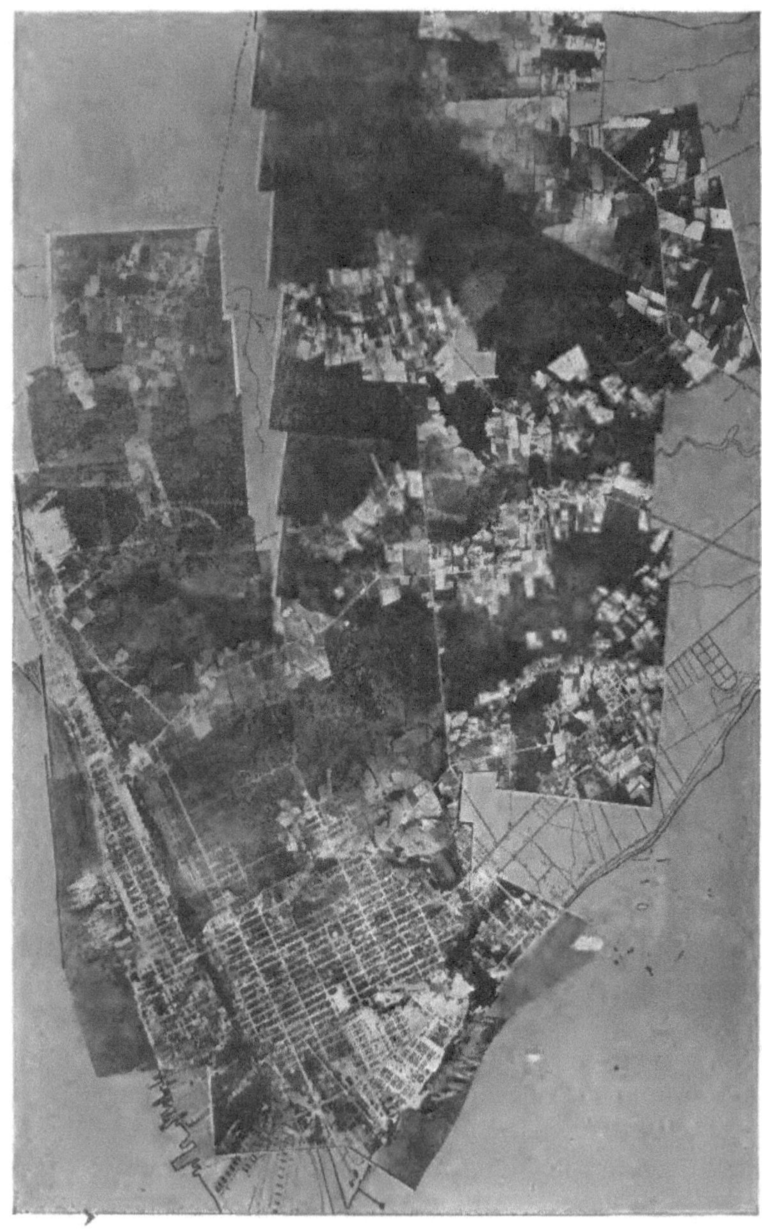

FIG. 143.—A partly completed map. Prints mounted over an outline sketch map to proper scale.

out of date. The war-time procedure in country largely unknown, such as Mesopotamia, was probably closely that which will be necessary in peace. Conspicuous points in the landscape were first triangulated from friendly territory, and from these the outline map was drawn, whose details were to be supplied by aerial photographs. Much of the "mapping" of cross country aerial routes so far done is frankly of a pictorial nature, showing conspicuous landmarks and good landing fields—extremely valuable and useful, but not to be confused with precision mapping. In assembling mosaics of this kind the elaborate procedure described above is not followed. The process is the simple one of juxtaposing adjacent prints as accurately as possible by visual examination. Errors are of course cumulative, but as long as exact distances are not in question this is no matter.

CHAPTER XXVII

OBLIQUE AERIAL PHOTOGRAPHY

Oblique views from the airplane are of very great value. While vertical views are more searching in many respects, they do nevertheless present an aspect of the earth with which ordinary human experience is unfamiliar. Consequently they are difficult to interpret without special training. They suffer, too, from the military standpoint, from the limitation that it is with vertical extension just as much as with horizontal that an army has to contend in its progress. Elevations and depressions of land show on an oblique view where they would be entirely missed in a vertical one. For illustration, study the picture of part of the outskirts of Arras (Fig. 144), presenting moat, walls and embankments, all of which would be serious obstacles, but would hardly be noticed on a vertical view. Pictures taken from directly overhead are eminently suited to artillery use, but oblique views of the territory to be attacked, taken from low altitudes, formed an essential part of the equipment of the infantry in the later stages of the war.

Pictorially, oblique views are undoubtedly the most satisfactory. The most revealing aspect of any object is not one side or face alone, but the view taken at an angle, showing portious of two or three sides. Best of all is that taken to show portions of front, side and top—the well-known but heretofore fictitious "bird's-eye view" (Fig. 145). This possibility is ordinarily denied the surface-of-the-earth photographer, but the proper vantage point is attained in the airplane.

Aerial obliques may be taken at any angle, although a

distinction is sometimes made between obliques of high angle and panoramic or low angle views (Fig. 146). In addition to ordinary obliques, a very beautiful development is the stereo oblique. Both kinds of oblique photography call for special instrumental equipment and technique.

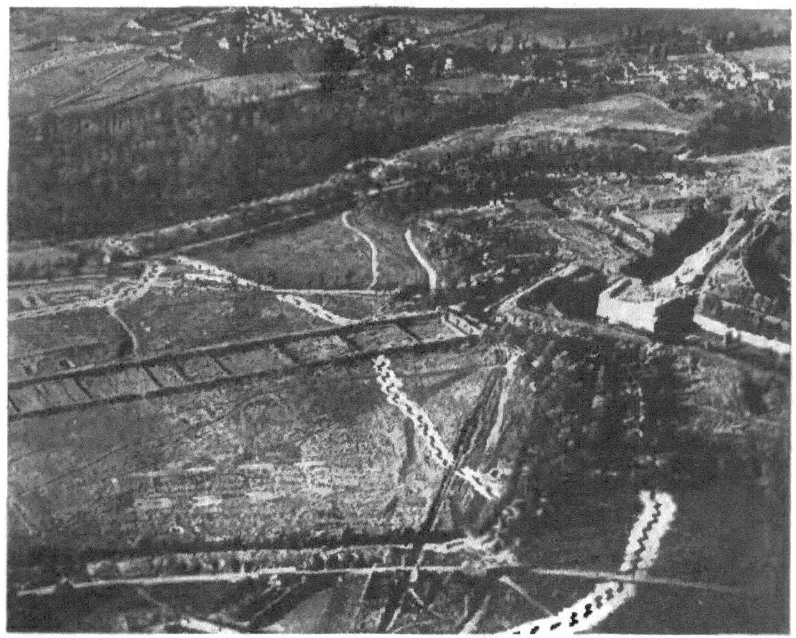

Fig. 144.—The outskirts of Arras. Low oblique showing contours.

Methods and Apparatus for Oblique Photography.—The simplest method of taking oblique pictures from a plane is to use a hand camera pointed at the desired angle. Its limitations are in the size and scale of the picture obtainable, and in the inherent limitations to the method of camera support. A step in advance of this is to mount the camera above the fuselage, on the machine gun ring or turret, in place of the gun. Considerably greater rigidity is thus

21

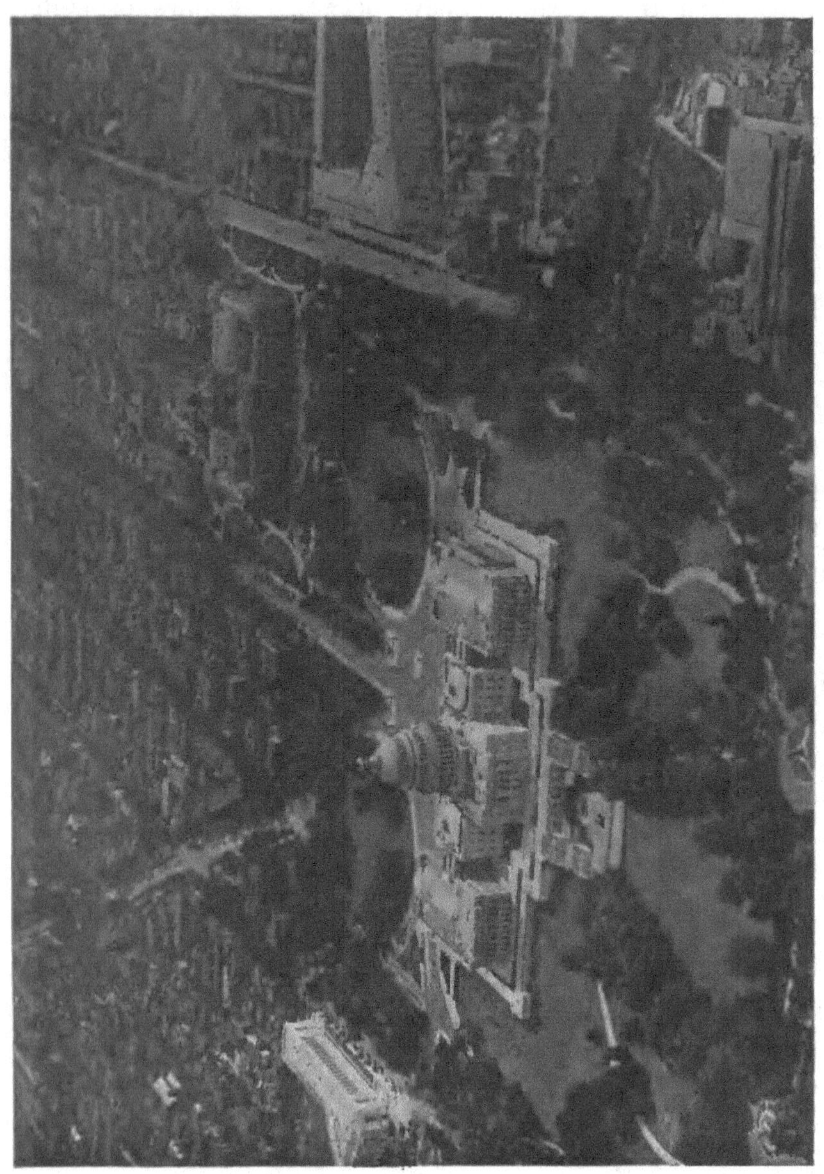

Fig. 145.—Oblique view of Capitol and Congressional Library, Washington.

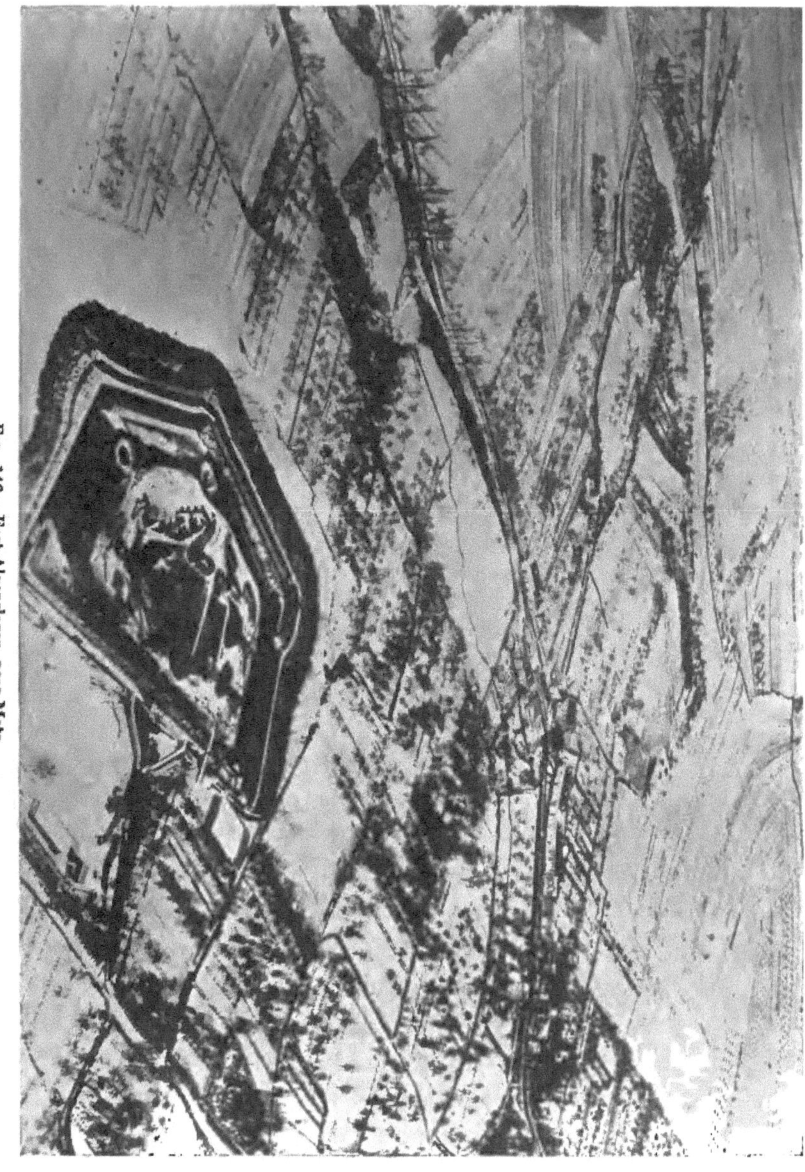

FIG. 146.—Fort Alvenslegen, near Metz.
Photo by Photographic Section A. E. F.

obtained, and heavier cameras can be utilized, although the wind resistance is a serious factor. Excellent obliques have been made in this way, even with 50-centimeter cameras, but the scheme is impractical in military planes, because of the removal of machine gun protection.

If the camera is fixed in the fuselage in its normal vertical position, obliques may be and have been taken by the simple expedient of *banking* the plane steeply. This is not to be recommended as a standard procedure, especially for taking a consecutive series of exposures.

The most satisfactory arrangements for taking obliques are two; first, to mount *the camera obliquely in the plane*, and second, to use a *mirror or prism*, in front or behind the lens of the vertically mounted camera. The first method has been employed chiefly by the French, the latter by the English, whose gravity fed cameras could not be mounted obliquely

Taking up first the oblique mounting of cameras, we find two ways of doing this: longitudinal mounting and lateral mounting. In longitudinal mounting the camera projects forward and downward, usually from the nose of a pusher or bi-motored plane. With this form of mounting (Fig. 147) it is necessary of course to fly directly toward the objective. If this is a portion of enemy trench, which must be photographed from a height of 400 or 500 meters, the plane will be directly on top of its objective a few seconds after the exposure is made, and be a conspicuous target, in imminent danger of destruction. Moreover, only a single short section of the trench would be obtained for each crossing of the line. The one case where resort to this method is practically forced is with the 120-centimeter cameras which simply cannot be slung athwart the plane. There is a slight advantage in this method of carrying in that the motion of the image is less if the objective is approached, instead of being passed at the side, and so longer exposures can be made. The longitudinal mounting has, however, been very generally superseded by the lateral.

Methods for mounting cameras obliquely for taking pictures through the side of the plane have been discussed in detail in connection with camera mountings and installations (Fig. 93). The chief difficulties are want of space, obstacles

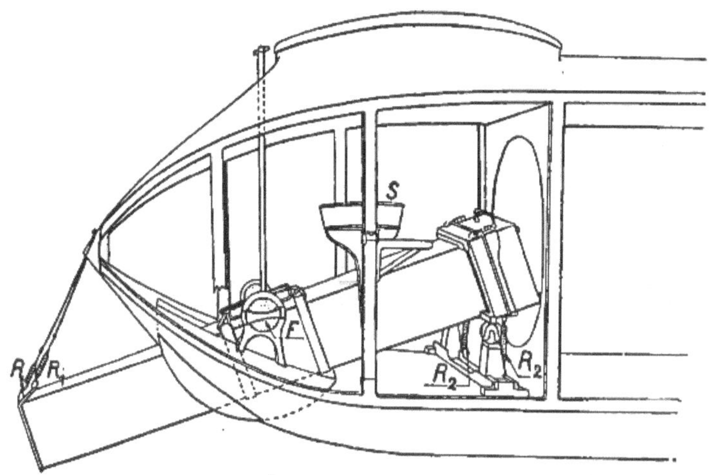

FIG. 147.—120-centimeter camera mounted obliquely in the fore-and-aft position.

at the side such as control wires and longerons, and failure of the camera to function properly at an angle. Even in the broad circular sectioned fuselage of the Salmson plane, quarters are so cramped that the French 50-centimeter camera when obliquely mounted cannot be used with the 12-plate magazine, and recourse is made to thin flat double plate-holders. Holes in the side of the fuselage should clear all wires and should command a view unobstructed by the wings—which often means that the camera must be carried behind the observer's cockpit, irrespective of the suitability of that space from other standpoints. Cameras dependent for their action on gravity, such as the deRam and English L type, are unsuited for oblique suspension.

For cameras which, because of their method of operation or shape cannot be slung obliquely, the only way to take obliques is to employ mirrors (Fig. 148) or prisms. These must be of the same optical quality as the photographic lens. They are both necessarily of considerable weight because they must be of large area of face to fill the entire aperture of an aerial lens. Mirrors are lighter than prisms, but must be quite thick to prevent distortion of the surface due to any possible strains to their mount. Right angle glass prisms have been used by the English with the 8 and 10 inch L cameras. The prisms were uniformly tilted to an angle of 12½ degrees from the horizontal.

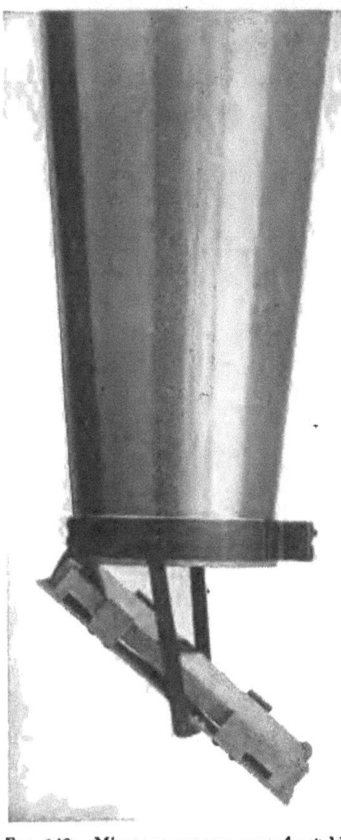

Fig. 148.—Mirror on camera cone for taking oblique views.

Glass mirrors can be silvered either on the rear or front surface. If on the rear, both surfaces must be accurately parallel, which means much greater labor and expense than if the front surface can be utilized. The difficulty with front surface mirrors is that the metallic coating is easily tarnished or scratched, especially if silver is used, which is almost imperative, since all the other metals have considerably lower reflecting powers. (Gold might serve both as mirror and color filter, because of its yellow color.) Placing the mirror inside the camera body in part obviates this trouble, but means the use of a special elbow lens cone. In any case the mirror or prism occasions at least a 10 per cent. loss of light. Pictures taken by reflectors of any kind are reversed, and must either be printed in a camera, or on transparent film which may be viewed from the back.

The most usual condition for making obliques is to fly very low (300 to 600 meters), with the line of sight of the camera from 12 to 45 degrees from the horizontal. This low altitude necessitates very short exposures, to avoid movement of the image. The picture may be taken either the long or the short way of the plate, depending on the character of the object and the information desired. It is to be noted that successive oblique pictures cannot be mounted to form a continuous panorama—this being possible with obliques only if they are taken from one point, as from a captive balloon. If successive views are made on a straight flight at intervals so as to exactly juxtapose in the foreground, they overlap by a large margin the middle, and a point on the horizon, if that shows, will be in the same position in every picture. Mosaics of obliques could be made only by some system of conical mounting.

Sights for Oblique Photography.—Any of the sights previously discussed for vertical work, such as the tube sights, are applicable to obliques. They must, however, be suited for mounting at an angle, in a position convenient for the observer. In addition, provision must be made for adjusting the angle so that the lines of sight of camera and finder are parallel. Mounting outside the fuselage is practically the only feasible way, and is less objectionable with oblique than with vertical sights, as oblique sighting does not require the observer to stand up and lean over the edge of the cockpit. Windows in the side of the fuselage, either of celluloid or non-breakable glass, are a great aid to oblique observation. Marks upon the transparent surface can be utilized for the rear points of a sight of which the front point is a single fixed bead or rectangle.

CHAPTER XXVIII

STEREOSCOPIC AERIAL PHOTOGRAPHY

One of the most striking and valuable developments in aerial photography has been the use of stereoscopic views. Pairs of pictures, taken with a considerable separation in their points of view and studied later by the aid of the stereoscope, show an elevation and a solidity which are entirely wanting in the ordinary flat aerial vista. Often, indeed, these attributes are essential for detecting and recognizing the nature of objects seen from above. Stereoscopic aerial photography has been justly termed "the worst foe of camouflage."

Principles of Stereoscopic Vision.—The ability to see objects in relief is confined solely to man and to a few of the higher animals in whom the eyes are placed side by side. When the eyes are so placed they both see, to a large extent, the same objects in their fields of view. Owing to the separation of the eyes the actual appearance of all objects not too far away is different, and it is by the interpretation of these differences that the brain gets the sensation of relief. Thus in Fig. 149 the two eyes are shown diagrammatically as looking at a cube. The right eye sees around on the right-hand face of the cube, the left eye on the left-hand face of the cube. The two aspects which are fused and interpreted by the brain are shown in the lower diagram.

Stereoscopic views or stereograms, made either by photography or, in the early days, by careful drawing, consist of pairs of pictures made of the same object from two different points. For ordinary stereoscopic work these points are separated by the distance between the eyes, approximately 65 millimeters or $2\frac{3}{4}$ inches. These two pictures are then so viewed that each eye receives its appropriate image from the proper direction, whereupon the object delineated stands out in relief.

Fusion of the two elements of the stereoscopic picture can take place without the assistance of any instrument, if the eyes are properly directed and focussed, but this comes only with practice. Holding the stereogram well away from the face the eyes are directed to a distant object above and beyond, in order to diverge the axes. Then without converging, the eyes are dropped to the picture, which should spring into relief. It is necessary in moving the eyes from the distant object to the near stereogram to alter their focus somewhat, depending on how near the stereogram is held; and the success of the attempt to fuse the images depends on the observer's ability to maintain the eyes diverged for a distant object while focussing for a near one. Near-sighted people (on taking off their glasses) fuse the stereoscopic images quite easily, since their eyes do not focus on distant objects even when diverged for them. Transparencies are easier to fuse than paper prints, but in any case where a stereoscope is not used the separation of image centers should not be more than that of the eyes.

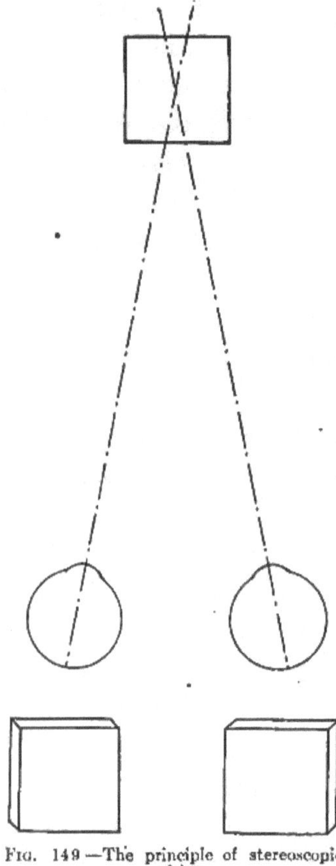

Fig. 149.—The principle of stereoscopic vision.

Stereoscopes.—The easier and more usual method of fusing the stereoscopic images is by a *stereoscope.* The simplest form consists merely of two convex lenses, one for each eye, their centers separated by a distance somewhat greater than that between the eyes. Their function is to bring the stereogram to focus, and, by the prismatic action of the edges of the lenses, to converge the lines of sight

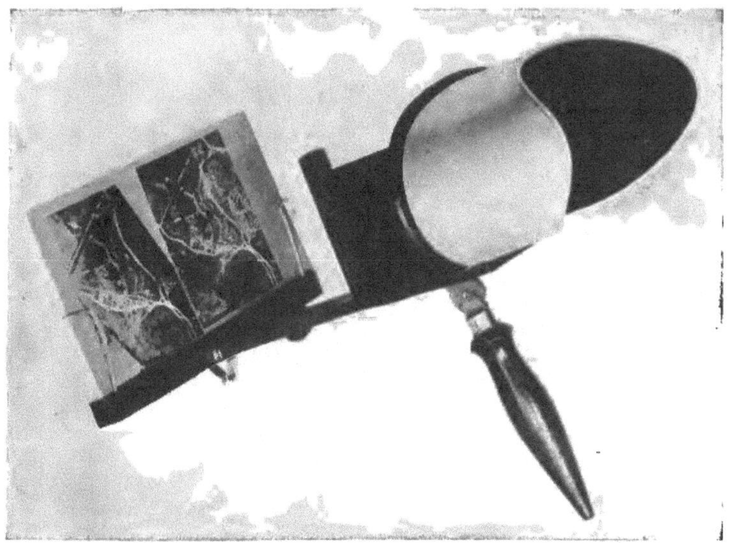

Fig. 150.—Common form of prism stereoscope.

which pass through the centers of the two pictures to a point in space in front of the observer. The two lenses should be mounted so as to provide for the adjustment of their separation to fit different eyes and print spacings. The most common form of stereoscope is that designed by Holmes, for viewing paper print stereograms (Fig. 150). It has prismatic lenses of an appropriate angle to converge pictures whose centers are three inches apart, instead of the lesser distance appropriate to stereograms intended for fusing without an instrument. No adjustment is provided for varying the lens separation, but the print can be moved to and fro for focussing.

Another form of stereoscope, one of the first produced, is the mirror stereoscope (Fig. 152), now used extensively for

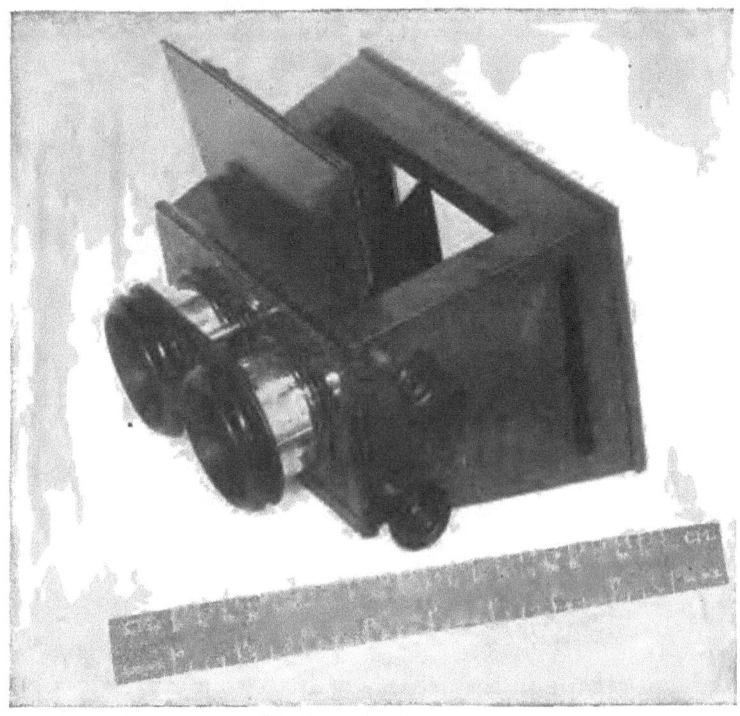

FIG. 151.—Box stereoscope.

viewing stereo X-ray pictures. It consists of two vertical mirrors at right angles to each other, with their edge of contact between the eyes. The two prints to be studied are placed to right and left, an arrangement that permits the use of prints of any size. The convergence point is controlled by the angle between the mirrors. The Pellin stereoscope

(Fig. 153) utilizes two pairs of mirrors in a way to permit the use of large prints. The prints are, however, placed side by side on a horizontal viewing table, which avoids certain difficulties of illumination met with in the simpler mirror form. The box form of stereoscope (Fig. 151) using either prisms or simple convex lenses, is particularly adapted for viewing transparencies, although the insertion of a door at the top provides illumination for paper prints. The Schweiss-

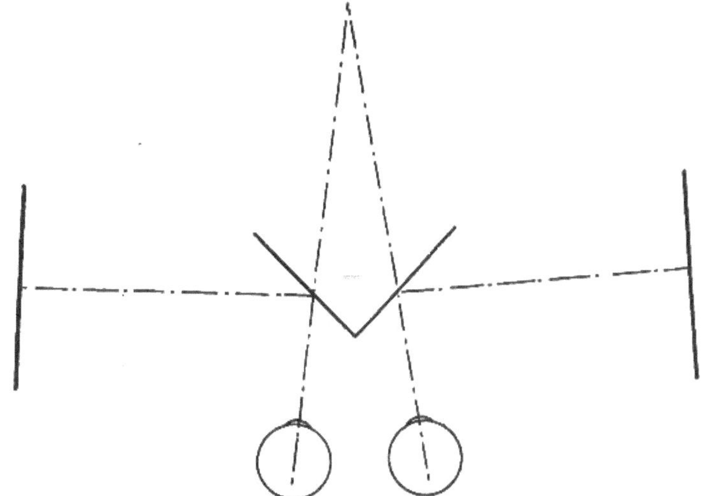

FIG. 152.—Diagram of mirror stereoscope.

guth design (Fig. 154) is intended primarily as an aid to selecting the portions of the prints to be cut out for mounting. The platform on which the pictures rest is composed of two long rectangular blocks, on which are plates of glass raised sufficiently to permit the prints to be slid underneath. The space between the blocks allows the unused portion of the photograph to be turned down out of the way. Prints of any size can thus be moved about until the proper portions for stereo mounting are found. Either block can be moved in its own plane and also to and from the eye, whereby two prints of somewhat different scales can be fused.

The Taking of Aerial Stereograms.—The normal separation of the eyes is altogether too small to give an appearance of relief to objects as far away as is the ground from a plane at ordinary flying heights. In order to secure stereoscopic pairs it is therefore necessary to resort to a method originally

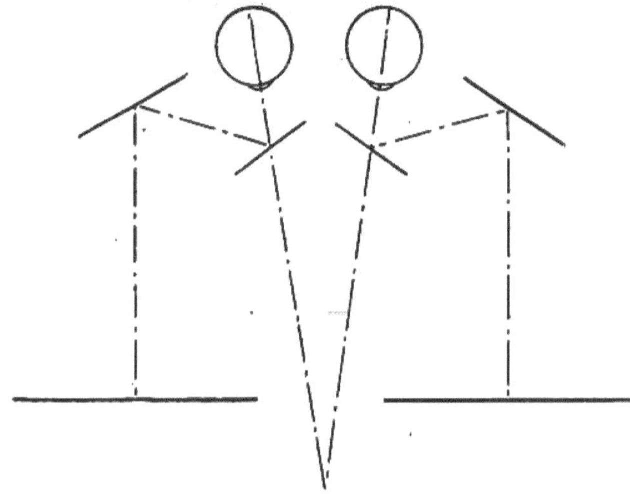

Fig. 153.—Pellin double mirror stereoscope.

employed for photographing distant mountains and clouds. This is to take the two pictures from points separated by distances much greater than the inter-ocular separation—by meters instead of millimeters—corresponding to the positions of the eyes on a veritable giant. In the airplane this is accomplished by making successive exposures as the plane flies over the objective, at intervals to be determined by the speed, the altitude and the amount of relief desired (Fig. 155).

An all important question which arises immediately is: What separation of points of view shall we select? If the exposures are too close together there will be little relief; if too distant the relief will be so great as to be unnatural, even offensive. Obviously we cannot here establish a criterion of *natural* appearance, since the natural appearance

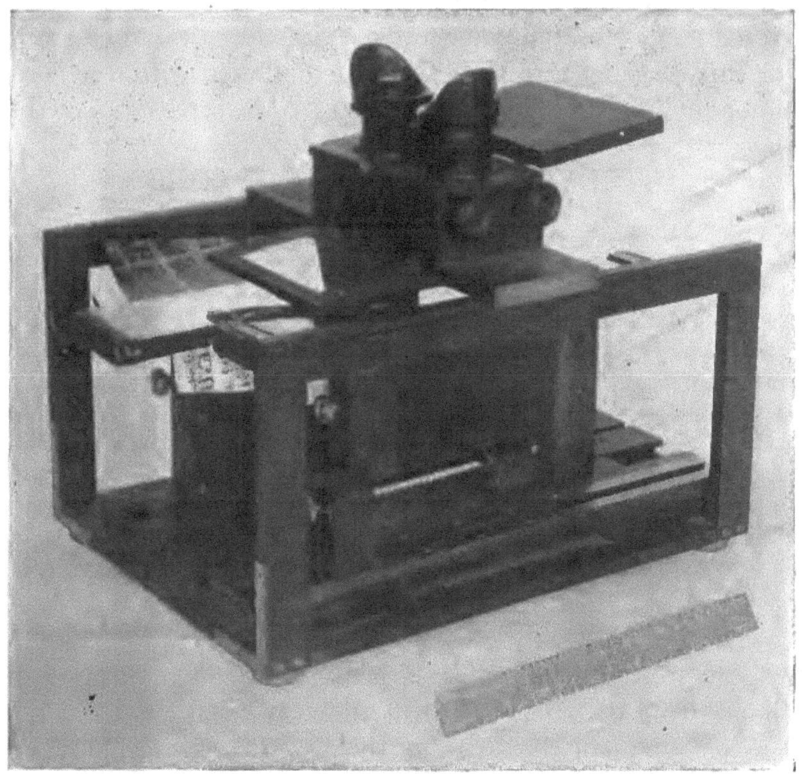

Fig. 154.—Schweissguth stereoscope, used for selecting portions of prints to be mounted.

to ordinary human eyes is devoid of relief. We may, however, define *correct relief* as that obtained when *the apparent height of elevated objects is right as compared with their extension or plan.*

In order to secure this condition it is necessary, first, that each element of the stereoscopic pair be correct in its perspective. This is fortunately an old photographic problem, already well understood. Its solution is to view the photograph from a distance exactly equal to the focal length of the camera lens. Since the normal viewing distance is not less than 25 centimeters, lenses of this focal length at least are requisite for correct perspective. Secondly, it is necessary for correct relief that the two views be taken with a separation equal, on the plane of the plate, to the separation of the

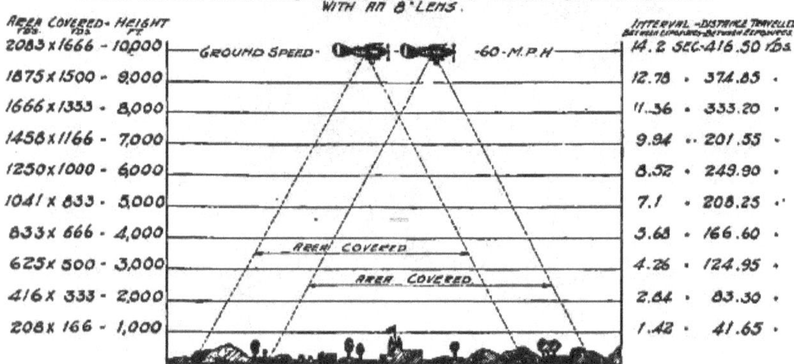

Fig. 155.—Method of taking stereoscopic pictures.

eyes, or 65 millimeters. If d is the interocular distance, a the viewing distance, identical with the focal length of the lens used, and A the altitude, then D, the distance between exposures, is given by the relation—

$$\frac{d}{a} = \frac{D}{A}$$

For $a=25$ centimeters, $\frac{D}{A}=\frac{6.5}{25}$, approximately $\frac{1}{4}$, or the interval between exposures must be a quarter the altitude. With a 50 centimeter lens this becomes $\frac{1}{8}$, and so on. These figures show the fallacy of the suggestion sometimes made that we take stereoscopic pictures by two cameras placed one at the extremity of each wing.

When lenses of more than 25 centimeters focal length are employed, the stereoscope should be one capable of throwing the convergence point farther away than the customary 25 centimeters. In the simple lens type of instrument we can do this by bringing the centers of the lenses closer together, and by making the focus agree with the convergence point by adjustment of the distance between lenses and stereogram. If enlargements are used they should be treated in all respects as originals made by lenses of the greater foci corresponding to the scale of the enlargement.

When all the conditions are covered, the appearance presented in the stereoscope is that of a *model* of the original object at a distance a, and $\frac{a}{A}$ times natural size. If pictures are made at exposure intervals less than those indicated for correct relief, they show insufficient relief. This does not, however, give an unnatural effect, because anything between no relief and "correct" relief appears natural with large objects which are not ordinarily seen in relief by eyes not Brobdignagian. Conversely, stereograms made with too large exposure intervals show exaggerated relief. Yet this is often no objection. It is indeed rather an advantage if we wish to bring objects of interest to notice. Consequently, so long as the exaggeration of relief is not offensive, the permissible limits of exposure interval are pretty large. Actually, the eye tolerates such great deviations from strictly normal conditions that satisfactory stereoscopic effects are obtained for pictures viewed at very different distances from the focal length of the taking lens, and with the axes of the eyes parallel or even diverging, although there is some strain whenever focus and convergence points differ. On the whole, therefore, it may be said that the conditions above laid down

22

for correct relief are only a normal, to be approximated as nearly as is practicable.

Having established the correct relation of taking points for stereos the next problem is how to determine these when in the plane. The simplest way is by means of a *stereoscopic sight.* This consists essentially of two lines of sight (fixed by beads, crosses, or other objects), inclined toward each other at the angle determined by the ratio of the ocular separation to the focal length of the lens. If the back sight is made a single bead or cross, the rest of the stereo sight will consist of two beads or crosses, separated from each other by the ocular distance of 65 millimeters, and distant from the back sight by the focal length of the lens (Fig. 157). The first picture is taken when the object is in line with the forward pointing line of sight, the second when it lies along the backward pointing one. Like other sights, the stereoscopic sight may be attached either to the camera, or if this is fixed in position, to any convenient part of the plane. A very simple sight for vertical stereoscopic photography consists of an inverted V painted on the side of the fuselage, so that the eye can be placed at the vertex and sighted along either leg.

The common method of determining the space between exposures is by the *time* interval. If V is the speed of the plane, and t the desired time interval, we have, from the last equation—

$$t = \frac{D}{V} = \frac{dA}{aV}$$

If $A = 2000$ meters, $d = 65$ millimeters, and $a = 25$ centimeters, and if the plane is traveling 200 kilometers per hour, the time interval must be—

$$\frac{.065 \times 2000 \times 3600}{.25 \times 200{,}000} = 9.4 \text{ seconds}$$

At 1000 meters altitude the interval will be half this, and so on in proportion. If the pictures are taken with a 50 centimeter focus camera, and are hence to be viewed at 50 centimeters convergence distance instead of at 25, the time will again be halved. These relations are clearly shown in the diagram (Fig. 156). Here the left-hand portion shows how

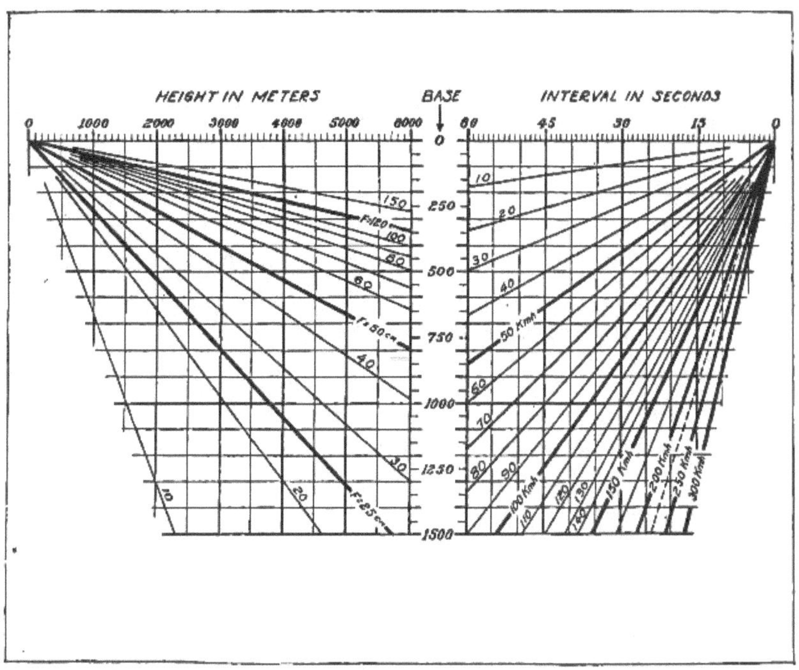

Fig. 156.—Chart for calculating intervals between exposures for stereoscopic pictures.

to find the stereoscopic base line at each altitude for each focal length; while the right-hand portion shows how to translate this into time interval for any plane velocity. The Burchall slide rule (Fig. 130) shows another way to arrange these data in form for rapid calculation.

Plates used for stereoscopic negatives should be at least twice as long as the ocular separation, if correct relief is desired, and the full size of the stereoscope field is to be utilized. This relation follows at once if we consider that we wish to cut from each negative a rectangle 65 millimeters wide, and that the image of the target has shifted 65 millimeters between exposures. If the plate is larger than this there is opportunity to select the view, or to pick several. If the plate is smaller the elements of the stereogram must be narrow strips. This, however, holds only for contact prints.

The ordinary English practice in making stereo negatives is to take successive pictures with an overlap of 60 to 75 per cent. This practice is probably dictated by the 4×5 inch plate, since 60 per cent. overlap on 4 inches means a separation of just over an inch and a half instead of 2¾, but it leaves 2½ inches of picture common to the two negatives. With ¾ overlap the common portion is 3 inches, which permits of cutting 2¾ inch prints, and allows some latitude for irregular motion of the plane or for chance error in calculation of intervals. Data on the basis of ¾ overlaps for a 4-inch plate are shown in connection with Fig. 155 which shows in diagrammatic form the variation of exposure interval with height, together with other points of interest.

Elevation Possible to Detect in Stereoscopic Views.—Can the actual difference in elevation be discovered by the use of stereoscopic views? An approximate idea may be obtained from the following considerations: Suppose we have two small point-like objects, one above the other, such as a street lamp globe and the base of the lamp pillar. In a view taken from directly overhead these will be superposed, and so will not be capable of separation. But, as the point of view is shifted sideways, the two objects separate, until a point is reached where they can just be distinguished as double. When this condition holds for either picture of the stereoscopic pair it will be possible to obtain stereoscopic relief.

Now the separation which can just be distinguished is commonly assumed to be one minute of arc. This angle corresponds to about $\frac{1}{3400}$ the distance from the eye to the object. If the object is assumed at a distance a from the face, and on a line with one of the eyes, which are separated by the distance d, then (all angles being small) the object must be of height $\frac{a}{d}$ times the horizontal distance which corresponds to one minute. For 25 centimeters' viewing-distance this quantity is about 4, so that the least perceptible elevation is $\frac{4}{3400}$ or about $\frac{1}{900}$. The stereogram having been made under conditions giving correct relief, this fraction is also the fraction of the altitude of the plane when the photograph was taken which may be detected. An object as high as a man (6 feet) should be visible as a projection in a stereoscopic view taken at $6 \times 900 = 5400$ feet. This relation—$\frac{1}{900}$—holds (irrespective of the focal length of the lens), as long as the conditions for correct relief are maintained.

Stereoscopic Aerial Cameras.—Cameras for aerial stereoscopic photography need in no way differ in construction from those made for mapping or spotting, provided only they permit exposures to be made at short enough intervals. The addition of special sights, as already discussed, constitutes the only real difference between single view and stereoscopic aerial cameras. But even without such sights ordinary aerial cameras are applicable to stereo work by the usual procedure of determining the exposure spacing by time.

One scheme employed for taking low stereos, where the interval is only two or three seconds, is to mount two cameras in the plane, exposing them one after the other at the correct interval. Another method which has been tried with success is the use of a double focal-plane shutter in a single lens camera (Fig. 157). The two shutters are side by side, with their slots parallel to the line of flight. To take a stereo negative we expose first the shutter nearer the tail of the plane, and then the other, after an interval which can be calculated from the speed and altitude, or, better, determined by a stereoscopic sight. The two views are thus obtained on a single plate. Prints from these negatives are transposed

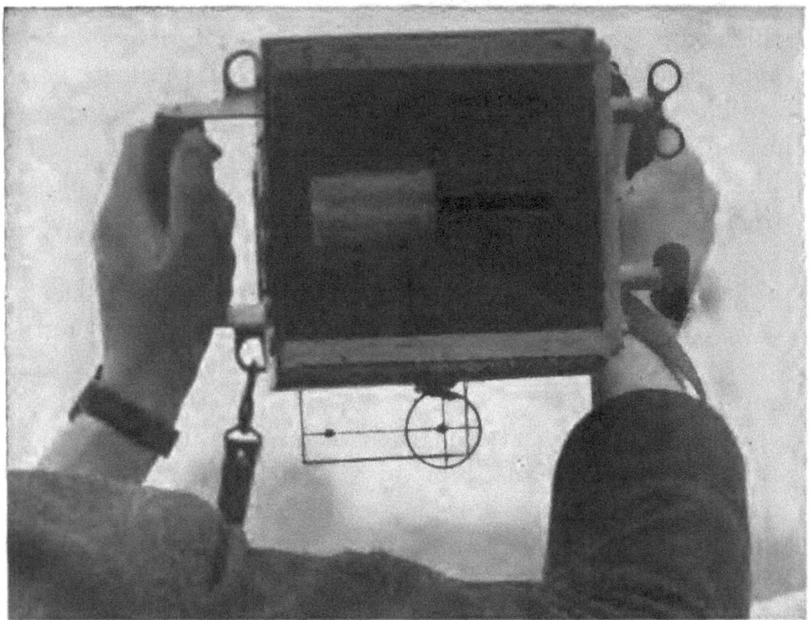

FIG. 157.—Aerial hand camera fitted with two complementary shutter slits and double sight, for stereoscopic photography.

right and left, and, if the prints are viewed in an ordinary stereoscope, have to be cut apart and transposed for mounting, or else this may be done to the negatives.

In this connection attention may be drawn to an alternative method of viewing stereograms, which may be used on transposed prints—a method which needs no instrument, and so has sufficient advantage to even warrant mounting ordinary stereoscopic pairs in the transposed position for observation. This method consists in crossing the optic axes, in the fashion illustrated in Fig. 158. A finger is held in front of the face in such a position that the left stereogram element and the finger are seen in line by the right eye; the right element and the finger by the left eye. The proper position is found by alternately closing each eye, and advancing or retracting the finger. Then both eyes are opened and converged on the finger tip, which is thereupon dropped, leaving the picture standing out in relief. An opportunity to try this method is afforded by Fig. 159.

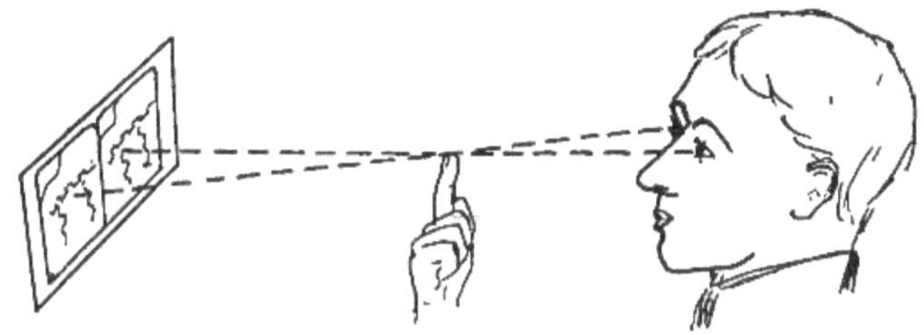

Fig. 158.—Method of fusing transposed stereoscopic images by crossing the optic axes.

Stereo Obliques.—The theory of making oblique stereo pictures is identical with that of other stereos. The only problem peculiar to obliques is that of making the exposures at short enough intervals apart. This problem is due largely to the fact that oblique views are ordinarily taken from low altitudes, for the purpose of "spotting" particular objects, rather than for mapping the gross features of an extended area. The same problem of how to secure a short exposure interval is met with when we attempt to take vertical stereos from a low altitude, but as already discussed, it is much preferable from the pictorial standpoint that pictures of definite small objectives be made obliquely.

Another reason for taking stereo obliques from points but

little separated is of some interest in connection with the discussion above given of "correct" and "natural" relief. When the relief is "correct" the object appears, as already stated, to be a small model in its true proportions, standing at the convergence distance. When the eyes are converged to a small object 25 to 50 centimeters away all objects beyond are hopelessly transposed and confused. This does not happen when we look at large distant objects, since their background is at a distance effectively but little beyond them. As a result, when a stereo oblique is made in "correct" relief of such an object as the Washington monument with buildings beyond, the confusion of the background presents an appearance entirely contrary to our visual experience with objects as large as the neighboring buildings are known to be. This effect may be avoided by choosing a uniform background such as grass, or by taking the pictures very much closer together, at the expense of "correct" but at a gain in "natural" relief.

Stereo obliques can of course only be made with any facility by laterally pointing cameras. From the calculations already given it appears that a "correct" stereo oblique of an object 500 meters away will mean exposures only two or three seconds apart, too short an interval for any of the ordinary plate-changing and shutter-setting mechanisms; and the case is even worse should less relief be desired. One solution of this problem has been the use, already mentioned, of two cameras mounted together, either side by side or one over the other, with separate shutter releases. Both releases may be controlled by the observer, using a sight, or else pilot and observer may work in harmony as has been recommended in the English service, where the pilot releases one shutter and the observer counts time from the instant he sees the first shutter unwind and releases the second.

A very satisfactory apparatus for the taking of stereo obliques consists of a 10-inch focus hand-held camera (Fig. 157), provided with a two-aperture focal-plane shutter. The right-hand half of one curtain aperture is blocked out, the left-hand half of the other. The first pressure on the exposing lever exposes one-half of the plate, the second the other. A

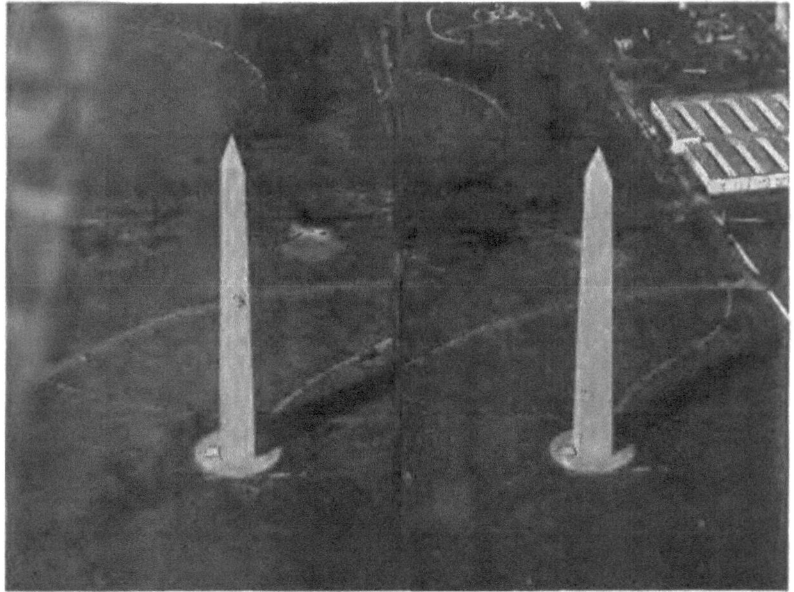

Fig. 159.—Oblique stereogram made with stereoscopic aerial camera (Fig. 157). To be viewed by crossing the optic axes (Fig. 158).

stereoscopic sight of the type already described is placed on the bottom. To make an oblique stereo negative the camera is held rigidly by resting the elbows on the top of the fuselage and the first exposure is made when the object comes in line with the rear sight and the leading front sight. The eye is then moved so as to look along the line of the rear sight and the following front sight, and when the object is again in alinement the second pressure is given the exposing lever.

Fig. 159 shows a stereo oblique made by this camera. The elements are transposed right and left, and the stereogram may be viewed by crossing the optic axes as shown in Fig. 158, or the two pictures may be cut apart and remounted.

The Mounting of Aerial Stereograms.—The first step in making the printed stereogram is to select two pictures taken on the same scale, but from slightly different positions. These may be two chosen from a collection made for other purposes, or else a pair taken at distances calculated to fit them for stereoscopic use. The next step is to mark the center of each picture, either with easily removed chalk or witn a pin point. They are then superposed, and afterward carefully moved apart by a motion parallel to the line joining their centers when superposed. The final step before mounting is to mark out and cut the two elements, their bases being parallel to the line of centers, their horizontal length the distance between the optic axes of the stereoscope (or as near this as the size of the prints will permit). They are then mounted on a card, with their centers separated by approximately 65 millimeters. The right-hand view is the one showing more of the right-hand side of objects, and *vice versa*. This process of arranging, cutting, and mounting is shown clearly in Fig. 160. In this case the stereoscopic elements lie symmetrically about the line joining the centers of the original prints. This is not necessary, as they may be selected from above or below this line so long as their bases are parallel to it. A simplification of this method consists in superposing the two prints, laying over them a square of glass of the size to which they are to be cut, then turning it so that a side is parallel to the line of centers, and cutting around it through both prints with a sharp knife. The principle and results are of course the same with both methods.

If large numbers of stereoscopic prints are required it is

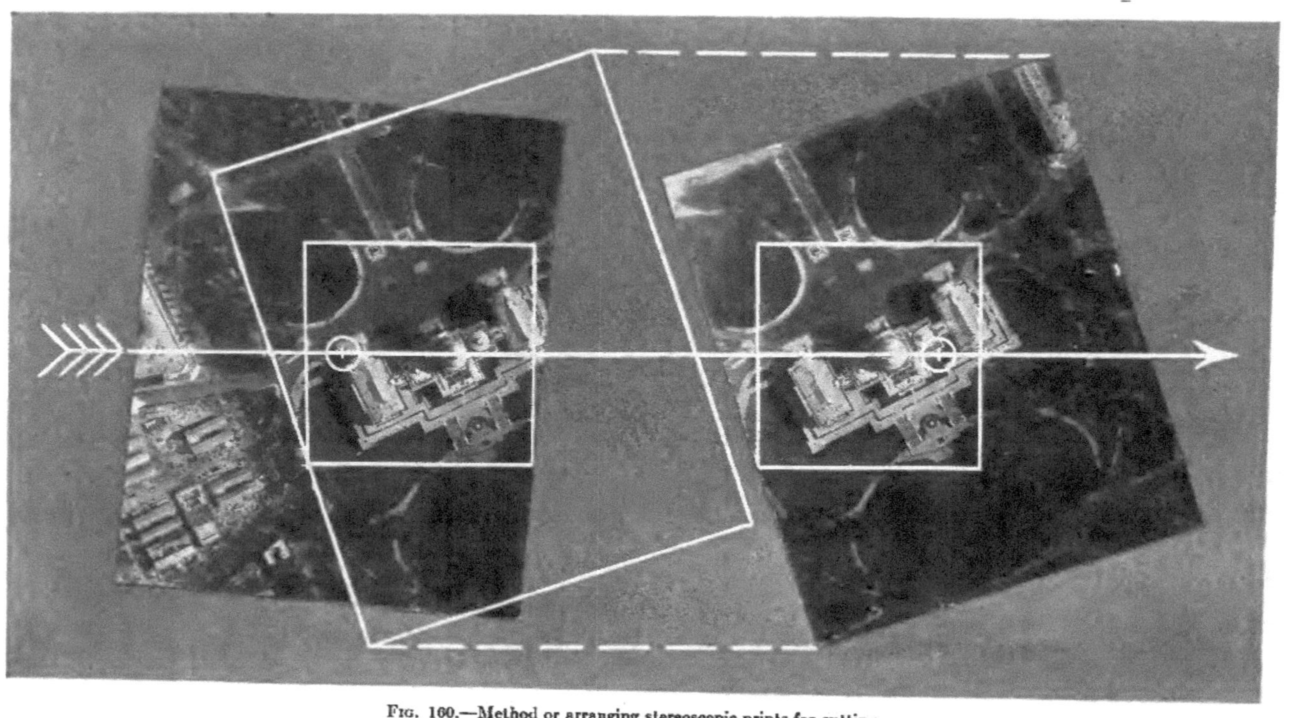

FIG. 160.—Method or arranging stereoscopic prints for cutting.

necessary, for economy of time, either to photograph a finished stereogram and make prints from this copy negative, or to set up special printing machines. Under the general discussion of printing devices a stereoscopic printer is described (the Richard) in which the two negatives are placed so that stereo prints can be got by two successive printings on one sheet of paper.

Uses of Stereoscopic Aerial Views.—Attention has already been called to the characteristic flatness of the aerial view. Neither the picture on the retina nor that on the photographic plate affords any adequate idea of hills and hollows. Unless shadows are well defined, small local elevations and depressions cannot be distinguished from mere difference in color or marking. Even in the presence of shadows it is often only by close study that differences of contour are noticeable. But with stereoscopic views these features stand out in a striking manner. Taking our illustrations from military sources, we may note the use of stereoscopic pictures to detect undulations of ground in front of trenches (Fig. 161). They reveal the hillocks, pits, small quarries, streams flowing behind high banks, and other features which make the attack hard or easy. Commanding positions are shown, the boundaries of areas exposed to machine-gun fire, and the defilades where the attackers may pause to reform. Concrete "pill boxes" are located in the midst of shell holes of the same size and outline, and can be differentiated from them.

Railway or road embankments and cuts can be detected and studied to extraordinary advantage in stereoscopic pictures. Thus what appears to be a mine crater on a level road, easily driven around, may be a gap blown in an embankment, a serious obstacle indeed. Bridges, observation towers and other elevated structures jump into view in

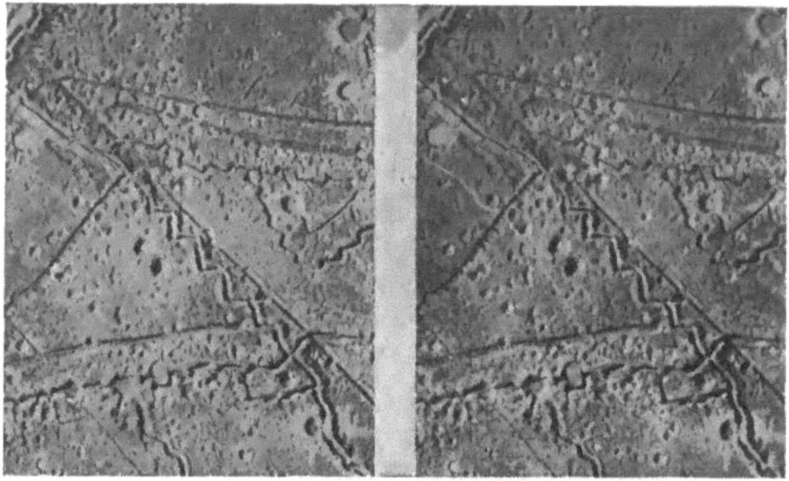

Fig. 161.—Typical stereogram of military detail. Fuse by looking at a distant object over the top of the page, and quickly dropping the eyes to the print.

the stereoscope when often they have entirely eluded notice in the ordinary flat picture. Once presented in relief, camouflaged buildings or gun emplacements, however carefully painted, are ridiculously easy to pick out.

Practical peace-time applications of stereoscopic views can easily be foreseen following the lines of war experience. Such, for instance, would be the study of proposed railway or canal routes. A series of stereograms would obviate the necessity of contour surveys, at least until the exact route was picked and construction work ready to start.

Apart from their utilitarian side, however, stereoscopic views have very great pictorial merit. Stereoscopic pictures of cathedrals, public and other large buildings, have often great beauty, and afford opportunities for the study of form given by no other kind of representation, short of expensive scale models. They may very well lead in the near future to a revival of the popularity of the stereoscope.

Impression of Relief Produced by Motion.—An appearance of solidity can be obtained in *moving pictures* by the simple expedient of slowly moving the camera laterally as the pictures are taken. As an illustration, if the moving picture camera is carried on a boat while structures on the shore are photographed, when these are projected on the screen they appear in relief, due to the relative motion of foreground and background. As relief of this sort is not dependent on the use of the two eyes, it demands no special viewing apparatus. This idea has been utilized to a limited extent in ordinary moving picture photography by introducing a slow to-and-fro motion of the camera, but this can hardly be considered satisfactory, since this motion is so obviously unnatural.

In moving pictures made from the aiplane the normal rapid motion of the point of view is ideal for the production of the impression of relief in the manner just described. For instance, in moving pictures of a city made from a low flying plane, the skyscrapers and spires as they sweep past stand forth from their more slowly moving background in bold and satisfying solidity. In fact, such pictures probably constitute the most satisfactory solution yet found of the vexing problem of "stereoscopic" projection. No better medium can be imagined for the travel lecturer to introduce his audience to a foreign city than to throw upon his screen a film made in a plane approaching from afar and then circling the architectural landmarks at low altitudes.

CHAPTER XXIX

THE INTERPRETATION OF AERIAL PHOTOGRAPHS

Oblique aerial photographs if on a large enough scale are even easier to interpret than are ordinary photographs taken from the ground, since they practically preserve the usual view, and add to it the essentials of a plan. With verticals, however, this is far from the case. In them all natural objects present an appearance quite foreign to the ordinary mortal's previous experience of them. This may be easily demonstrated by taking any aerial view containing a fair amount of detail and trying systematically to identify each object. A necessary preliminary to doing this accurately is acquaintance with and study of the ground photographed, or of similar regions, and of objects of the same character as those likely to be included.

The interpretation of military aerial photographs is of such importance, and has become such an art, that it is the function of special departments of the intelligence service. Extended courses in the subject are now given in military schools. This instruction must cover more than the interpretation of aerial photographs as such. General military knowledge is essential, so that not only may photographed objects be recognized, but the significance of their appearance be realized. Whether attack or retreat is indicated; whether a long range bombardment is in preparation, or a mere strengthening of local defences.

The natural difficulties of interpreting aerial views are enormously increased by the unfamiliar nature and frequently changed character of the military structures, and particularly by the attempts made to conceal these from aerial observation by selection of surroundings and by camouflage. The small scale of the photographs, in which a machine gun shows as a mere pin point, adds to the uncertainty, with the net result of making interpretation a task of minute study and deduction worthy of a Sherlock Holmes.

Little detailed information on interpretation can be profitably written in a general treatise, partly because the illustrations available are of a highly technical military character, partly because original photographs instead of half-tone reproductions are practically imperative for purposes of study. Nevertheless some general instructions, applicable to any problem of interpretation, may be given, as well as a few illustrations, drawn from military sources, which will serve to show the detective skill necessary.

First of all it is important that the print or transparency be held in the right position. The shadows must always fall toward the observer; otherwise, reliefs will appear as hollows and hollows will show as hills. The reason for this is that the body ordinarily acts as a shield, preventing the formation of shadows except by light falling toward the beholder. Thus in Fig. 162 the slag heap looks like a quarry when the shadows fall away from one. The necessity for proper direction of shadows is, it may be noted, in conflict with the ordinary convention for the orientation of maps—at least in the northern hemisphere. A city map, made by sunlight falling from the south, presents its shadows as falling away from the observer, when it is mounted with its north point at the top, as is customary. As a consequence buildings in aerial photographic mosaics of cities occasionally look sunken instead of standing out.

The relation between the shape of the shadow and the object casting it must be well learned. This is a part of the

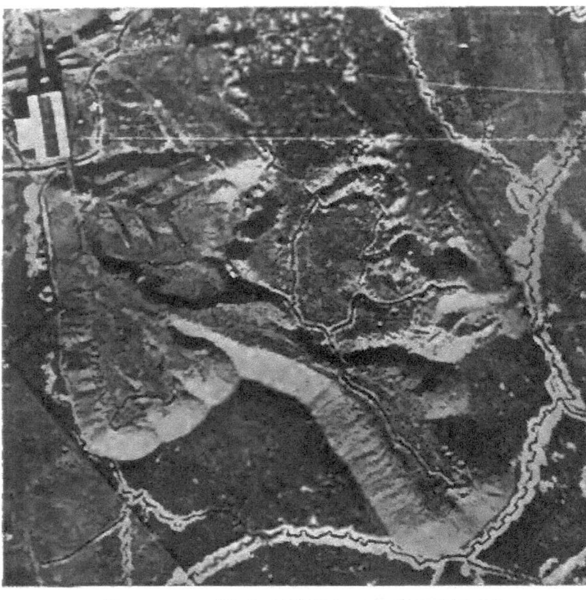

Wrong way. Shadows falling away from observer.

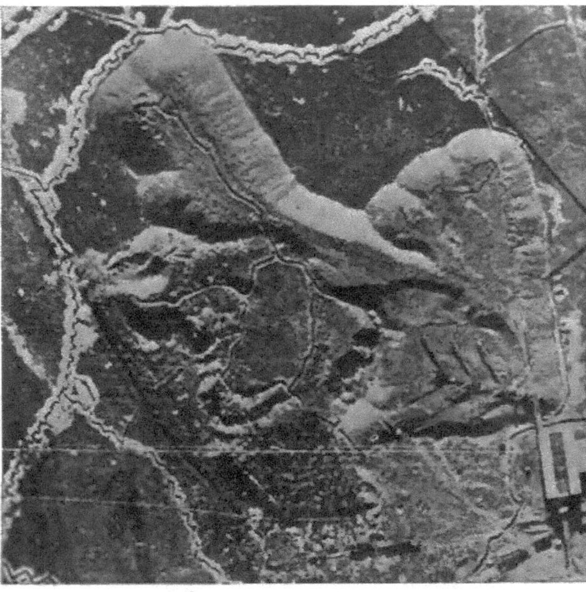

Right way. Shadows falling toward observer.

Fig. 162.—The wrong way and the right way to hold a photograph for interpretation.

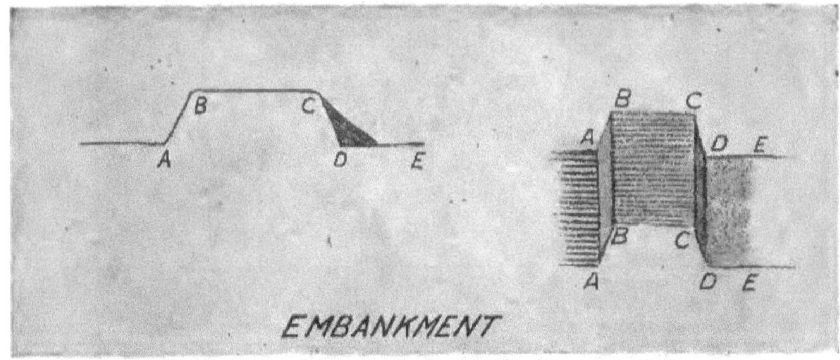

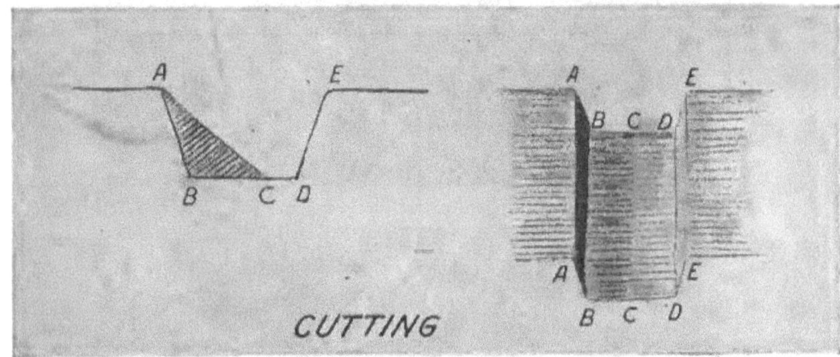

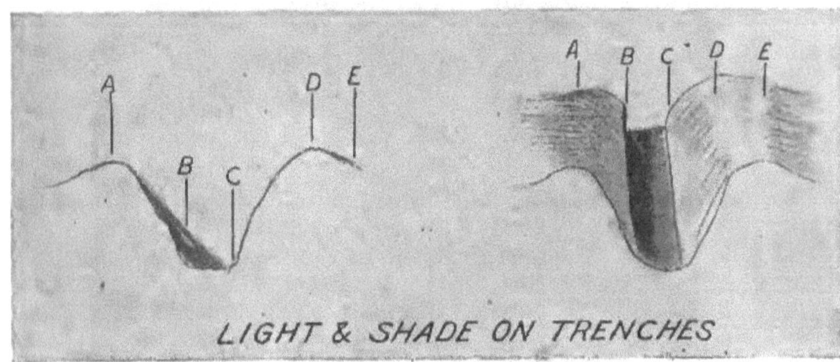

FIG. 163.—Guide to interpretation of trench details.

training of every architectural draftsman, but the appearance of shadows from above has not heretofore been a matter of importance. The difference between high and low

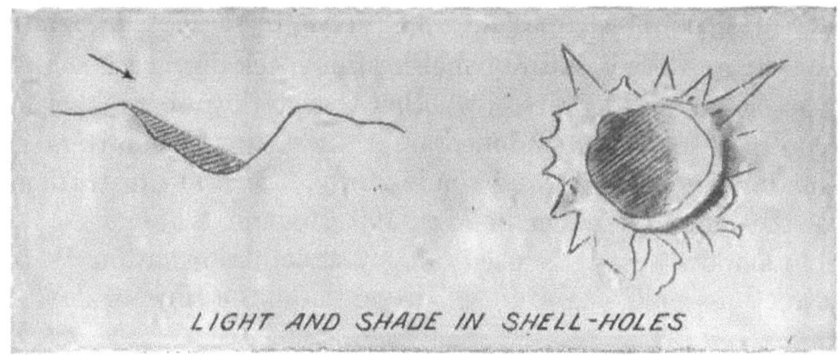

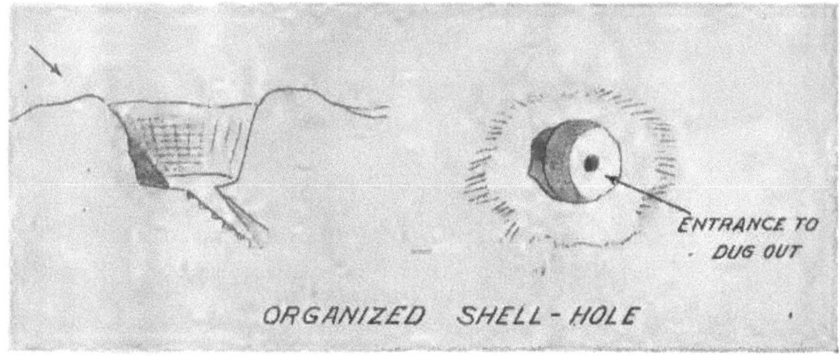

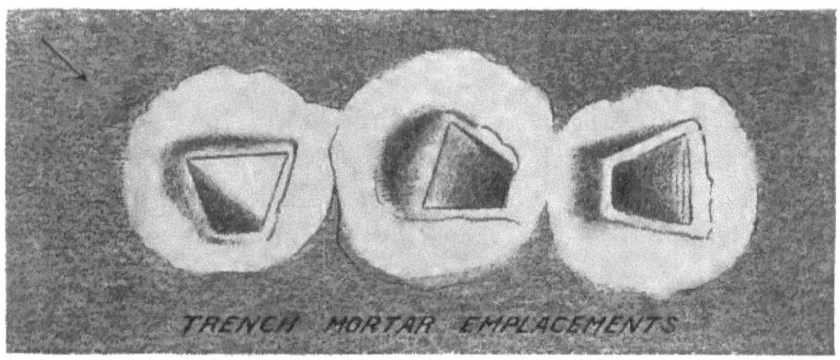

FIG. 164.—Guide to interpretation of shell holes and other pits.

trenches, between cuttings and embankments, between shell holes, occupied or unoccupied, and "pill boxes," must be detected largely from the character of the shadows. Which elevations and depressions are of military and which of merely accessory nature, whether this black dot is a machine gun or a signaling device, whether that dark spot is an active gun port or an abandoned one—these are all matters of shadow and of light and shade study. Several illustrations of these points appear in Figs. 163, 164 and 165.

Shadows may be used to get exact information as to directions and magnitudes. If we know the time of day at

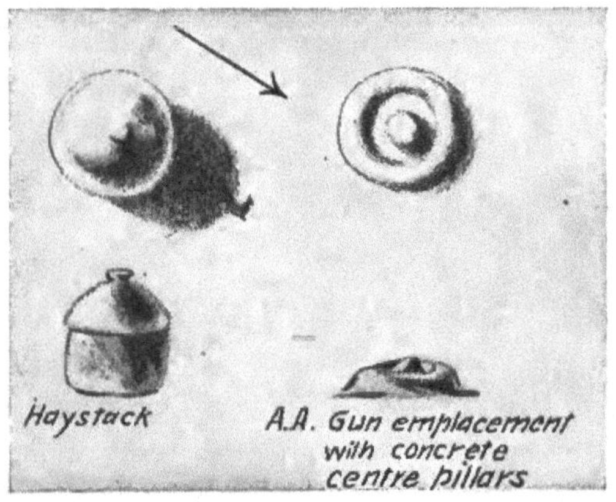

FIG. 165.—Illustrating the importance of distinguishing between objects of similar appearance but different military importance.

which a picture is taken, the direction of the shadows will give the points of the compass. A chart for doing this is shown in Fig. 166. The length of a shadow is a measure of the height of the object casting it, and the exact relation between the two dimensions is determined by the day and hour. Fig. 167 embodies in chart form the values of this relationship for all times of the year and day, while Fig. 168 shows the kind of picture in which shadow data could be utilized to great profit.

Minute changes, both in light and shade and in position, must be watched for with great care. Naturally growing foliage and the cut branches used for camouflage differ in color progressively with the drying up of the leaves. Hence a mere spot of lighter tone in a picture of a forest, especially

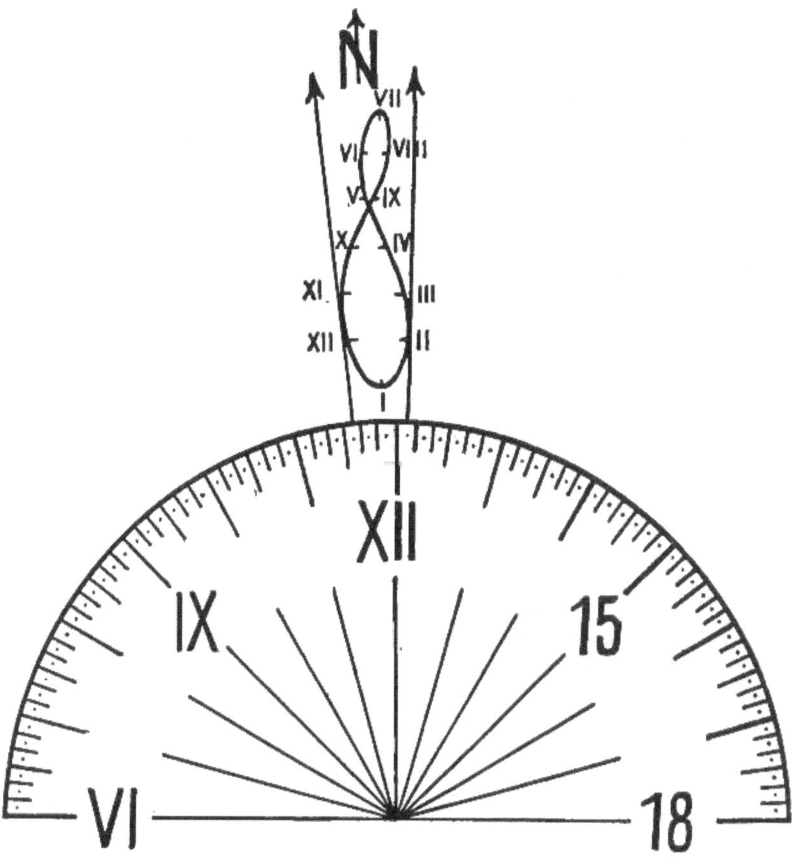

Fig. 166.—Location of true north from direction of shadows. Place the dial on the photograph, the hour line corresponding to the time it was taken being pointed in the direction of the shadows. North lies between the two arrows, the exact direction being obtained by joining the center of the dial to the point on the figure of eight corresponding to the date on which the picture was taken. (Numbers on figure of eight represent the 1st of the month).

if the picture is taken through a deep filter, becomes instant object for suspicion. The complete study of any position calls for photographs of all kinds—verticals, obliques, and stereos. Stereoscopic views are the worst foe to camouflage. A bridge painted to look like the river beneath is labor thrown away if the stereo shows it to be a good ten feet above the real river!

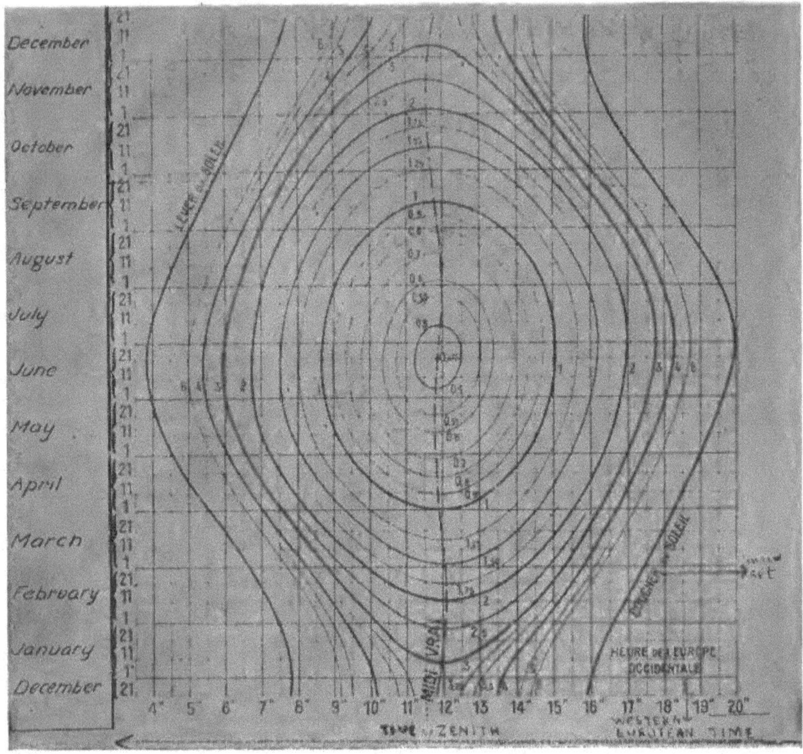

FIG. 167.—Length of shadow of object one meter high, at different times of the day and year, for latitude of Paris.

A few illustrations of the more ordinary and obvious objects whose detection is the subject of aerial photography are shown in accompanying figures. Fig. 169 pictures a typical trench system, with barbed wire. The trenches show as narrow castellated lines, from which run the zigzag lines of communicating trenches, saps, and listening posts. The

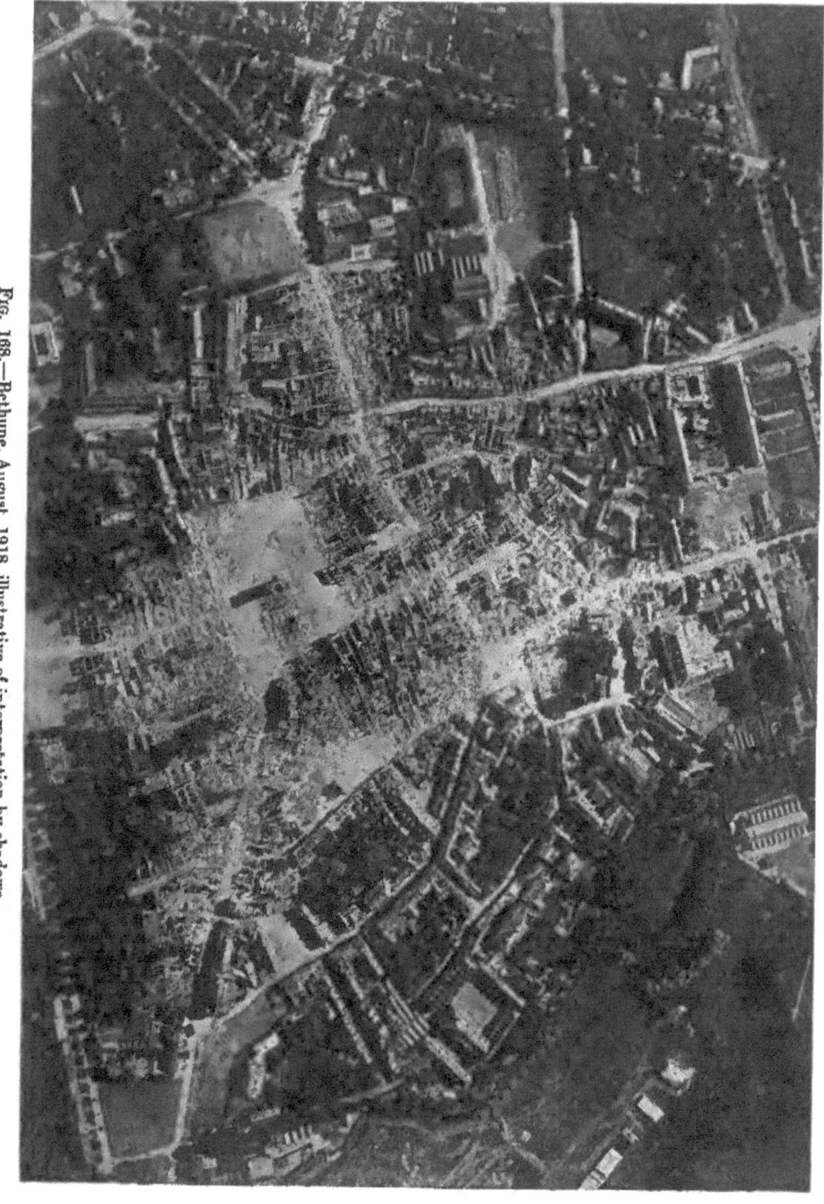

FIG. 168.—Bethune, August, 1918, illustrative of interpretation by shadows.

minute pockmarks behind the main trench lines are shell holes and machine gun pits. The barbed wire shows as double and triple gray bands, intricately criss-crossed at

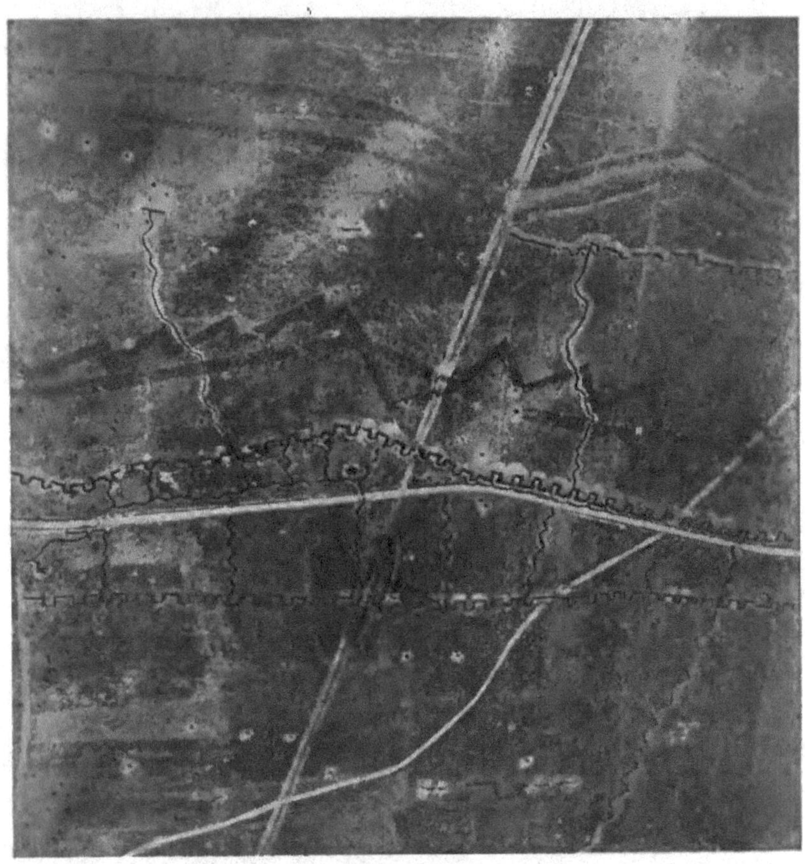

FIG. 169.—Typical trench photograph showing first and second lines, communicating trenches, listening posts, machine gun emplacements, and barbed wire.

strategic points. Another form of defence, intended for the same purpose as the barbed wire of the western front, is that furnished by overthrown trees in forest regions. Fig. 170 reveals a mountain fortress surrounded by a zone of felled trees, and indicates in striking manner the value of the information a single aerial photograph may furnish to an attacking force. Fig. 123 shows on a comparatively large scale opposing trench systems in which a natural obstacle—a river—separates the adversaries. Nicks and dots indicate machine guns to the skilled eye, and several rectangular

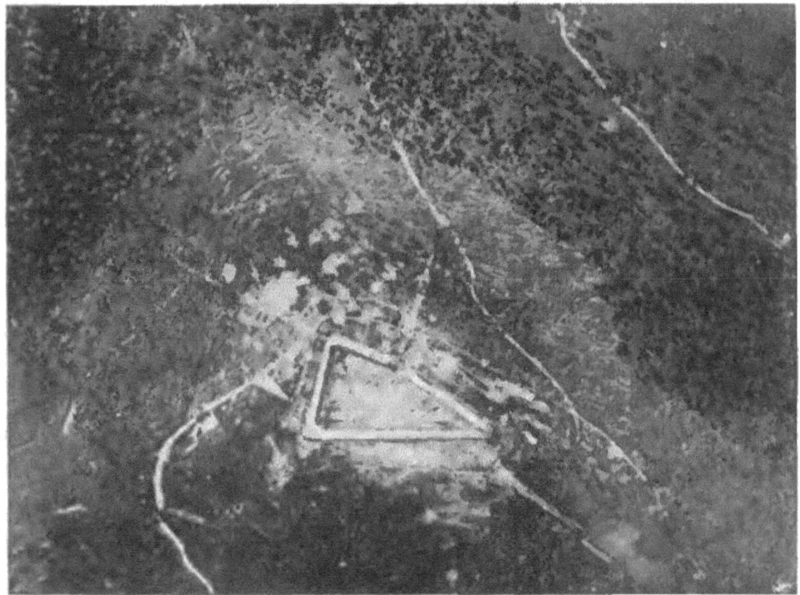

Fig. 170.—A mountain fort surrounded by felled timber.

structures are revealed as concrete buildings which have survived unscathed the shell fire which has obliterated, and caused to be rebuilt, nearly every other element of the trench system.

Isolated battery emplacements (Fig. 171) must be carefully studied to learn if they are in use. The chief indication is given by the paths the men make in going and coming; these show as fine light lines, obliterated by growing vege-

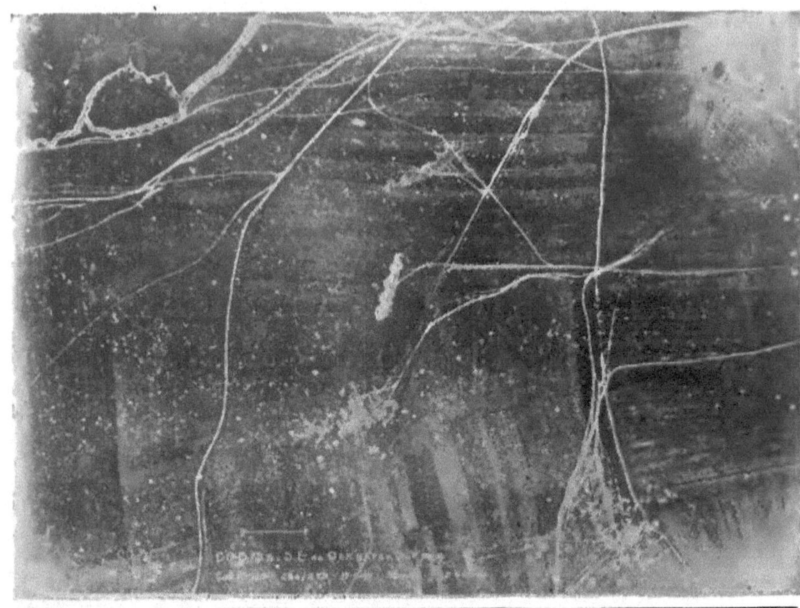

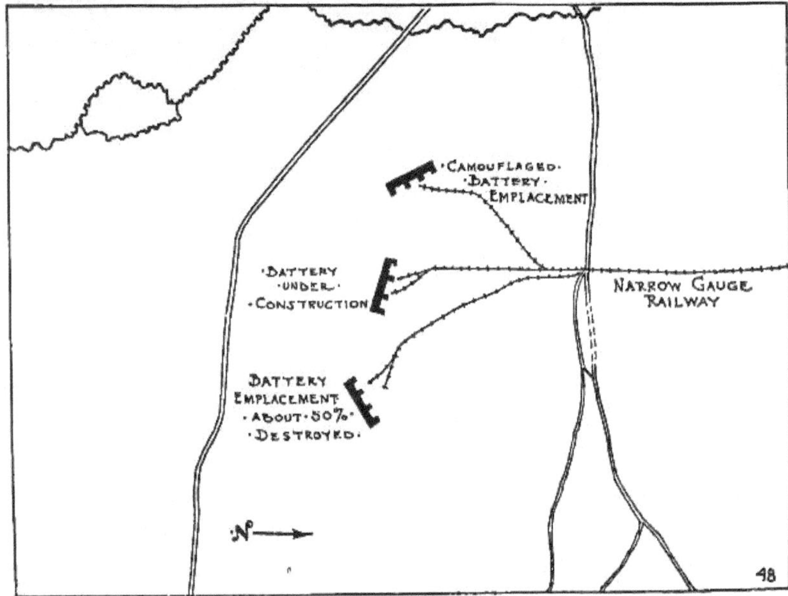

FIG. 171.—Three stages in the life of a battery.

tation if long disused. Another indication is the blast marks in front of the gun muzzles; occasionally the sensitive plate will catch the actual puff of smoke as the gun is discharged.

Railways of various gauges show as thin lines, crossed by ties, and exhibiting the characteristic curves and switches. They are particularly important to detect because they naturally lead to guns or supplies of importance. Abandoned railways from which the rails and ties have been removed leave their marks on the ground and must be carefully distinguished from lines in actual operation.

Aviation fields are easily recognized by the hangars, often with "funk hole" trenches alongside for the men to take shelter in during air raids (Fig. 172). Other characteristic features are the "T" which shows the direction of the wind to the returning pilot, and of course the planes themselves, standing on the ground. But the field may be inactive, and the planes merely canvas dummies, so that to pierce the disguise, all paths, ruts, and other indications of activity must be minutely studied.

Overhead telegraph and telephone lines are revealed when new by a series of light points (Fig. 174), where the posts have been erected in the fresh turned earth. Later, when the fields through which they pass are cultivated, the post bases show as islands left unturned by the plow. In winter the wires reveal their position by black lines in the snow caused by drippings. Buried cables are indicated while building by their trenches, and for some time afterward by the comparatively straight line of disturbed earth.

Just as the detective of classic story makes full use of freshly fallen snow to identify the footprints of the criminal, so does the aerial photographer utilize a snowfall to pierce the enemy's attempts at deception. Tracks in the snow show which trenches or batteries are in actual use. Melting

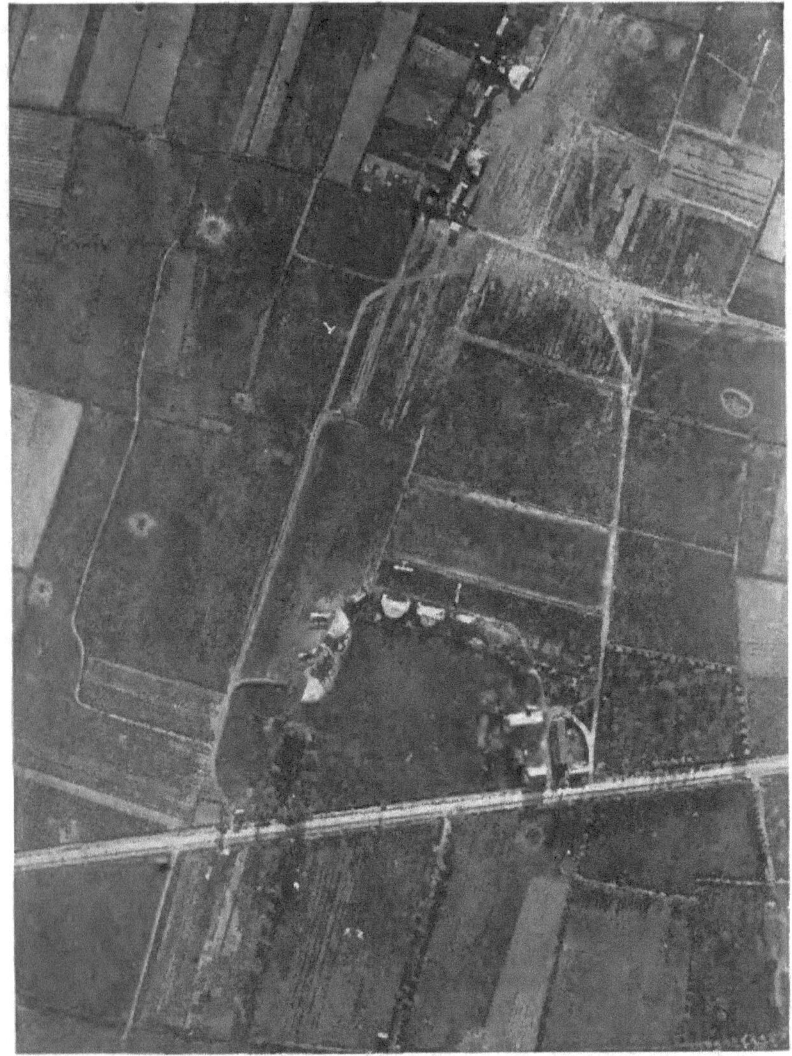

Fig. 172.—Aviation field, showing hangars, planes, landing "T" and refuge trench.

of the snow in certain places may mean fires in dugouts beneath. Black smudges in front of trench walls show where guns are active. Guns, wire and other objects, however

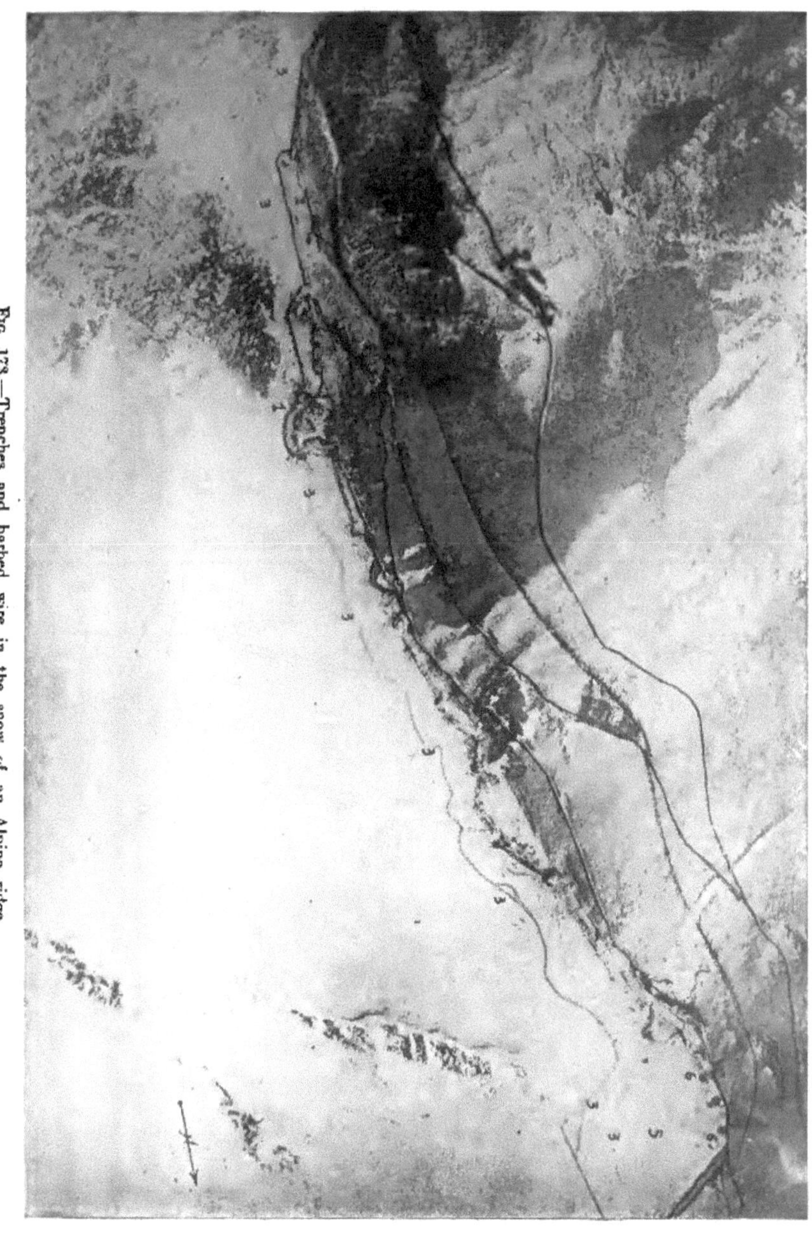

Fig. 173.—Trenches and barbed wire in the snow of an Alpine ridge. Italian Air Service photograph.

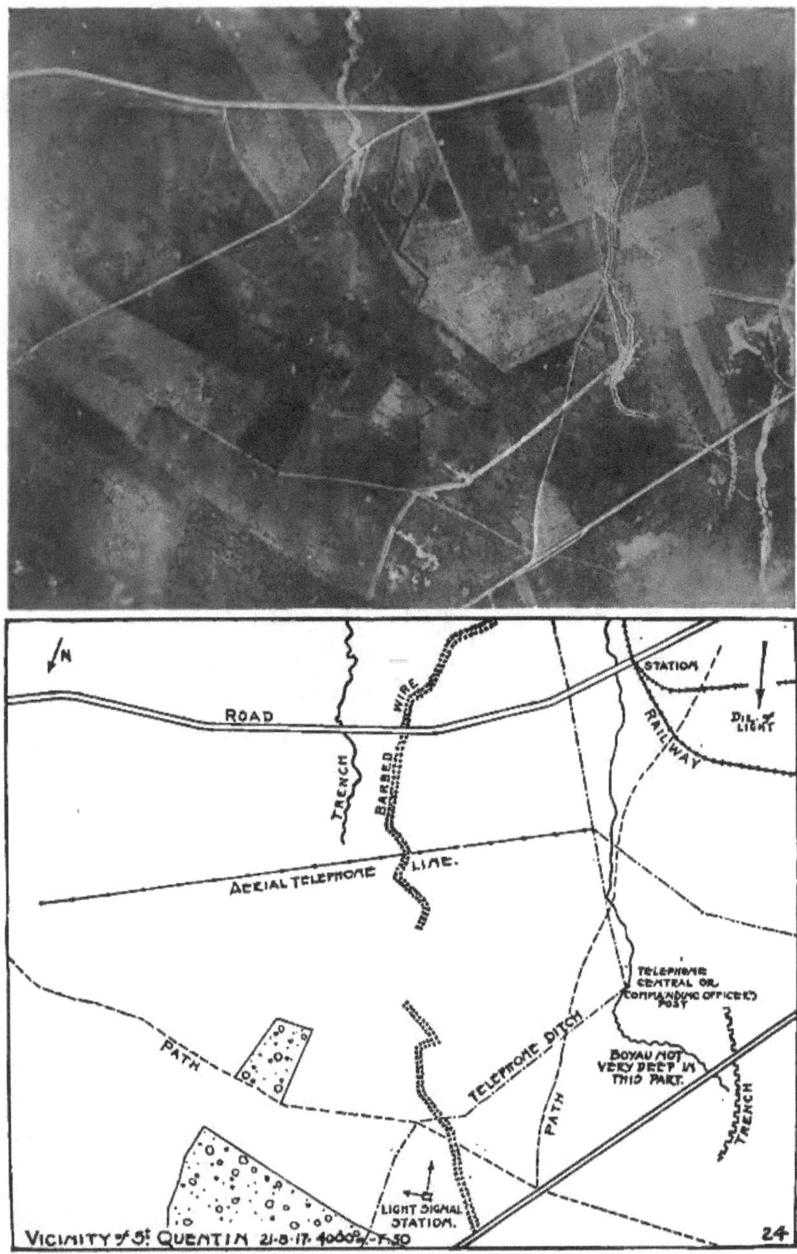

FIG. 174.—A fully interpreted aerial photograph.

carefully painted to match the gray-green earth, stand out in violent contrast to this new white background (Fig. 173).

After the aerial photograph has been interpreted the results of the interpretation must be made available to the artilleryman or the attacking infantryman. This may be done by legends marked directly on the photograph. Another method is to mount over the photograph a thin tissue paper or oilskin leaf, with the interpretation marked on it. A yet more elegant method consists in outlining all the chief features of the photograph in ink, writing in the points of importance in interpretation, and then bleaching out the photograph with potassium permanganate solution. Photographic copies of the resultant line drawing are then mounted side by side with the original photograph. Fig. 174, which shows a fully interpreted photograph, is an example of this kind of mounting.

CHAPTER XXX

NAVAL AERIAL PHOTOGRAPHY

The problems of naval aerial photography are qiute different from those of military aerial work, and on the whole they are more simple. At the same time, photography has played a considerably less important part in naval aerial warfare than in land operations. Photography as a necessary preparation for attack has not figured in naval practice nearly so much as have the record and instruction aspects. To some extent this is due to the nature of the naval operations in the Great War, to some extent to the limitations of ceiling and cruising radius of the naval aircraft.

A photographic reconnaissance, preceding and following a bombardment of shore batteries; a photographic record of the ships at anchor, as at Santiago; a photograph of the forts defending a channel, as at Manila; photographs, quickly developed and printed, of an approaching fleet—all these are possibilities of great usefulness in naval warfare between contestants both of whom "come out" and carry the struggle to the enemy's gates. But in the recent war the use of the submarine, operations under cover of fog, the striving for "low visibility," and the considerable distances to be traversed to reach the enemy lairs, have conspired to limit the development of photography as a major aid to naval combat. Probably when the whole history of the conflict is told we shall learn that the Zeppelins which cruised over the North Sea, keeping the Allied fleet under observation, had a regular routine of photographic work. In the Italian zone, where much of the enemy territory and several important naval centers lay at only short distances over the Adriatic, the naval photographic service more nearly rivalled that of the army than in the English, French and American zone of activity in the Channel and North Sea.

The majority of the photographs made in the British service were obliques, taken by short focus (6 to 10 inch)

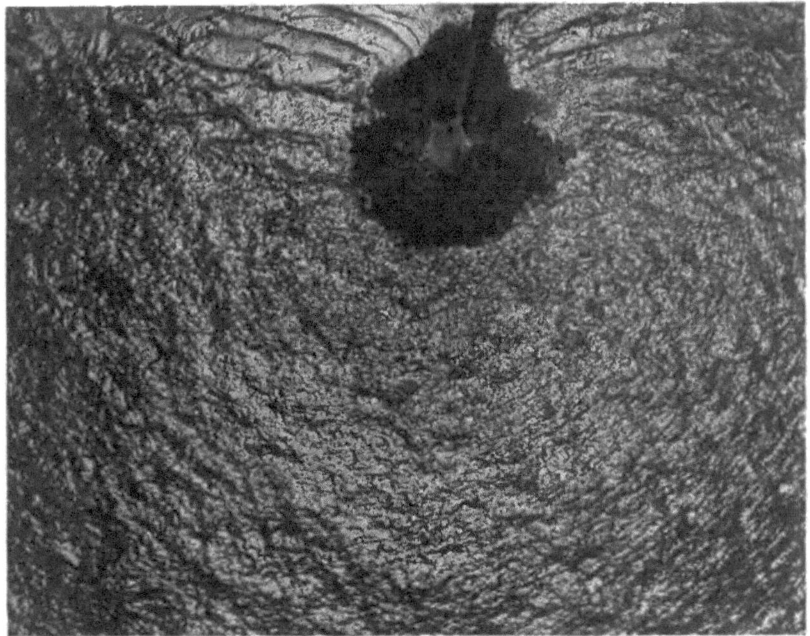

Fig. 175.—A lighthouse, as the naval flier sees it.

hand-held cameras. This type was employed partly because of difficulties to be noted presently, in using other forms of cameras, but more especially because such pictures sufficed for the kind of information desired. A hand-held camera formed part of the outfit of each flying boat and dirigible, but, unlike land reconnaissance, planes ascending primarily for picture taking were unknown in their naval service. Usually no photographic objective was predetermined—photographs were made only if objects of interest were come upon. Mapping also formed no part of the seaplane's work. Four plates would be carried, instead of as many dozens in the land machine, and often these would come back unexposed. There were of course some photographic flights planned out beforehand, for the purpose of photographing lighthouses and other landmarks whose

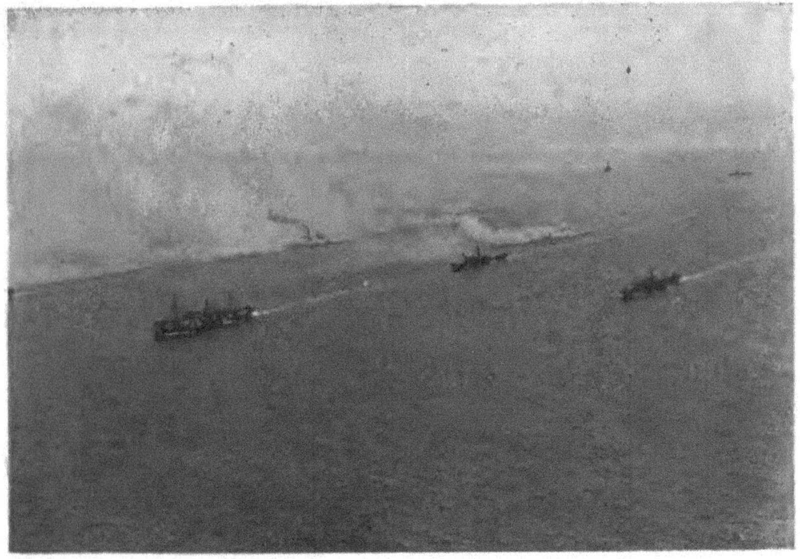

Fig. 176.—A threatened submarine attack. Throwing out a smoke screen to protect a convoy. British official photograph.

appearance from the air should be known to the naval aviator (Fig. 175). Among the accidental and record types of photograph come convoys (Fig. 176), whose composition and arrangement were made a matter of record, particularly if any ship was out of its assigned position. Photographs of oil spots on the sea surface, or other results of bomb dropping, were necessary evidence to establish the sinking of a submarine (Fig. 179). Pictures of all types of ships friendly, neutral, and where possible, enemy—were a much needed part of naval equipment, in particular pictures of friendly destroyers and submarines, which should not be bombed by mistake. For safe navigation it was essential to have photographs of uncharted wrecks (Fig. 181), of buoys out of place and of ships failing to return signals or otherwise

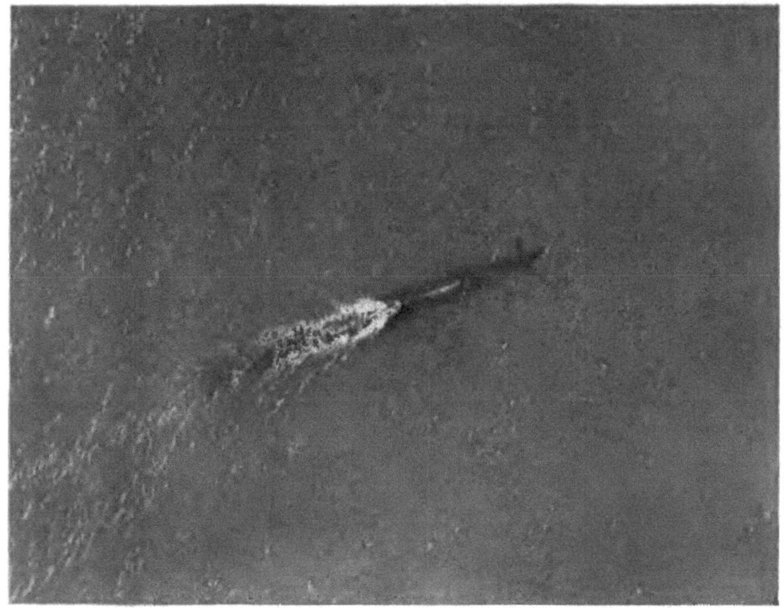

Fig. 177.—Submarine coming to the surface.
U. S. Naval Air Service photograph

to comply with rules. The great majority of the pictures were taken from altitudes of not more than 300 meters.

Hand-held cameras for naval work have practically the same design as those for land work. In view of the smaller number of pictures taken on naval trips, and the consequent absence of any need for great speed in changing plates, the ordinary two-plate dark slide has been found satisfactory in

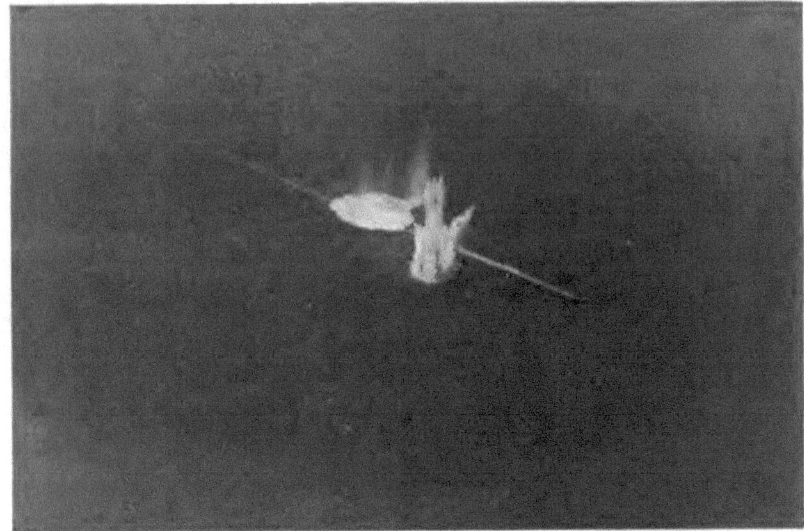

FIG. 178.—Dropping depth bombs.

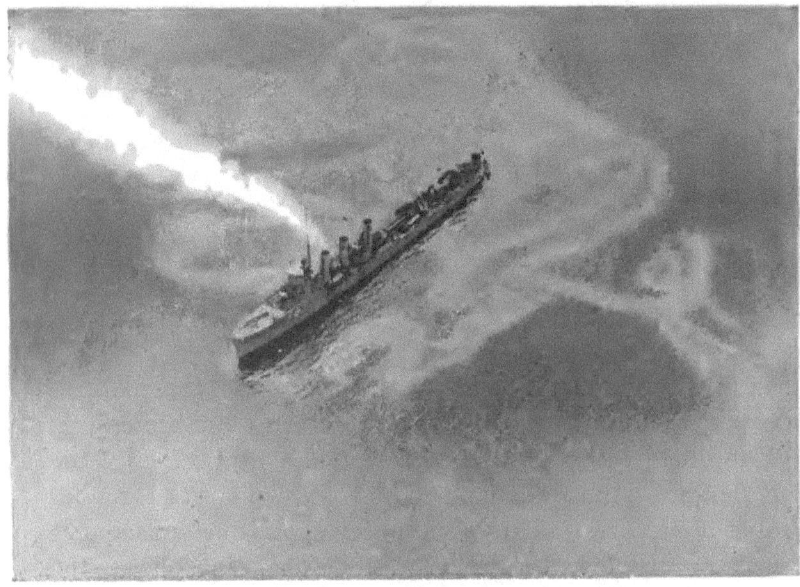

FIG. 179.—The submarine destroyed. Destroyer on tell-tale oil patch.
British official photographs.

the English service. But these are much less convenient than the bag magazines used in the U. S. Naval hand camera (Fig. 31). The sights on the naval hand camera are preferably of the rectangular, field indicating type, especially useful

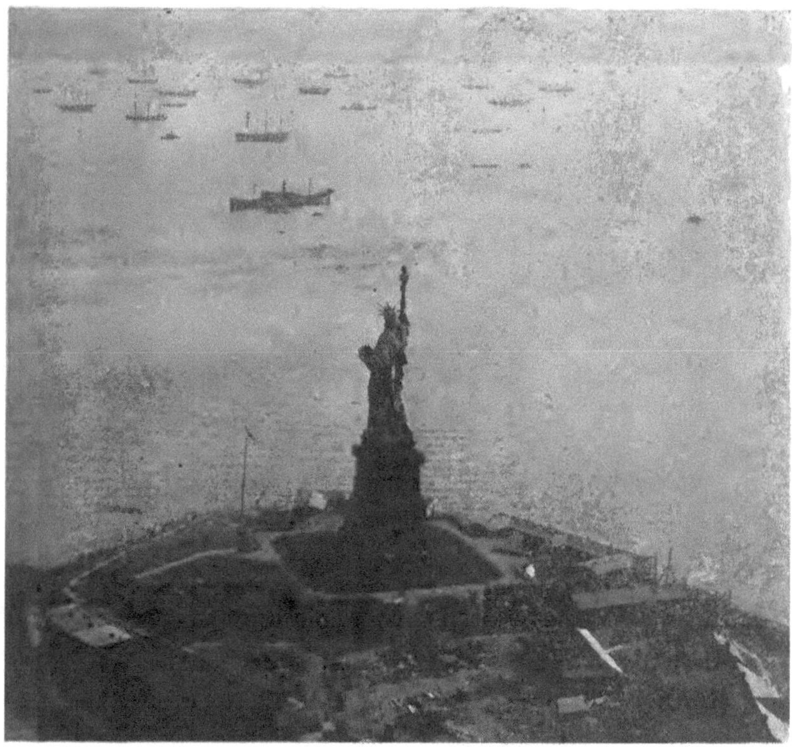

FIG. 180.—A convoy at anchor in port.

in photographing extended objects such as convoys. As the flying boat travels comparatively slow, it is easy for the observer to stand up to take pictures, and the sight is conveniently placed on top. But if held out over the side for verticals the sight must be on the bottom. Rectangular sights in both positions are provided in the English camera

(Fig. 186). Naval cameras should be immune from moisture, which means doing away with all wooden slides or grooves. A praiseworthy practice is to carry the camera in a waterproof bag.

Cameras other than of the hand-held form have been little used in sea planes, owing to the difficulties of installa-

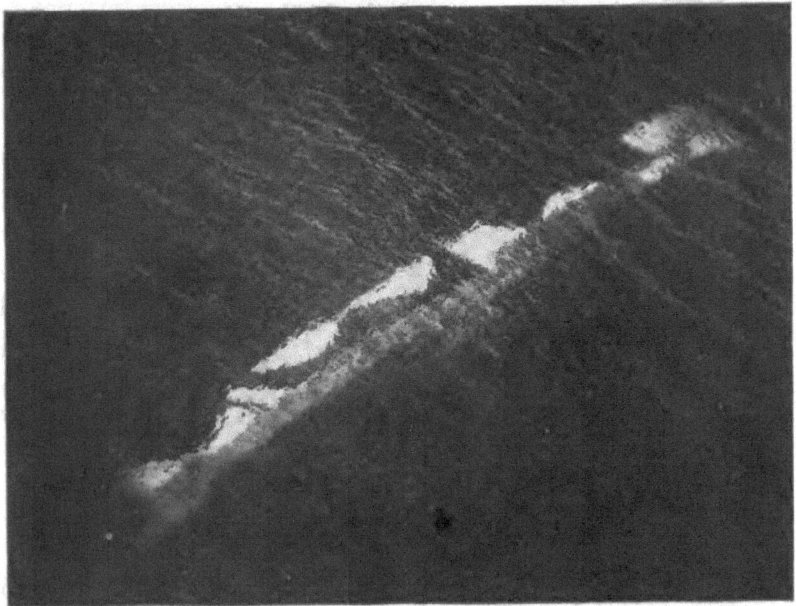

Fig. 181.—Airplane photography as an aid to salvaging. Position of wrecked merchantman twelve fathoms down revealed by photograph from the air.
Photograph by British Air Service.

tion. The hydro-airplane, consisting of an ordinary airplane fuselage mounted on two pontoons (Fig. 182), can carry the same kind of photographic equipment as the land machine. But if it has a single central pontoon this is not feasible. The hydro-airplane is, however, largely superseded by the flying boat (Fig. 183), whose fuselage, of boat form, rests directly on the water. In this type of sea plane, views taken

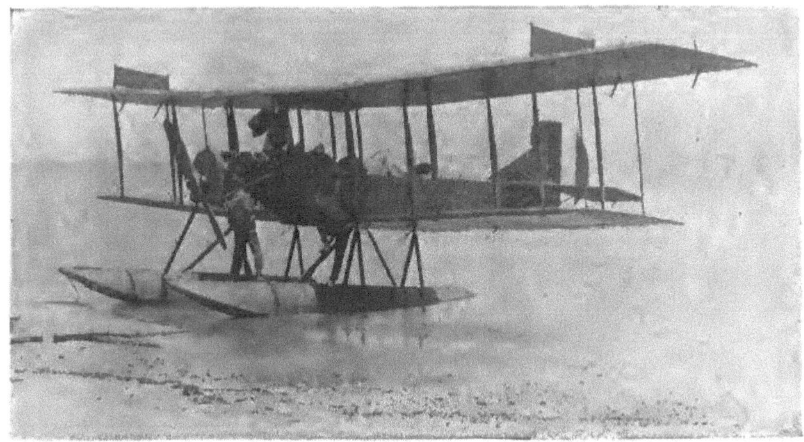

Fig. 182.—A sea plane.

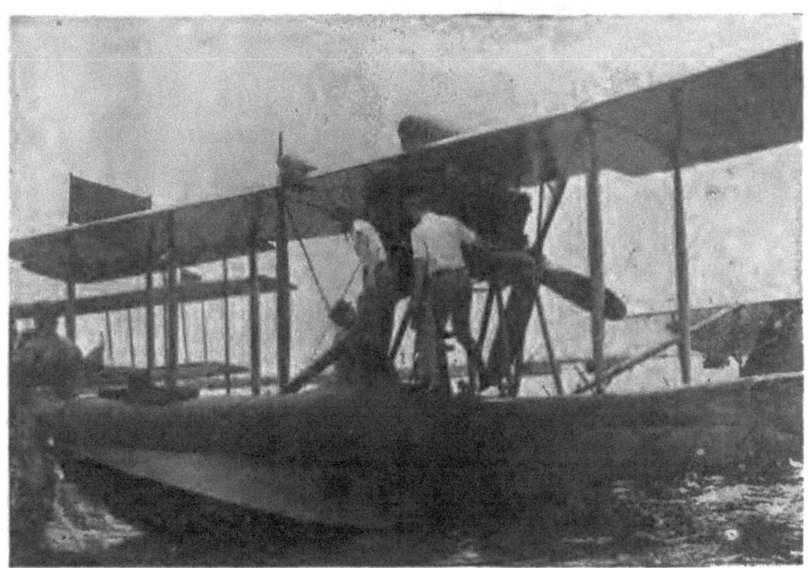

Fig. 183.—A flying boat.

vertically downward are not easy to make. In the larger flying boats the hull projects out horizontally a matter of several feet beyond the side of the cockpit. An ordinary

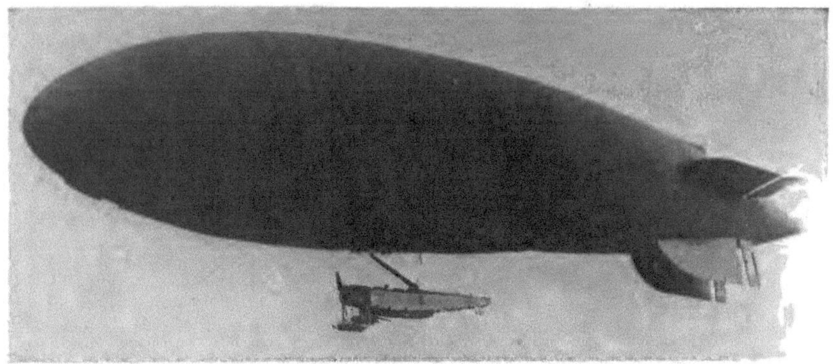

Fig. 184.—A dirigible or "blimp"—possibly the photographic aircraft of the future.

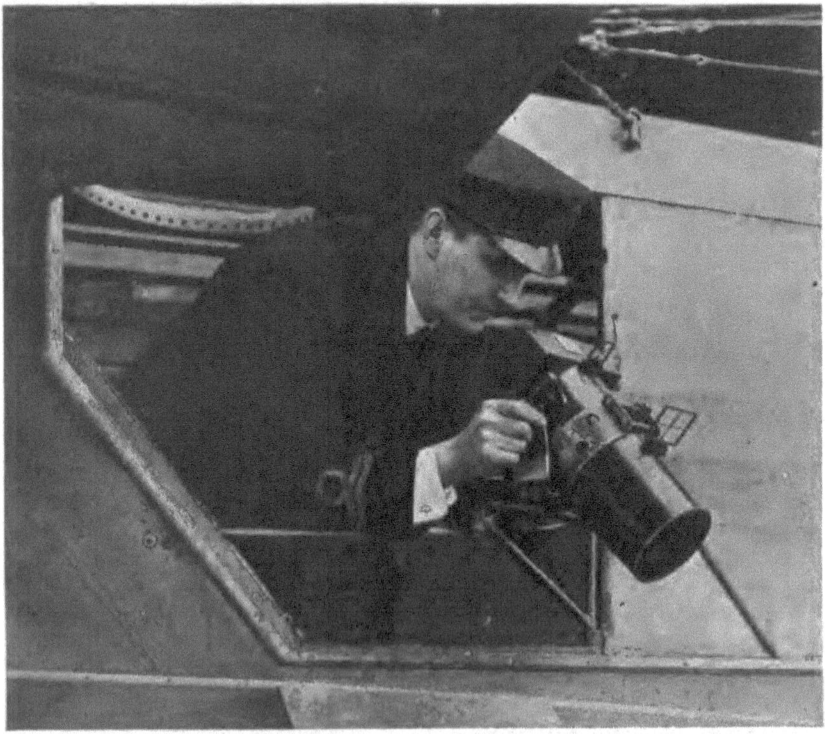

Fig. 185.—English "Type 18" hand camera on bracket for exposing through side window of flying boat. British official photograph.

outboard mounting is therefore out of the question. The camera must either be held out at arm's length or else mounted on a long bracket (Fig. 186). The usual place for carrying the camera is in the front cockpit with its magnificent all-round view. Obliques can, too, be taken in great comfort from the side windows behind the wings, as shown

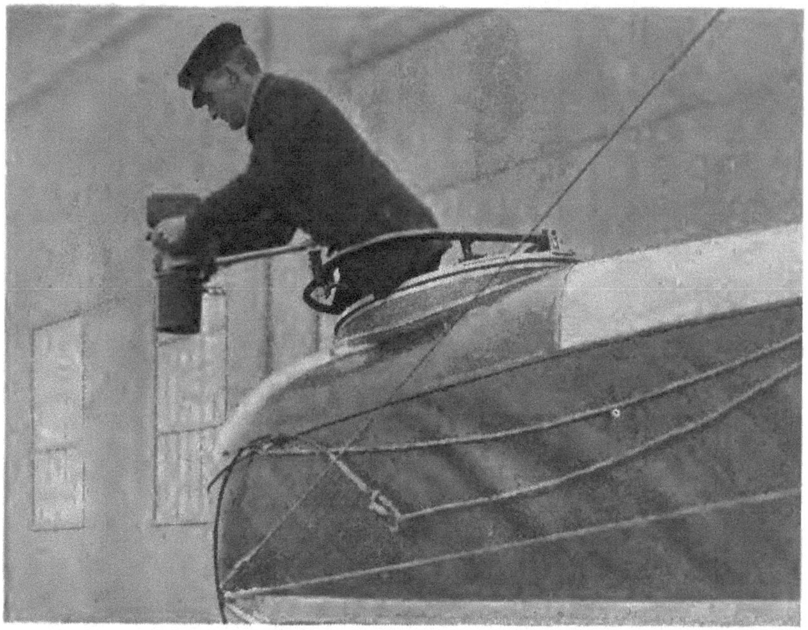

FIG. 186.—Camera mounted in bracket from forward cockpit of flying boat. British official photograph.

in Fig. 185. The possibility of cutting a hole in the bottom of a flying boat to take care of a vertical camera is not entertained in British and American naval cricles. Nevertheless it is the regular practice in the Italian service, with their small high ceilinged flying boats. In them a round hole is cut in the floor, stopped with a plug and rubber gasket. After the boat rises into the air the hole is opened, and the regulation Italian camera is set securely in a frame on the floor over the hole (Fig. 187). Photographs are taken to the capacity of the camera, and if it is desired another camera is put in its place, till all its plates have been exposed, and then even a third. Before coming down the hole must of course be closed again. Sliding doors have been designed to close this aperture, but have not proved sufficiently

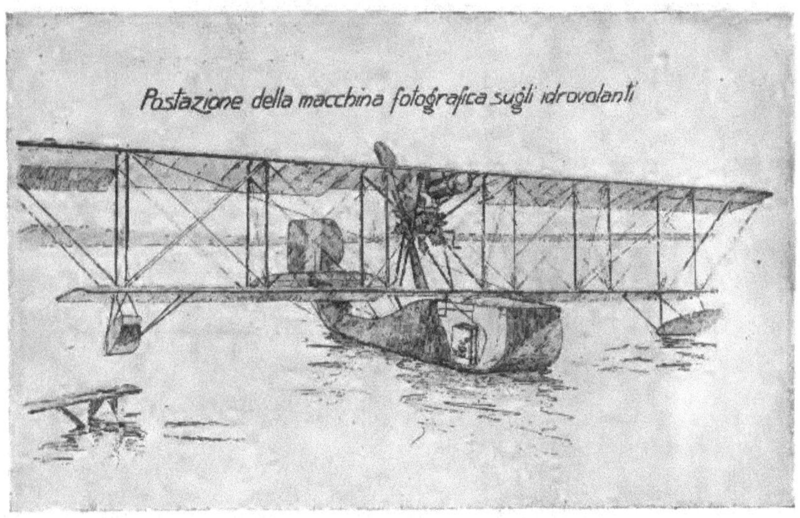

Fig. 187.—Italian flying boat with camera mounted on the floor.

water-tight, although such a device could undoubtedly be worked out.

With its space for five or more passengers, and with its low speed, the modern flying boat affords an excellent craft for photographic work. There is ample room for any size of camera, and for any style of mounting, if we assume that there is no objection to an opening in the bottom. The low ceiling of these ships, however, prevents their use for certain forms of aerial photography which should be of the greatest importance. Operations against shore stations—harbors, docks, shipyards, ships at anchor, and fortifications—cannot be undertaken for fear of anti-aircraft guns and hostile land planes. The solution of the problem of carrying and launching fast high flying planes from ships will immediately extend the usefulness of aerial photography to coastal work. In the recent war, such of this as was done, along the Belgian coast—the shore batteries, and the results of naval operations at Zeebrugge and Ostend—was done by land planes from territory held by the Allies. The photographic equipment of sea planes of the type suggested will of course present special problems, but the apparatus used will be apt to approximate closely to that of the land planes.

VII

THE FUTURE OF AERIAL PHOTOGRAPHY

CHAPTER XXXI

FUTURE DEVELOPMENTS IN APPARATUS AND METHODS

Prophecy is an undertaking that always involves risk. The prophet's guess of what the future will bring forth is based only on the tendencies of the past, the most urgent needs of the present, and the activity of his imagination. He may easily—and he usually does—entirely overlook certain possibilities which may arise apparently from nowhere and which profoundly affect the whole trend of development. Conditions which dominate at the present time—such as military necessity—may happily drop into the background and free the science from some of its severest restrictions. With this caution, some future possibilities in apparatus and methods may be presented along the lines already used in discussing the present status of aerial photography.

Lenses.—From the military standpoint the next steps in lens design would be toward telephoto lenses on the one hand, and on the other toward lenses of short focus and wide angle. The telephoto lenses used for spotting would be of long equivalent focus—a meter and more—but of handy size, that is, not more than 50 centimeters over all working distance. The wide angle short focus lenses would be designed for low flying reconnaissance or quick mapping work. They would also be demanded for peace-time mapping projects, where the largest possible amount of territory should be covered in a single flight. Both types of lens should be pushed to the extreme in aperture, for short exposures and the maximum of working days will always be demanded.

Cameras.—Peace-times will give the necessary opportunity to develop self-contained and therefore simply installed cameras. They will at the same time be made very completely automatic but simple to operate in spite of their complexity. Such cameras have, during the war, been the ideal of all aerial photographers, but the time has been too short since the necessary conditions have been understood for that lengthy development work and those complete service tests which are so necessary to develop all automatic apparatus. Several designs which are now being perfected may be counted on to take us a long way toward this ideal.

On the other hand, that military ideal which leaves the camera operator the greatest possible freedom for other activities, is apt to be entirely reversed in peace. The camera operator can now be required to be an expert, who will be free to change plates or filters and to estimate exposures, instead of giving his best efforts to the problem of defence. For him a simple and reliable hand-operated or semi-automatic camera is entirely satisfactory, and the great expense of complicated automatic apparatus has no longer its former justification.

Camera Suspension.—Perhaps the most pleasing prospect before the aerial photographer as he turns from war to peace work is that of having planes built for and dedicated primarily to photography. Instead of his camera being relegated to an inaccessible position, picked after the plane design has been officially "locked;" instead of yielding first place to controls, machine gun and ammunition; instead of being jealously criticised for the space and weight it takes up, the camera can now claim space, weight, and location suitable for any likely aerial photographic need. High speed no longer will be vital, and slower planes, permitting longer exposures in inverse ratio to their speed, will be chosen for photographic purposes.

A development which is sure to intrigue many investigators is the gyroscopically controlled camera. This has its chief *raison d'être* in precision mapping, whose possibilities from the air will undoubtedly be intensively studied at once. With the automatically leveled camera will come renewed attention to indicators of time, altitude, and direction, with the ultimate goal of producing aerial negatives that show upon their face the exact printing and arranging directions necessary to put together an accurate map.

Sensitive Materials.—Manufacturers of plates and films will direct efforts toward producing emulsions of good contrast, high color sensitiveness and high effective speed, especially when used in conjunction with the filters necessary for haze penetration. Exposure data will be accumulated and exposure meters appropriate for aerial work will be developed.

Color Photography.—Color photography from the air by any of the screen-plate or film-pack methods is probably out of the question because of the long exposures required. The screen-plates are unsuitable also because of the relatively large size of their grain compared to the detail of the aerial photograph. Ordinary three-color photography, using three separate negatives, is always subject on the earth's surface to the difficulty that the three negatives must be exposed from the same point of view, either in succession or by means of some optical arrangement which is costly from the standpoint of light. In photographing from the air this difficulty of securing a single point of view for the three photographs is absent. Three matched cameras, side by side in the fuselage, have identical points of view as far as objects on the earth below are in question. Consequently, three-color

25

negatives are entirely possible, and indeed will be simple to make as soon as plates of adequate color sensitiveness and speed are available. Probably the new Ilford panchromatic plate has the necessary qualities.

Night Photography.—The searching eye of photography was so omnipresent in the later stages of the Great War that extensive troop movements and other preparations had to be carried out either in photographically impossible weather or else at night. The natural reply to the utilization of the cover of night is to "turn night into day" by proper artificial illumination. At first thought it might well appear that the task of illuminating a whole landscape adequately for airplane photography is well-nigh hopeless by any artificial means. On one hand we have the very short exposures alone permissible; on the other the fact that the intensity of daylight illumination is overwhelmingly greater than those common in the most extravagant forms of artificial illumination.

Toward the close of the war, however, actual experiments made with instantaneous flashes of several million candlepower showed that if proper means were provided to insure the flash going off near the ground, and if its duration were made no longer than about $\frac{1}{50}$ second, interpretable photographs were obtainable on the fastest plates. It appears, therefore, merely a matter of manufacturers perfecting the technique of flash production, and of inventors providing the launching and igniting devices to push this kind of photography to the practical stage. The achievement of night photography cannot fail to have an enormous effect on future tactics.

The technique of night photography may take either of two directions. On one hand we may develop flashes of the requisite intensity to give all their light in $\frac{1}{100}$ second; on the other hand, it may prove more feasible to use flashes of longer duration and to arrange for the camera shutter (of the between-the-lens type) to be exposed synchronously with the middle of the flash. One way, frequently suggested, to use these longer flashes would be to trail the charge on a long wire, through which the ignition is effected electrically. This is not likely to be satisfactory, however, for the resistance of a wire is so great that when the plane flies at any practical height, the trailed flash, if it reaches near the ground, will be forced to a very great distance behind. Probably the best solution will involve accurate synchronizing of the fuse in the freely dropped sack of flash powder with the exposing mechanism in the camera.

CHAPTER XXXII

PICTORIAL AND TECHNICAL USES

Aside from their element of novelty, aerial photographs have undouted qualities of beauty and utility. The "bird's-eye view" has always been a favorite for revealing to the best advantage the entire form and location of buildings and

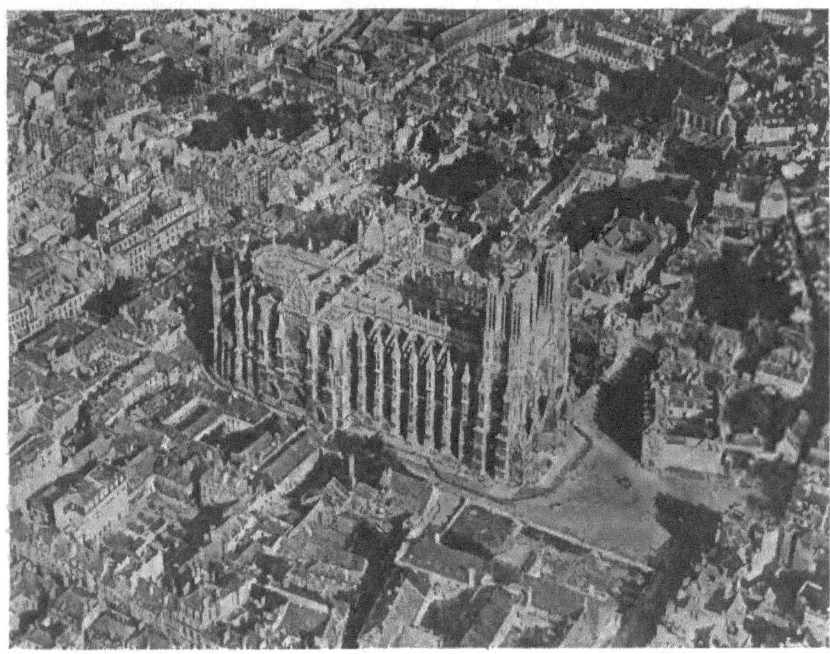

Fig. 188.—Rheims Cathedral.

of other large objects. Heretofore such views have usually had to be drawn by an imaginative artist.

Aerial oblique views possess the virtues both of pictures and of plans. They are destined to be extensively used in the study of architecture (Fig. 188). Cathedrals, castles,

388

town halls, particularly those still in their cramped medieval surroundings where they can never be seen in their entirety from the ground, come forth in all their beautiful or quaint proportions from the airman's point of vantage. Stereoscopic aerial views are destined to occupy a valuable position also. Stereo prints of the famous buildings of Europe, taken from

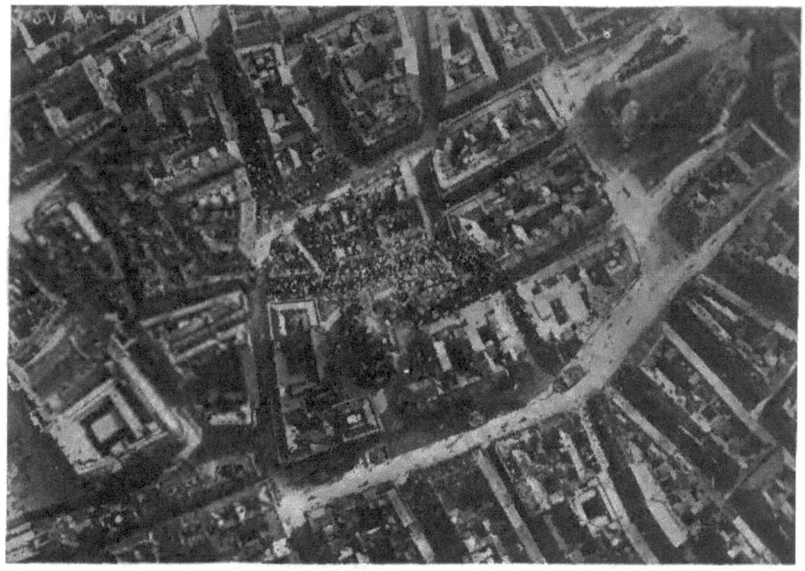

FIG. 189.—A portion of Vienna seen from the air, during a "propaganda raid." Italian official photograph.

the air, will give to the prospective traveler or the arm-chair tourist a many fold more accurate idea of their construction than will any number of mere surface views.

A vertical aerial photograph is most closely akin to a map, but has advantages over any ordinary surveyor's product. As a guide it is infinitely superior to the best draftsman's diagram, for it provides a wealth of detail whereby the traveller may definitely locate himself. At a

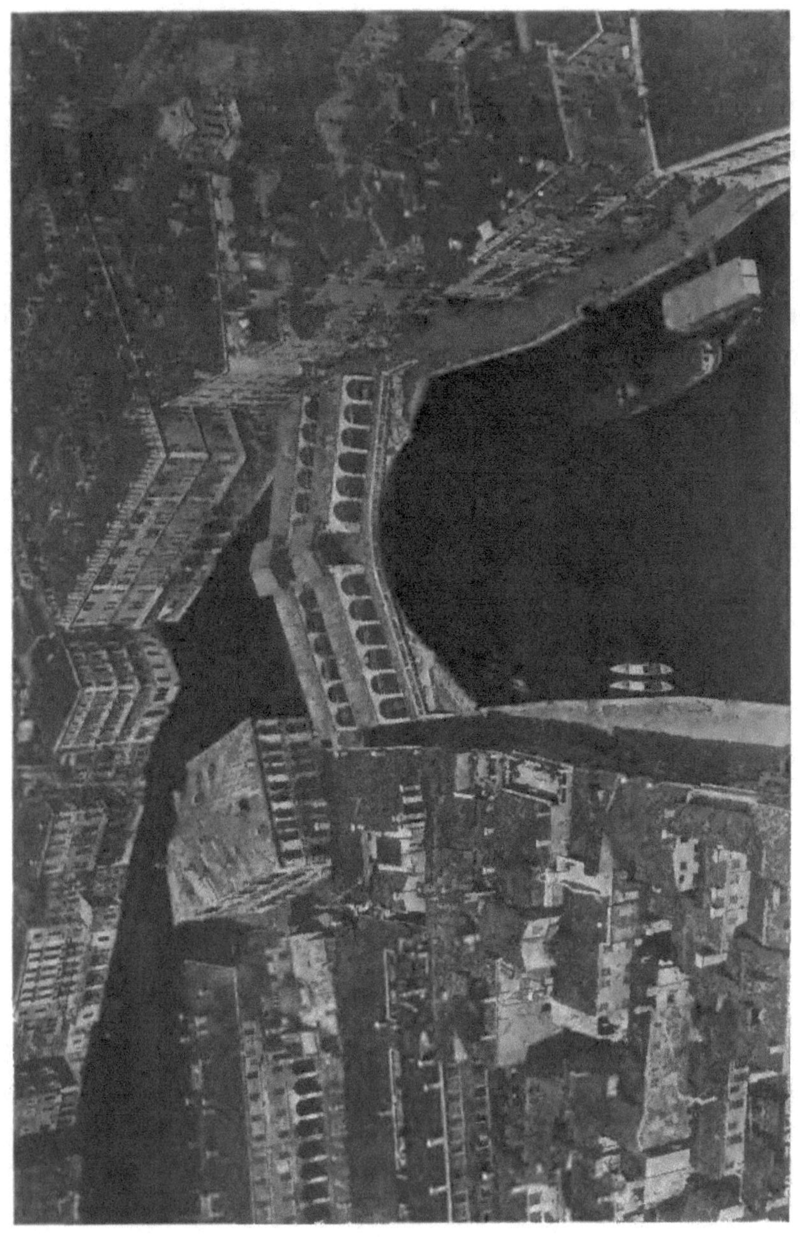

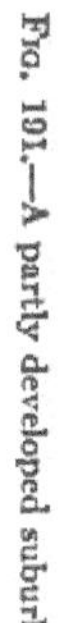

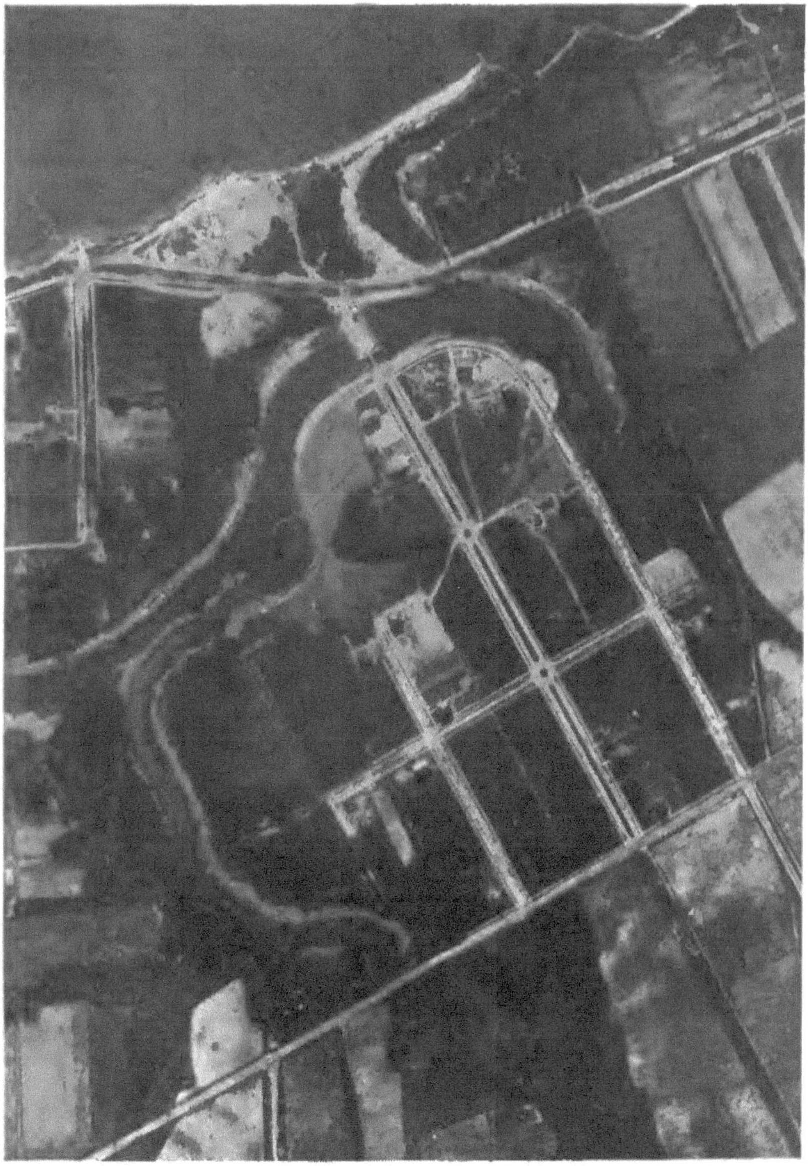

FIG. 191.—A partly developed suburb.

single glance he notes the objects of interest within his radius of easy travel. The guide-book of the future will therefore be incomplete without numerous aerial views, both vertical and oblique. As an illustration of the peculiar merit of the view from the air, consider the photograph of Vienna made during d'Annunzio's "propaganda" bombardment. Or the

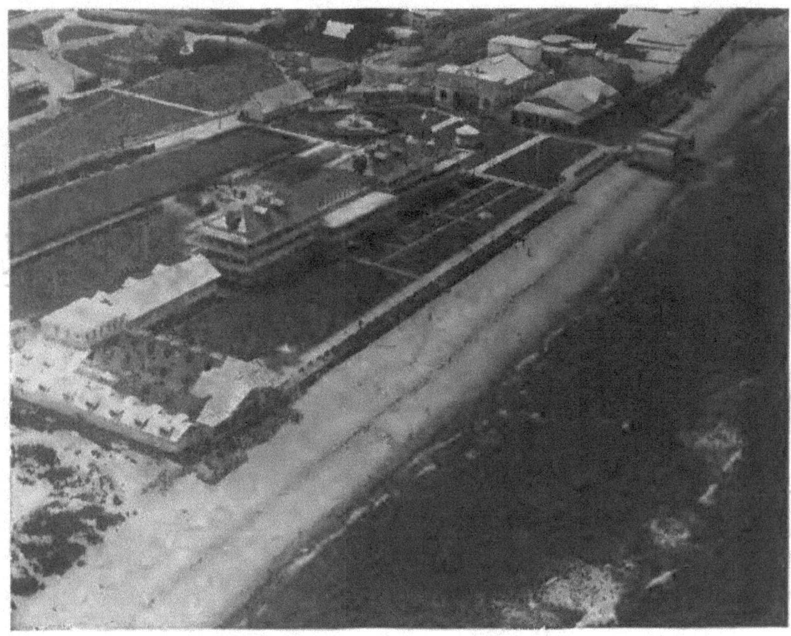

Fig. 192.—A sea-side resort.

picture of the Rialto bridge (Fig. 190). No ordinary photograph from land or water suggests the central roadway and no map shows the beauty of its elevation. Both are shown here, as well as an intimate view of the arched and pillared courtyard of the Fondaco de' Tedeschi to the right.

Airplane photographs will undoubtedly be widely used in certain fields of advertising. Architects and real estate agents may be expected to display their wares by the aid of aerial views. A well-planned country estate or golf course, or a suburban development (Fig. 191), can be shown with a completeness, both as to environment and stage of progress which no other form of representation can approach. A seaside resort can now show the extent and grouping of its

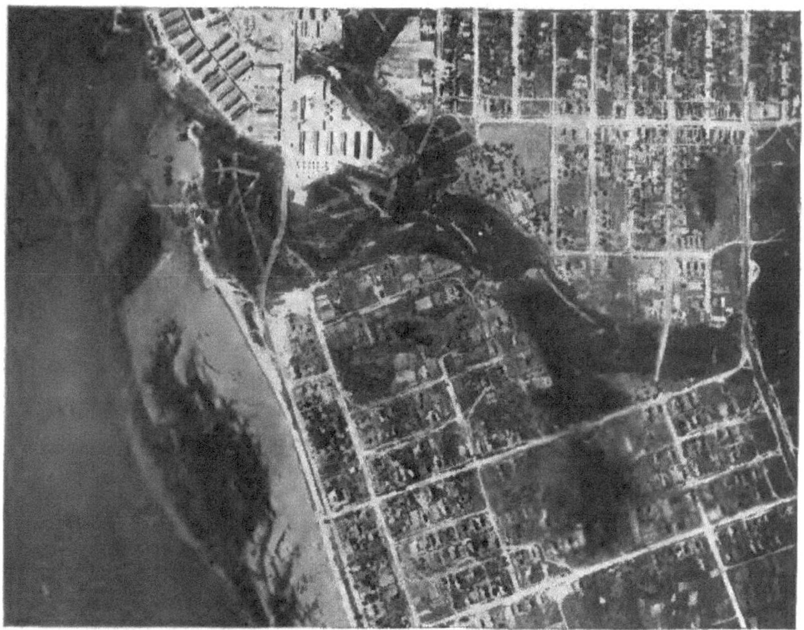

Fig. 193.—A bathing beach seen from the air.

natural and artificial amusement features in a single picture (Fig. 192). Even the extent of its bathing beach under water is revealed to the aerial photographer (Fig. 193). Real estate agents can utilize aerial photographic maps of cities to great advantage. On these their properties can be pointed out, with the nature of their surroundings shown at a glance, together with their relation to transportation, schools,

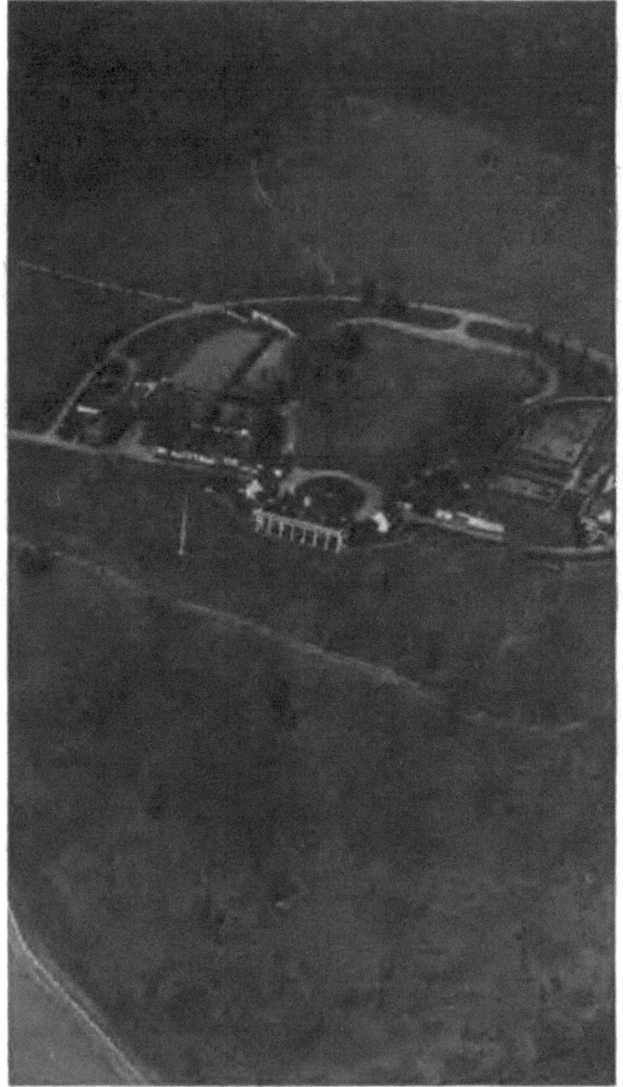

FIG. 194.—Mt. Vernon from the air.

churches, shopping districts, parks, or factories. The future purchaser of lots in a distant boom town will no longer be satisfied with a map outlining the streets with high-sounding names. He will demand an authentic aerial photograph, showing the actual number of houses under construction, the streets, gutters and sidewalks already laid, the size and planting of trees.

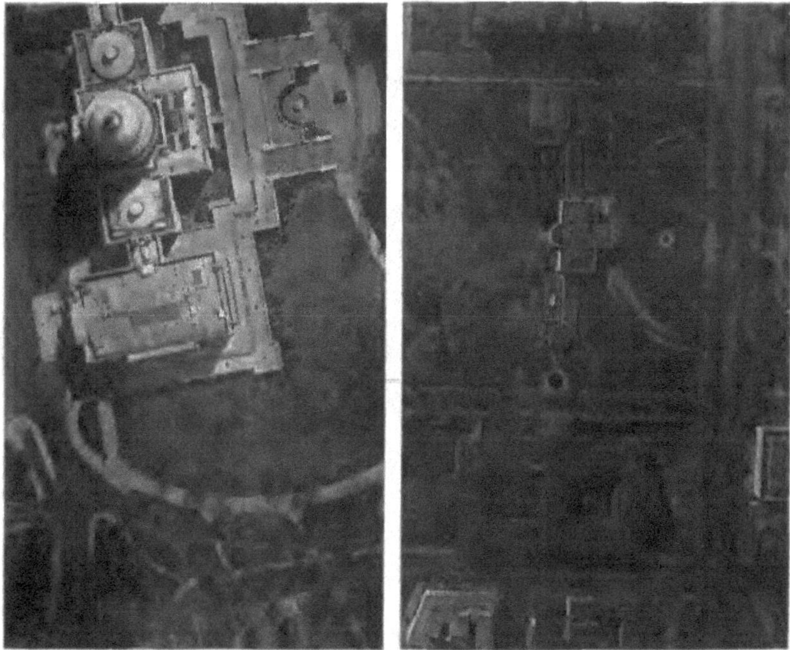

FIG. 195.—A contrast in roofs. The Capitol retains its individuality, while the White House loses all character when seen from above.

The study of landscape gardening is another field for which the aerial photograph is peculiarly fitted. A collection of oblique pictures of the châteaux and palaces of Europe showing their approaches and grounds, or of the historic estates of our own South, (Fig. 194), will be worth more to the prospective designer of a country estate than maps and ground pictures can ever be. Closely allied to landscape gardening is city planning, for which the aerial map will be quite indispensable. The appearance of a city from the air may indeed become a matter of pride to its inhabitants, and not only the arrangement of streets and parks, but even the character of the roofs of the buildings, be the subject of study (Fig. 195).

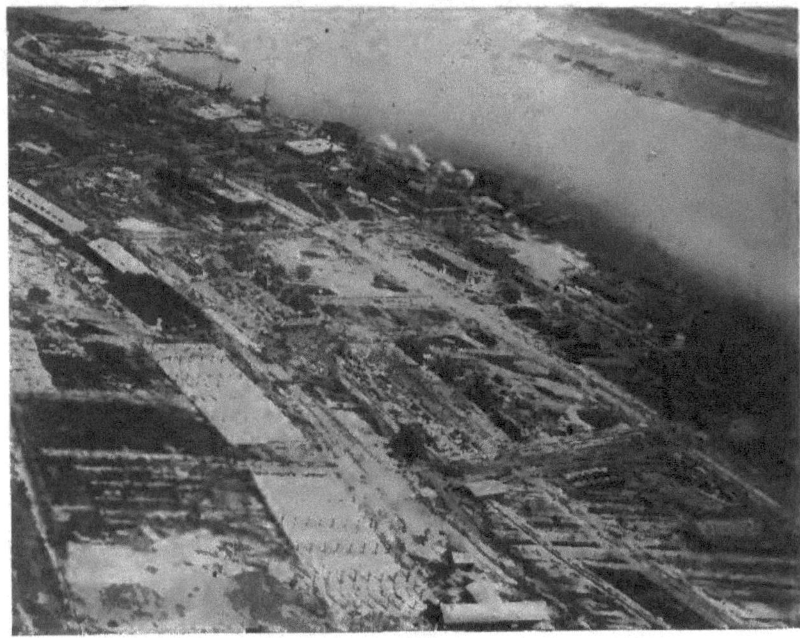

FIG. 196.—An aviation field under construction; early stage.

Engineers and constructors will depend more and more on preliminary photographic surveys as a basis for locating their operations. At the later stages of their work they will use aerial photographs for recording progress. Periodic photographs of buildings in process of construction, such as are now made from the ground, are much more illustrative when made from the air. Only from above is it possible to obtain in a single picture the progress of the complete project, such as the construction of an aviation field (Figs. 196 and 197) or of a shipyard. The building of large structures—bridges, hotels, ships on the stocks—particularly demands aerial views if the foreground is not to eclipse the center of real interest.

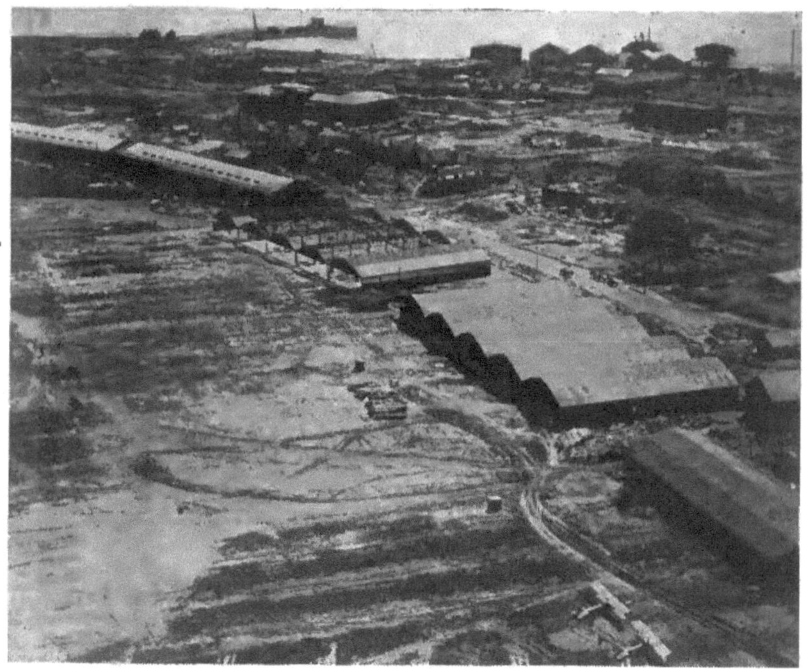

FIG. 197.—An aviation field under construction; later stage.

News events will soon call for an aerial photographic service. Already we are seeing newspapers and magazines featuring aerial photographs of the entry into conquered cities and the parades of returning fleets. Accidents, fires, floods and wrecks, of either local or national interest, can best be represented by this newest form of photography.

The photographing of wrecks, fires and floods suggests the importance of aerial views to insurance underwriters,

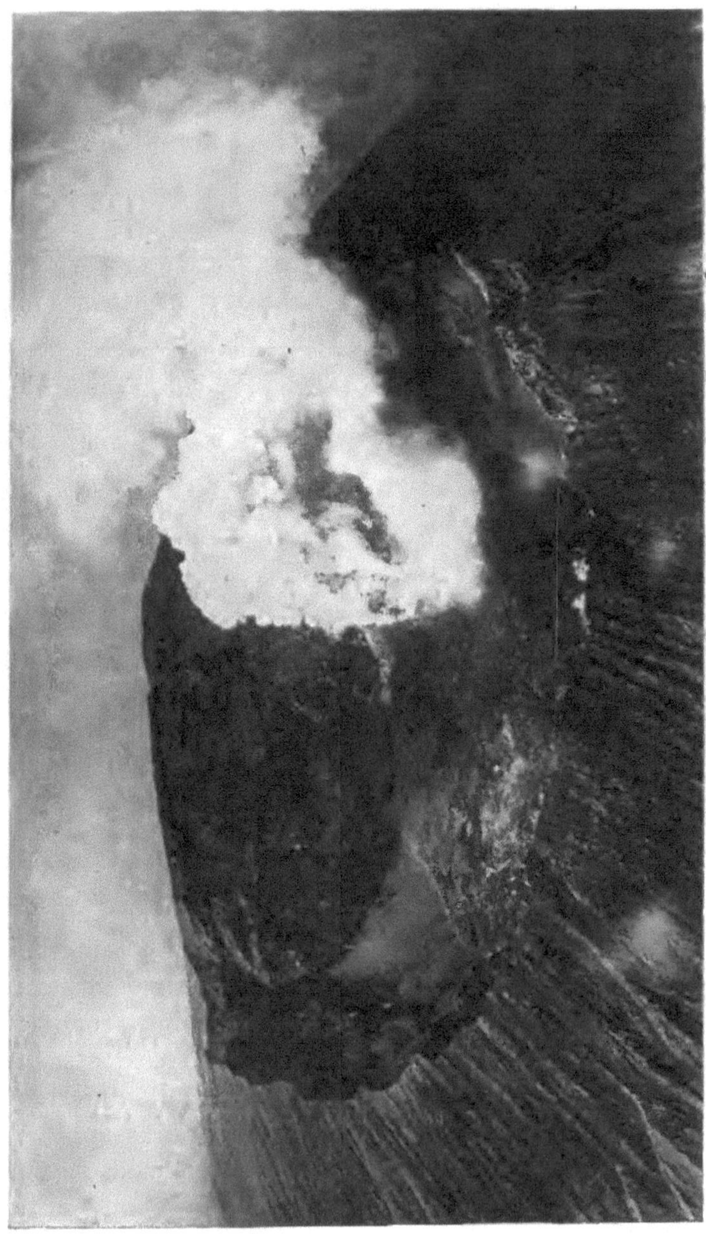

Fig. 198.—The crater of Vesuvius.
Photograph by Royal Italian Air Service

who require the most minute information on the characteristics of buildings in every neighborhood, and on the extent and nature of damage done. Marine insurance companies might with profit use the airplane camera to help estimate

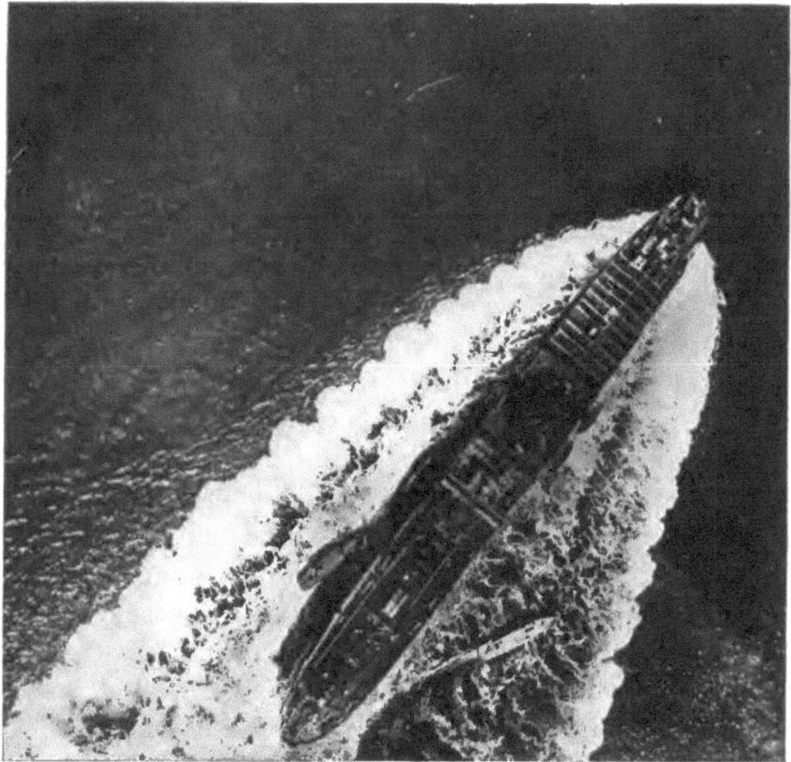

FIG. 199.—Waves set up by a ship—of interest to the naval architect.

the chances of salvage of a stranded ship or a vessel foundered in shallow waters (Fig. 181).

Numerous scientific uses for aerial views seem likely. Prominent among these is their use in geology, for the study of the various forms of earth sculpture. Pictures from the air of extinct volcanoes will give information as to their configuration that would otherwise require months of painstaking survey to obtain. Aerial photographs of active volcanoes (Fig. 198), showing the results of a succession of outbursts—one obliterating the other—would prove of the greatest value, especially when studied in conjunction with other scientific data, the whole making a record unobtainable by any other means.

In earthquake regions—notably Southern Italy and Japan—the changing coast lines, shallows and safe harbors, could be promptly ascertained after the subsidence of each fresh shock, with a consequent keeping open of trade routes and often the saving of life. River courses, glacier formations, cañons, and all the larger natural formations which man usually sees only in minute sections, and which he must build up in his mind's eye or by models, are today quickly and accurately recorded by the camera in the air. Such formations as coral reefs, whose configurations can now be accurately learned only by laborious surveys of a limited number, could be studied in quantity and with heretofore unknown satisfaction as the result of a single expedition with a ship-carrying seaplane and aerial camera.

Another scientific field—probably one of many similar ones—lies in the study of the waves set up by ships (Fig. 199). These are of extreme importance in the realm of naval architecture, but before the day of the airplane could never be easily studied in full scale.

CHAPTER XXXIII

EXPLORATION AND MAPPING

Aerial photographic mapping in war-time has been almost entirely confined to inserting new details in old maps. For such work some distortion or a lack of complete information on altitude and directions is not a serious matter, because the known permanent outlines serve as a basis. Furthermore, in so far as outline maps are concerned, as distinguished from pictorial maps, these have been drawn on the ordinary scales, and with the ordinary conventions of engineering map practice.

Aerial photography may be used in the future in practically the same way, as an aid to the quick recording of those minute details which would ordinarily consume an enormous amount of labor to survey directly. The region shown in Fig. 200 affords a good illustration. A discouraging amount of time and effort would be required to map this section of Virginia by the usual methods, while the smallest curve of creek and shore is instantly and completely recorded on a single photographic plate. But there are other possibilities, diverging from this application both toward greater and lesser requirements for precision.

Pictorial maps, in which the actual photographs figure, promise to be an essential part of the airman's equipment, whether he be pilot or passenger, mail carrier or sportsman. Without any pretention to detailed accuracy of location, these maps will show, in strip or mosaic form, the general appearance of the country to be traversed, with particular reference to good landing fields and other points of interest to the aviator. Vertical pictorial maps may be supplemented by obliques giving the view ahead, whereby the pilot may direct his ship. Thus the Washington monument as seen by the pilot from Baltimore is a truer guide than is the country beneath him. The crossing of mountain ranges is another case where the oblique picture will be more useful than the vertical (Fig. 201).

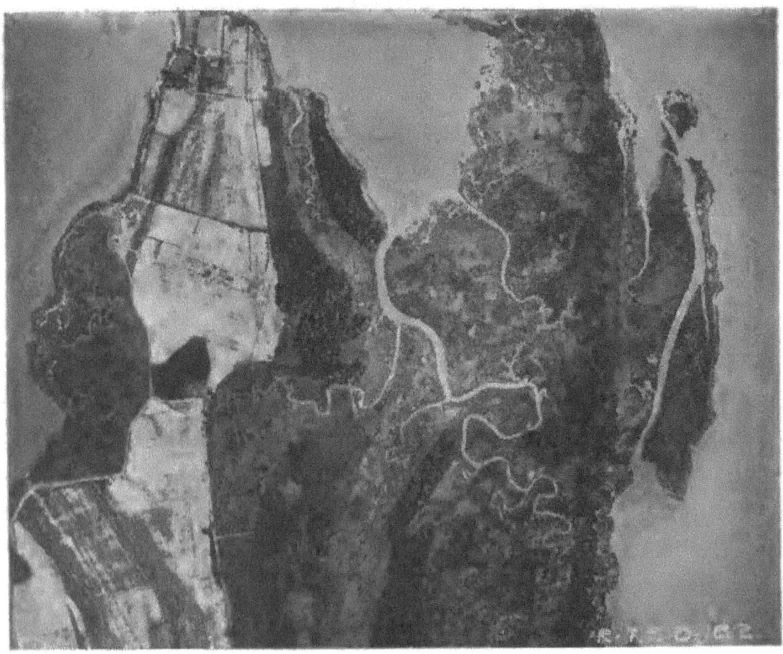

Fig. 200.—An aerial photographic survey of ground difficult to cover by ordinary surveying methods.

Contrasted with the merely pictorial maps will be precision surveys. Whether it will prove practical to make these entirely from the air is still an open question. It is to be assumed that cameras can be constructed with lenses having negligible distortion of field, with between-the-lens shutters to obviate the distortions due to the focal-plane type, with auxiliary devices for indicating compass direction,

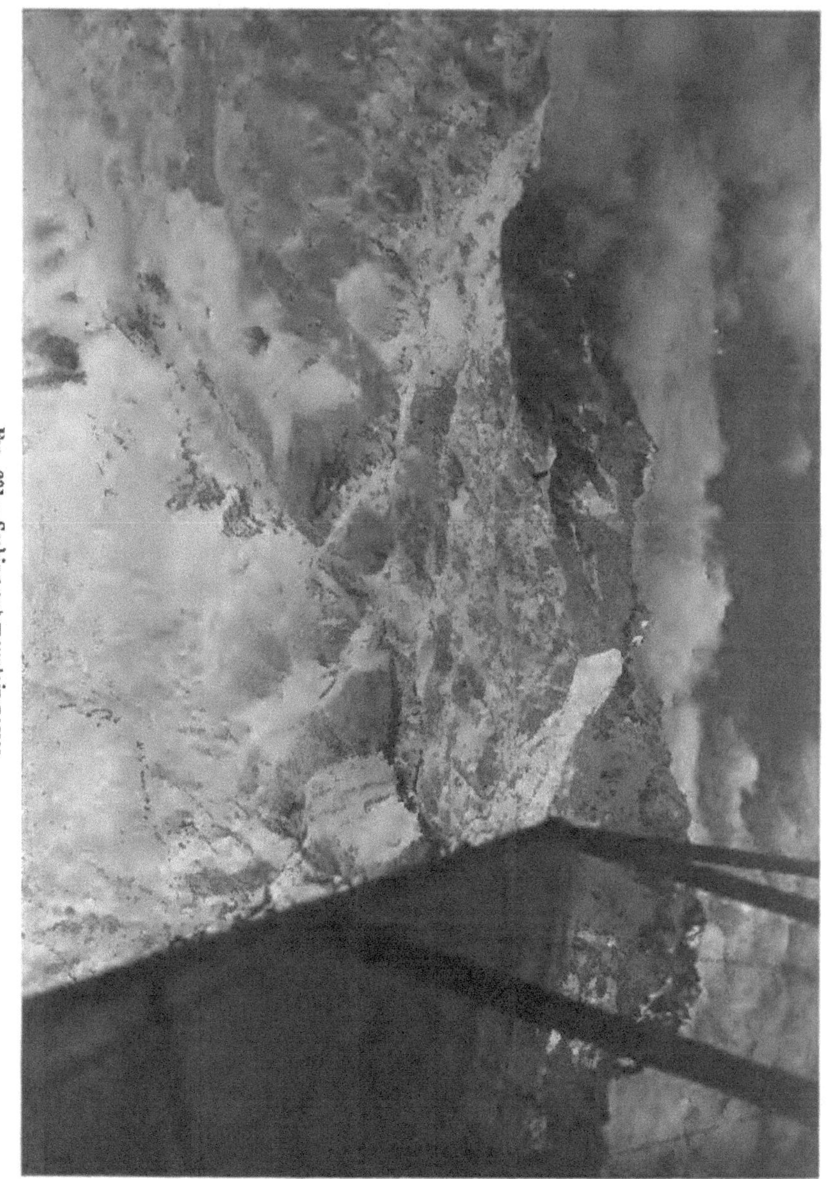

Fig. 201.—Seeking out mountain passes.

altitude, and inclination, or with gyroscopic mounting so that an inclination indicator is unnecessary. The application of aerial photography to precision mapping will depend upon the perfection which such cameras attain, as estimated by the permissible errors in this form of mapping. Entire dependence on photography, as in uncharted regions, is likely to be worked up to slowly, beginning with a stage of rather complete triangulation of natural or artificial points—say three in each constituent picture—then through several stages each successively employing fewer and fewer well determined points. The photographic mapping of some of our Western States will be greatly facilitated by the 100-yard squares into which the land is divided and already marked in a manner which shows clearly in aerial photographs.

A theoretical possibility is the plotting of contours from stereo-aerial pictures. Given two elements of a stereoscopic pair, taken from points whose separation is known, the position of any point in space shown in the stereoscopic view can be determined by the use of the stereocomparator. This is an instrument already employed in mountain photosurveying, which consists essentially of a compound stereoscope in whose eye-pieces are two points movable at will so that the relief image formed by their fusion can be made to coincide with any chosen part of the landscape. The chief difficulty in the application of this idea to aerial work is to fix the base line. This problem may be met in some cases by using stereo obliques, and getting the base line by simultaneously made vertical photographs of well surveyed territory beneath. Possibly also methods can be developed by which photographs from two or more known altitudes may furnish the requisite data.

City mapping is a field for which aerial photography is

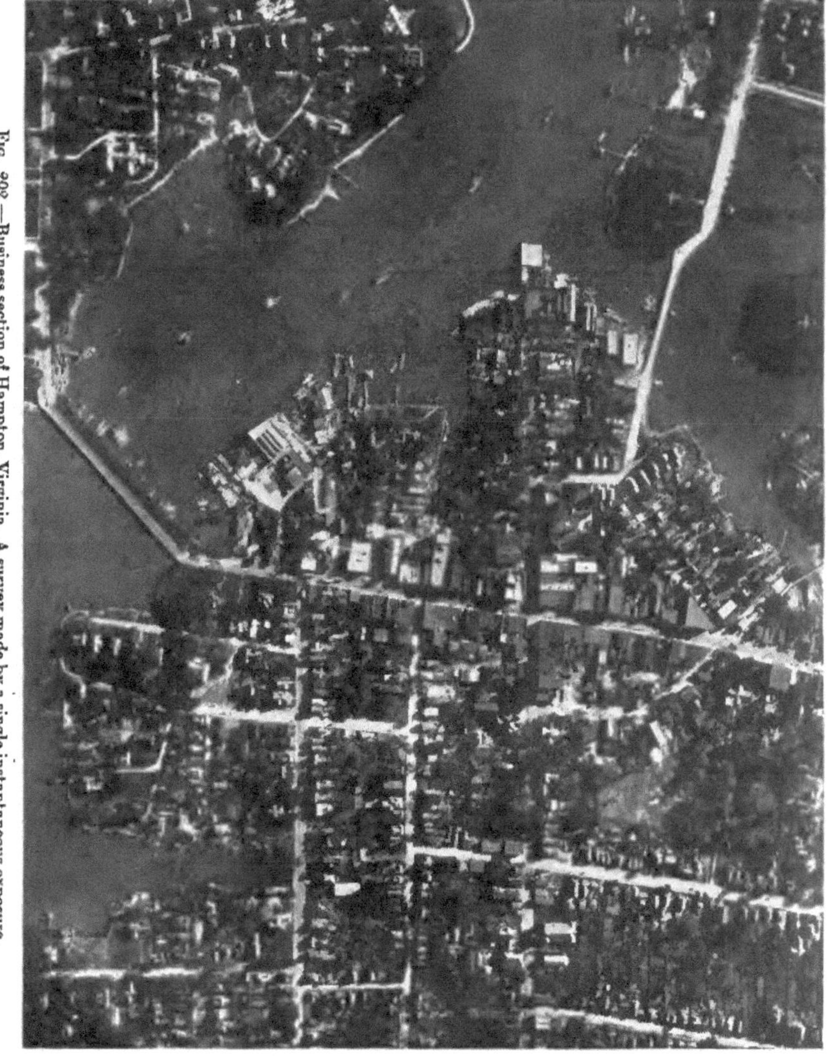

Fig. 202.—Business section of Hampton, Virginia. A survey made by a single instantaneous exposure.

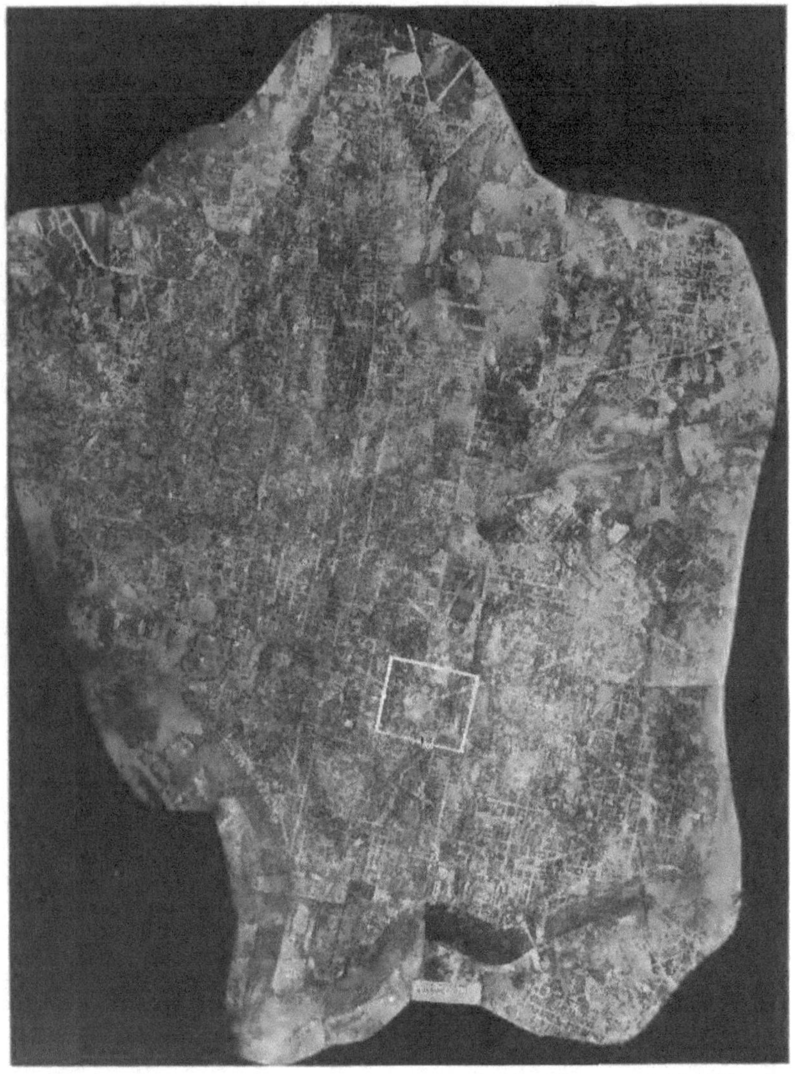

Fig. 203.—Mosaic map of the City of Washington. White rectangle shows portion included in next figure.

peculiarly fitted (Fig. 202). A complete map of a large city is a labor of years. In fact, a modern city is always

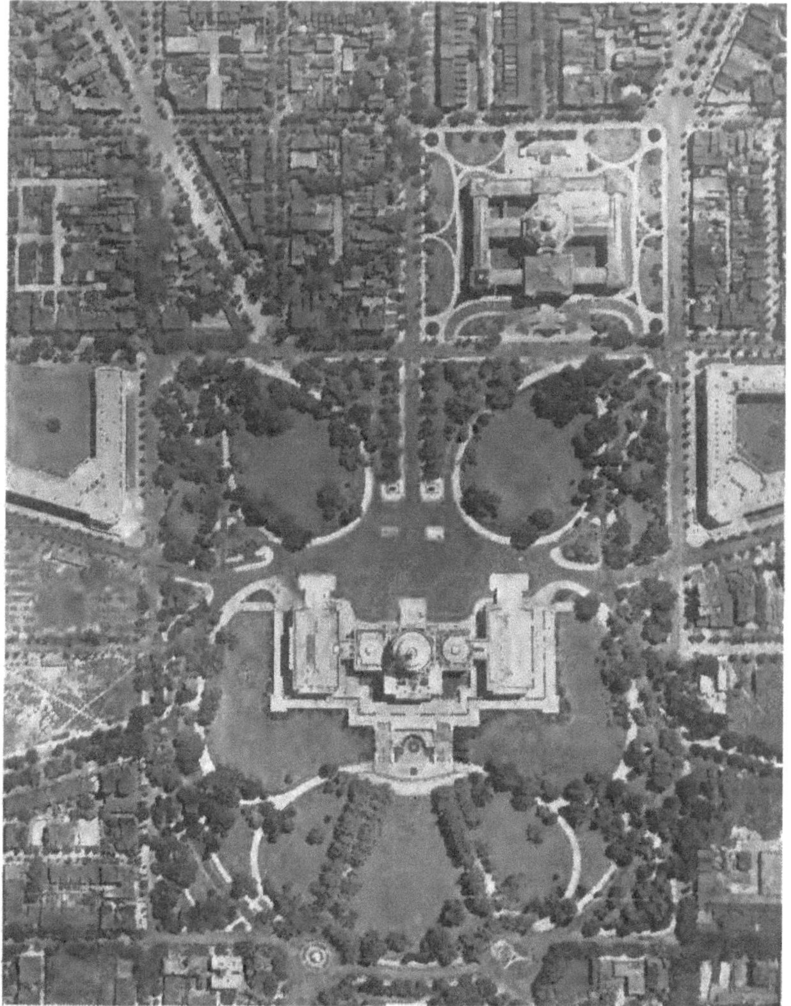

FIG. 204.—Portion of Washington mosaic, full size.

dangerously near to growing faster than its maps. An aerial map, on the contrary, can be produced in a few hours. Paris was mapped with 800 plates in less than a day's actual flying. Washington was completely mapped in 2½ hours, with less than 200 exposures. The entire map is shown, on a greatly reduced scale, in Fig. 203, while Fig. 204 shows a small portion of it in full size, from which can be obtained an idea of the dimensions of the original. These maps, while not accurate enough for the recording of deeds and mortgages, yet serve the majority of needs. There is indeed no reason why with long focus cameras, given several accurately marked points, the photographic map of a piece of real estate should not be made with all the accuracy needed, still leaving the whole process of partial surveying helped out by photography an enormously simpler one than the usual method.

Rougher types of surveying, in open country, offer a most promising opportunity. Railway surveys, showing the character of the country: passes through mountain ranges: the available timber and other materials of construction. Canal routes, with the available sources of water supply, and the best choice of course to avoid deep cuttings and aqueducts. Irrigation projects, with the natural lakes, river courses and valleys, which may be dammed to form storage basins. Coast, river and harbor surveys are possible by aerial means with a promptness and frequency which should entirely revolutionize the making of maps of waterways. Shifts in channels and shallows, even of considerable depth, stand out prominently in the aerial photograph. The actual bottom, if not more than three or four meters down—as in a bathing beach—shows in the aerial photograph (Fig. 193), while the varying surface tints caused by light reflected from the bottom at far greater depths are readily differentiated by the camera from the air. An instantaneous photograph will thus perform the work now done by a week's soundings. Fig. 205, taken near Langley Field, shows how the aerial photograph may be used to chart natural channels, while Fig. 206 shows the dredged chan-

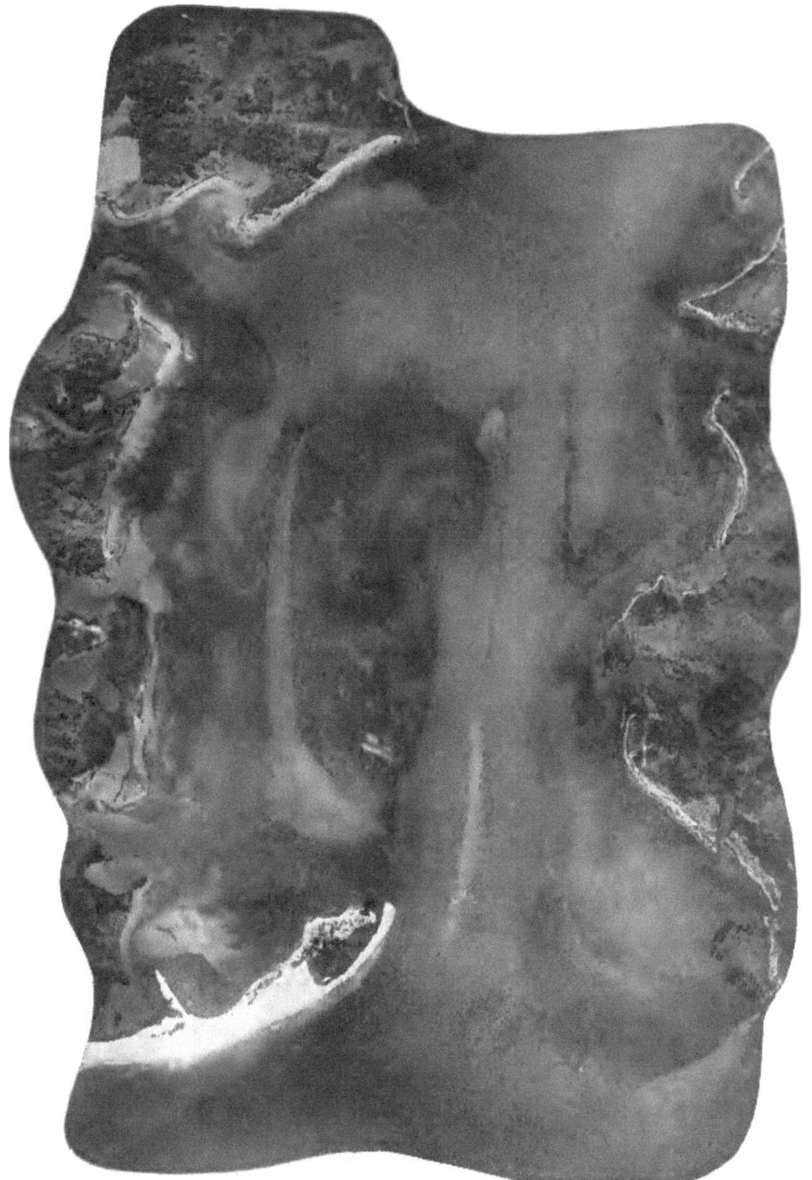

FIG. 205.—Shallows and channels revealed by the aerial photograph.

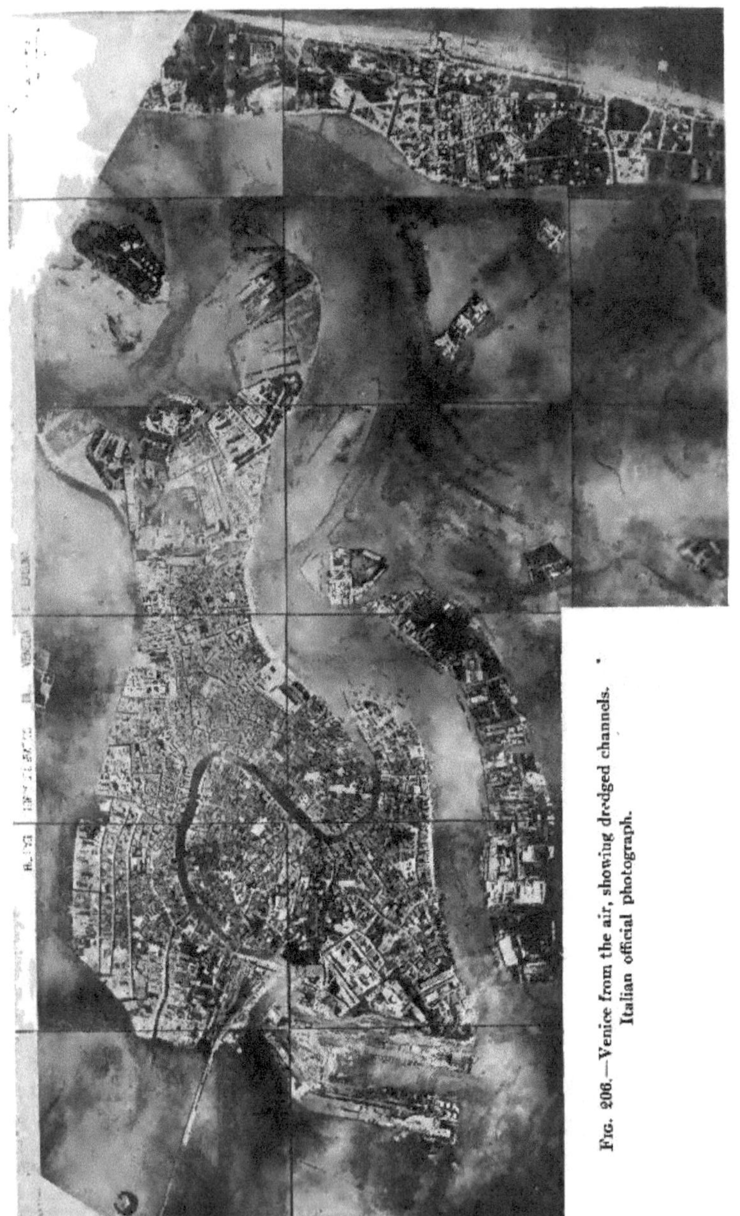

Fig. 206.—Venice from the air, showing dredged channels. Italian official photograph.

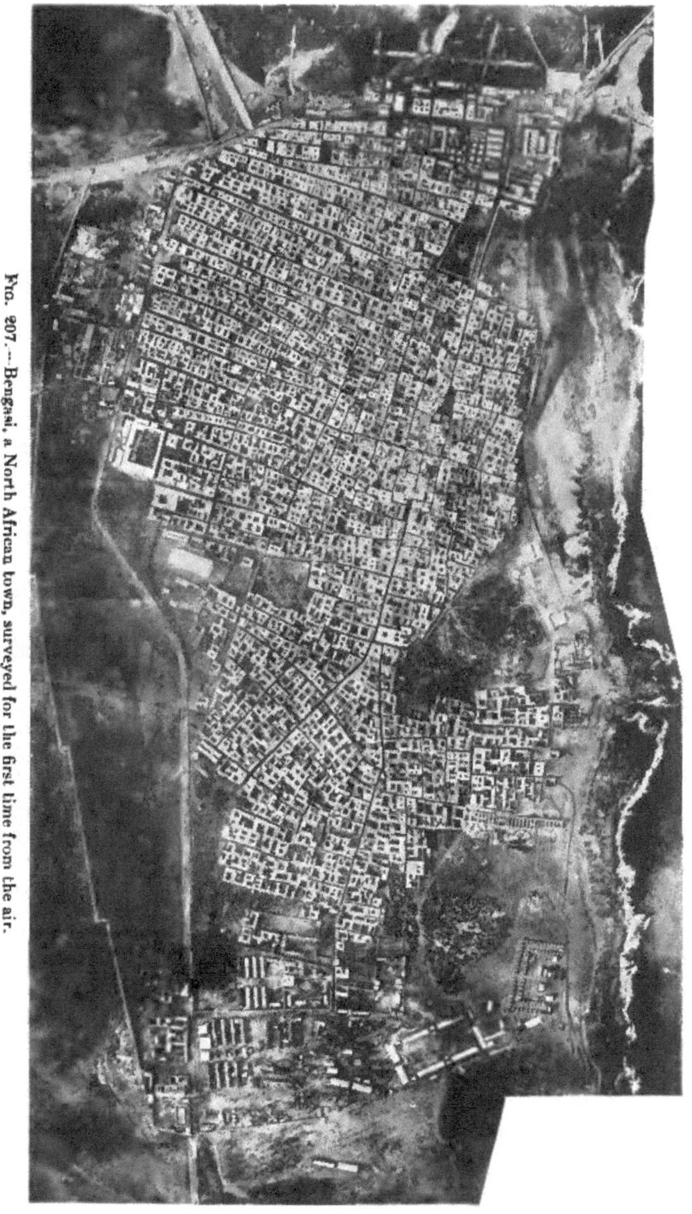

Fig. 207.—Bengasi, a North African town, surveyed for the first time from the air. Italian official photograph.

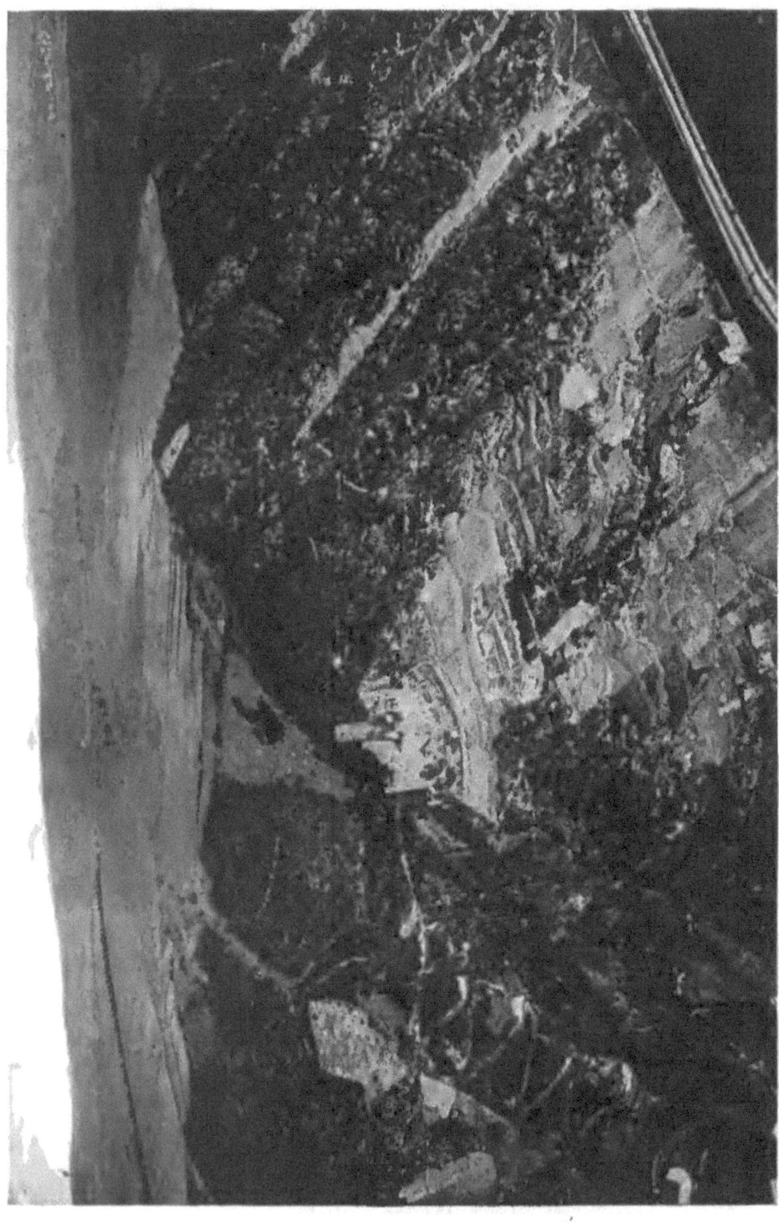

Fig. 208.—Thurnberg on the Rhine.
Photograph by photographic section A. E. F.

nels of the port of Venice. Navigation of such a river as the Mississippi with its shifting bars may come to be guided by monthly or even weekly aerial photo maps.

Among other uses for aerial photography will be the location of timber. As one illustration, may be taken the discovery of mahogany trees. Their foliage at certain times of the year is of characteristic color. This may be recorded on color sensitive plates with a scientifically chosen filter, and the cutting expedition sent out with the photograph as a guide. In this as in other cases where rough or unexplored country is to be covered, it is a question whether the airplane will after all be the most feasible craft, on account of its necessarily rapid rate of travel, and its need for known landing fields. The dirigible of large cruising radius, which can seek its landing field at leisure, is probably indicated for this kind of work. It may indeed, as already hinted, prove to be the chief photographic aircraft of the future.

Archæological surveys offer a fascinating opportunity for airplane or dirigible balloon photography to render scientific service. Buried in desert sands or overgrown with tropical vegetation the ancient cities of Asia Minor, of Burma, and of Yucatan evade discovery, and even when found remain unmapped for decades. Discovery and mapping can now go hand-in-hand. The topography of barbaric or colonial towns and villages, whose importance could not warrant elaborate surveys, but which should nevertheless be a matter of record, will be quickly and easily plotted by photography (Fig. 207). To this day who knows how the streets run in Timbuctu, and how, save from the air, can we ever map the teeming cities of China? He who would follow in the footsteps of Haroun-al-Raschid can even now explore the by-ways of Bagdad by the aid of the Royal Air Force photographic map!

INDEX

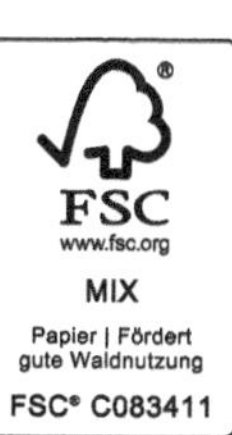

Zeitfracht Medien GmbH
Ferdinand-Jühlke-Straße 7
99095 Erfurt, Deutschland
produktsicherheit@kolibri360.de